OCCUPATIONAL HEALTH AND SAFETY FOR THE FIRE SERVICE

Joseph V. Bruni Jr. M.Ed.

St. Petersburg Fire & Rescue Division Chief

St. Petersburg College Associate Professor of Fire Science

St. Petersburg, FL

PEARSON

Boston Columbus Indianapolis New York San Francisco Upper Saddle River Amsterdam
Cape Town Dubai London Madrid Milan Munich Paris Montreal Toronto Delhi
Mexico City São Paulo Sydney Hong Kong Seoul Singapore Taipei Tokyo

Publisher: Julie Levin Alexander
Publisher' Assistant: Regina Bruno
Senior Acquisitions Editor: Stephen Smith
Associate Editor: Monica Moosang
Development Editor: Anne Marie Masters
Editorial Assistant: Samantha Sheehan
Director of Marketing: David Gesell
Marketing Manager: Brian Hoehl
Marketing Specialist: Michael Sirinides
Marketing Assistant: Crystal Gonzalez
Managing Production Editor: Patrick Walsh
Production Liaison: Julie Boddorf
Production Editor: Lisa S. Garboski, bookworks editorial services
Senior Media Editor: Amy Peltier
Media Project Manager: Lorena Cerisano
Manufacturing Manager: Alan Fischer
Creative Director: Jayne Conte
Cover Designer: Karen Salzbach
Cover Photo: Joseph Bruni
Composition: S4Carlisle Publishing Services
Printing and Binding: R.R. Donnelley/Willard
Cover Printer: Lehigh-Phoenix Color/Hagerstown

Credits and acknowledgments borrowed from other sources and reproduced, with permission, in this textbook appear on appropriate pages within the text. Unless otherwise stated, all photos have been provided by the author.

Library of Congress Cataloging-in-Publication Data

Bruni, Joseph (Date)
Occupational health and safety for the fire service / Joseph Bruni.
p. ; cm.
Includes bibliographical references and index.
ISBN-13: 978-0-13-513808-3
ISBN-10: 0-13-513808-6
1. Fire fighters—Health and hygiene. 2. Fire fighters—Safety measures.
3. Emergency medical technicians—Health and hygiene. 4. Emergency medical technicians—Safety measures.
I. Title.
[DNLM: 1. Accidents, Occupational—prevention & control—United States.
2. Emergency Medical Technicians—United States. 3. Fires—United States.
4. Emergencies—United States. 5. Occupational Health Services—organization & administration—United States. 6. Safety Management—methods—United States. WA 487.5.F4]
RC965.F48B78 2012
363.37—dc23

2011020696

10 9 8 7 6 5 4 3 2 1

ISBN 10: 0-13-513808-6
ISBN 13: 978-0-13-513808-3

DEDICATION

I would like to dedicate this book to two very important people in my life.

First, I am dedicating this book to my mother, who passed away after a long and difficult battle with cancer during the writing of this text. She was one of the few family members who provided me with support and encouragement when I started in the fire service. For a short period, my mother was one of the first women in the volunteer fire department where I started. She continued to stay strong all the way to the end of her life. She understood that an individual's character and integrity are of great importance. She encouraged each of us in the family not only to exercise forgiveness every day but also to help others through difficult circumstances. She also encouraged me to avoid anger by helping me to understand its destructive forces. She was the rock that I could turn to for wisdom and insight in any situation. I will continue to give thanks for the opportunity for her to be not only my mother but also my dearest friend.

Second, I am dedicating this book to FDNY Lieutenant Andrew Fredericks, who was killed in the North Tower of the World Trade Center on September 11, 2001. Andy was not only a great firefighter but also one of the best instructors and authors the modern-day fire service has ever had. I had the pleasure of meeting Andy when I was conducting fire stream and nozzle research in my own department in St. Petersburg, FL. As part of a nozzle panel discussion board at the Fire Department Instructors Conference (FDIC), Andy encouraged me over lunch to write a sidebar to one of his nozzle articles in *Fire Engineering Magazine*. I was reluctant at the time, as I had not considered myself to be an accomplished writer, even though I was attending college to obtain my undergraduate degree. I completed the article, and our friendship grew as Andy continued to provide support and be a powerful mentor as I kept writing. Little did I know that just 4 short months after Andy and his family came to Florida for a visit with my family he would be taken from his family and the family we know as the fire service. Andy's spirit continues to live on in those of us he provided true leadership to, as we move forward to make the fire service and the world a better place.

CONTENTS

Chapter 5 Risk Assessment and Safety Planning 89

Chapter 6 Fire Emergency Safety 117

Chapter 12 Data Collection and Information Management 246

Chapter 13 Specific Issues Concerning Occupational Safety Services 263

Chapter 14 Lessons Learned from Incidents 275

PREFACE

Year after year the number of emergency responders injured and killed in the line of duty continues to remain the same, with little variation. Many have identified occupational health and safety in recent years as a key factor concerning injury and death of emergency workers. It is hoped that this textbook will aid emergency services managers and leaders in reducing these statistics. The world of emergency services is a continually changing one. Yet amid this changing environment, responders must stay focused on current trends such as terrorism, technology, and, of course, health and safety. Many standards and regulations have been implemented to enhance health and safety, and as an effective way for managers and leaders to identify where their focus should remain. The continued updating of standards and regulations is a proactive approach addressing the health and safety of responders. Technology enhances and supports the health and safety of emergency responders, and will continue to do so well into the century.

The concepts of health and safety have been an industry standard for many years in the private sector, yet in the public sector, they constitute a relatively new discipline that is gaining in popularity and focus. Many leaders and managers have identified the need for accountability in the area of health and safety. It is necessary to have an increased emphasis on education and training to be successful in this important area of health and safety. Both fire and emergency medical services must realize the accountability of safe practices. Many groups and symposiums are taking place across the nation in an effort to remain focused on the issues and problems related to the health and safety of responders. The Fire and Emergency Services Higher Education (FESHE) program strongly recommends the topic of health and safety as a part of the associate's degree program. It is hoped that a focus in this area will save many lives and reduce injuries of the nation's emergency responders.

Throughout the text, the icon of the National Fallen Firefighter's Foundation is used appropriately to promote the 16 life safety initiatives that ensure that "Everyone Goes Home."

ResourceCentral

At the beginning of each chapter, we prompt readers to visit Resource Central at **www.bradybooks.com** for interactive learning exercises on the topics discussed. Students will find quizzes, web links, and more to supplement classroom learning. Answers to end of chapter review questions as well as answers to the case study questions may also be found here. Through Resource Central, this text also offers instructors a full complement of online supplemental teaching materials such as test banks and PowerPoint lectures to aid in the classroom.

ACKNOWLEDGMENTS

The acknowledgments for this book could very well be a chapter of their own. Let me begin by thanking my wonderful wife, Tammy, and my two beautiful daughters, Melanie and Autumn. They put up with countless hours of family neglect as I sat conducting research and typing away at the keyboard. Without their continued support, I would not have been able to complete this project.

I would also like to acknowledge my mother for her continued encouragement over the course of a lifetime and a career. At one point during the writing of this text, I had to take a great deal of time off to help my sister support our mother and take care of her needs as she withered from that terrible disease known as cancer. As firefighters, we have come to know this disease all too well. There is not a member of the fire service that cannot relate to someone, be it family member, friend, or coworker, who has been affected by it. I have lost two brother firefighters and mentors to this disease over the course of my career. My mother supported me throughout the course of my pursuit of a career in the fire service, and I was there to help her throughout her battle with cancer. My family has been my rock throughout life and my career. There is no such thing as a successful firefighter without the support of strong family members. I, like many others, have been fortunate to have had both my immediate family and my firefighting family as a strong foundation in my life. My uncle and mother, the only other volunteer firefighters in my family, set a great example for me to follow and encouraged me a great deal as a young firefighter. Unfortunately, I had to perform cardiopulmonary resuscitation on my uncle on my seventh wedding anniversary. I was in school at that time to become an emergency medical technician, and that experience moved me forward with my decision to leave Pennsylvania and become a career firefighter in his honor. My parents' relocation to Florida several years earlier made the transition from Pennsylvania to Florida easier on my family and I. I only regret my uncle was not able to see how far I have come on this path as a firefighter.

The individuals who mentored and had an impact on me throughout my career as a firefighter are too many to list. Many of them have passed on, but their spirits spur me on as I strive to honor them by living and working in the ways they taught me. My remaining colleagues know who they are and can all rest assured that every day I use something that I have learned in each of the three departments of which I have been a member. As a young volunteer firefighter, we were continually trained and encouraged to attend as many classes as possible to learn the craft of emergency services work. I wish to thank those who tirelessly drove us to classes and made sure we operated at the emergency scene as safely as possible. We learned very early on that continual knowledge, experience, and physical fitness all play a major role in keeping us from getting injured or possibly killed. I wish to honor those who continue to stay focused on the well-being and health of firefighters across the nation. It is true that a strong focus on risk management and safety in the fire and other emergency services will reduce injuries and deaths.

I also wish to thank my development team, including Stephen Smith, Monica Moosang, Anne Marie Masters, Julie Vitale, and Lisa Garboski, for their willingness to take a chance on me to continue this project on my own. Their patience and support provided the encouragement I needed to keep moving forward on those long days and nights when I sat at the keyboard. I had never envisioned how difficult the writing and research process would be to complete this project. These professionals provided me with patience, understanding, and friendship as I worked through the process. I would also like to thank all of the production staff at Brady/Pearson for their hard work.

Last, I would like to thank the following reviewers whose suggestions and comments helped to make this a better book.

John P. Alexander
Adjunct Instructor
Connecticut Fire Academy
Windsor Locks, CT

Lawrence T. Bennett, Esq.
Program Chair
Fire Science and Emergency Management
University of Cincinnati
Cincinnati, OH

Leo DeBobes, MA (OS&H), CSP, CHCM, CPEA, CSC, EMT
Assistant Professor/Fire Protection Technology
Program Coordinator
Suffolk Community College
Selden, NY

Randall Griffin
Lieutenant
DeWitt Fire District
DeWitt, NY

Christian M. Hartley
Lieutenant
Houston Fire Department
Houston, AK

Les Hawthorne, BA, NREMT-P
Coordinator
Paramedic Program
Southwestern Illinois College
Belleville, IL

Bryan T. Haywood
Founder and CEO
SAFTENG.net LLC
Milford, OH

David Marshall
Program Director
Fire Science
Yavapai College
Clarkdale, AZ

Nathan Sivils
Director
Fire Science
Blinn College
Bryan, TX

Thomas Y. Smith, Sr.
Fire Science Program Chair
West Georgia Technical College
LaGrange, GA

Gene Wilkerson
Deputy Chief
Burlington Fire Department
Burlington, IA

ABOUT THE AUTHOR

Joe Bruni is the division chief of safety and training for the St. Petersburg, Florida, Fire & Rescue Department. This department protects 250,000 people in a 63-square mile area of Pinellas County, Florida; is internationally accredited by the Commission on Fire Accreditation International; and operates out of 12 fire stations with services that include fire suppression, fire prevention, public education, advanced life support emergency medical response, hazardous materials, technical rescue, and water rescue/dive operations. The author is a 36-year veteran of the fire service, having begun his career with the Robinson Township #1 Fire Department in the suburbs of Pittsburgh, Pennsylvania. In the State of Florida, he has also worked for St. Petersburg Beach Fire Department and the City of St. Petersburg Fire & Rescue Department. He has worked through the ranks of firefighter, driver/engineer, lieutenant, captain, and division chief.

His education includes an associate's degree in Fire Science Administration from St. Petersburg College, a bachelor's degree in Organizational Studies with a concentration in Public Leadership from Eckerd College, and a Master's degree in Education with a concentration in Public Administration from Troy State University. He is a Florida certified Fire Officer I, Florida certified Fire Instructor III, and Florida certified Emergency Medical Technician. The author has performed as a field training officer, a hazardous materials technician, and a member of the St. Petersburg Fire & Rescue Technical Rescue Team. He also serves as a member of the Pinellas County Training Officers Group and the St. Petersburg College Fire Training Center Advisory Committee.

As an instructor, the author teaches various fire science classes at St. Petersburg College, including building construction for the fire service. He is also a lead instructor for the St. Petersburg College Fire Academy, and teaches continuing education classes through the Pinellas Technical Education Center.

Fire and Emergency Services Higher Education (FESHE) Grid

The following grid outlines Occupational Health and Safety for the Fire Service course requirements and where specific content can be located within this text:

Course Requirements	1	2	3	4	5	6	7	8	9	10	11	12	13	14
Describe the history of occupational health and safety.	X													
Identify occupational health and safety programs for industry and emergency services today.	X			X	X	X	X				X			
Compare the differences between standards and regulations.		X						X						
List and describe the components of risk identification, risk evaluation, and incident management.		X	X	X	X	X	X	X			X			
Describe the relevance of safety in the workplace, including the importance of PPE.	X	X	X		X	X	X	X		X				X
Apply the knowledge of an effective safety plan to pre-incident planning, response, and training activities.			X	X	X		X	X		X				X
Explain the components of an accountability system in emergency services operations.						X		X						
Discuss the need for and the process used for post-incident analysis.			X						X					X
Describe the components and value of critical incident management programs.					X				X					

Describe the responsibilities of individual responders, supervisors, safety officers, incident commanders, safety program managers, safety committees, and fire department managers as they relate to health and safety programs.		X	X		X	X		X	X	X		X	X	X
Describe the components of a wellness/fitness plan.		X	X		X					X				
Identify and analyze the major causes involved in line-of-duty firefighter deaths related to health, wellness, fitness, and vehicle operations.					X	X		X		X		X		X

knowledge enables emergency responders to develop "situational awareness" concepts, without which they will continue to incur a high rate of injury and death. The occupation requires a mastery of specific skills and the ability to apply them as needed. For those who have the ability to master these skills psychologically and physically, the career is one of the most rewarding jobs in the world.

Fire Service History and Culture

One must be aware of the history and culture of the fire service to appreciate what is coming in the future. In most cases, every type of emergency will require a response from the fire service. In the United States, the fire service responds to a variety of emergencies including vehicle accidents, fires, medical emergencies, building collapses, hazardous materials incidents, technical rescue incidents, weather-related incidents, civil disturbances, and terrorist incidents. Today's fire service reflects the society and community in which it serves. Women and minorities reflect a large portion of the fire service today. No matter the makeup of the fire service organization, firefighters work together, live together, and train together. The closeness and interaction between firefighters and other emergency responders is unmatched in any other type of organization. Emergency responders come to rely on each other and become members of an integral team.

The general public has become accustomed to emergency responders being able to evaluate any emergency or nonemergency situation and mitigate the situation. Emergency responders are ordinary people who have been trained to place themselves in extraordinary situations where they will work through the problems to successful completion. There is a high level of motivation among emergency responders, who, in many cases, are provided with effective training and the necessary equipment. Unfortunately, in some cases, training and equipment are tragically deficient. Hence, the best efforts of response personnel may not be enough in certain situations to solve confronted problems and issues, and the public must—although in many cases it does not—understand this reality.

In every disaster there are lessons to be learned. Wherever buildings are in close proximity, there has been a problem when fire has broken out. Fire has the capability to jump from building to building, thus destroying property and placing lives in harm's way. Emergency responders continue to stay proactive in their ability to adapt and overcome. Fire service responders' long history has demonstrated their ability to work tirelessly to contain and extinguish fire.

In the cities of Europe, it was common for citizens to be required to have not only a ladder long enough to reach the roof of their structure but also containers to keep water on hand to extinguish any uncontrolled fire. These measures were required up until the early 20th century. As humans evolved from nomadic tribes into an agrarian society, structures were built in close proximity to one another for various reasons. Many of these early buildings were built with some type of masonry walls; however, they often had thatched roofs that would burn. Many of these structures also contained some type of fireplace for heating and cooking. It was common for sparks and embers to land on thatched roofs, causing ignition. Unfortunately, fire suppression had not evolved beyond the "bucket brigade" where people stood side by side to pass buckets of water along a line in an attempt to put out uncontrolled fires. See Figure 1–1.

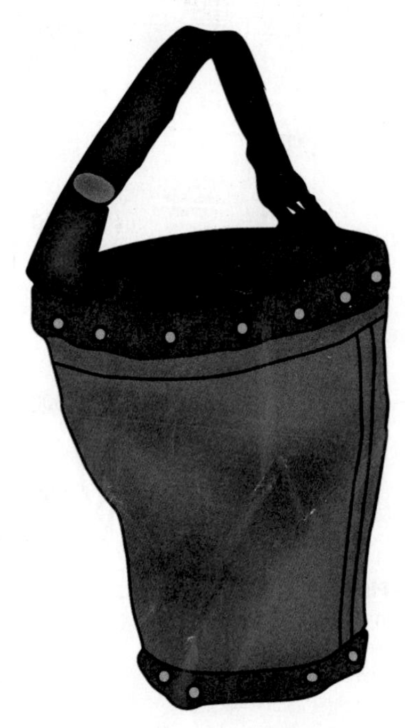

FIGURE 1–1 The early fire bucket was a valuable tool for the bucket brigade.

In some cases, this bucket brigade method worked to suppress fires. In other cases, the roofs would burn and fire would consume what was inside.

FIGURE 1–2 Benjamin Franklin as the founder of the Union Fire Company.

Many of the newcomers and immigrants to the United States brought with them the firefighting methods from their native countries. Because of this, the results were the same as in the European nations. Many organizations evolved from social and fraternal clubs dedicated to fighting fires. Benjamin Franklin founded the first fire company called the Union Fire Company. See Figure 1–2.

In many cases today, those willing are accepted as members of a fire service organization prior to receiving training or certification as firefighters. Many then enroll in basic firefighting courses through local community colleges or vocational technical institutions to meet the training and eligibility requirements. In present-day society, most emergency responders fall into the categories of career, paid-on-call, combination, and volunteer departments. Many firefighters also serve in federal, military, and private fire departments. A good number of the nation's firefighters are found in volunteer fire departments; however, it has become increasingly common for paid-on-call emergency responders to respond from their home or workplace and then receive compensation for every emergency response. In the United States, 29% of the fire departments are made up of career firefighters and 71% are volunteer firefighters. It seems that due to the efforts of the fire service in the area of occupational health and safety, the number of fireground injuries has declined over a 28-year period. Many firefighter injuries occur at non-fire-related events; however, it is important to note that the highest number of firefighter injuries happens while operating on the fireground. See Figure 1–3.

Although it is good news that the number of firefighter injuries occurring on the fireground has declined, the number of firefighter injuries per 1,000 fires has remained relatively the same within the same time frame. See Figure 1–4.

Modern Fire Apparatus

Internal combustion fire engines first saw service in 1907. During the early years, internal combustion vehicles were used either as pumping apparatus or as tractors to pull the equipment and apparatus. The Chicago, Illinois, fire department became the first

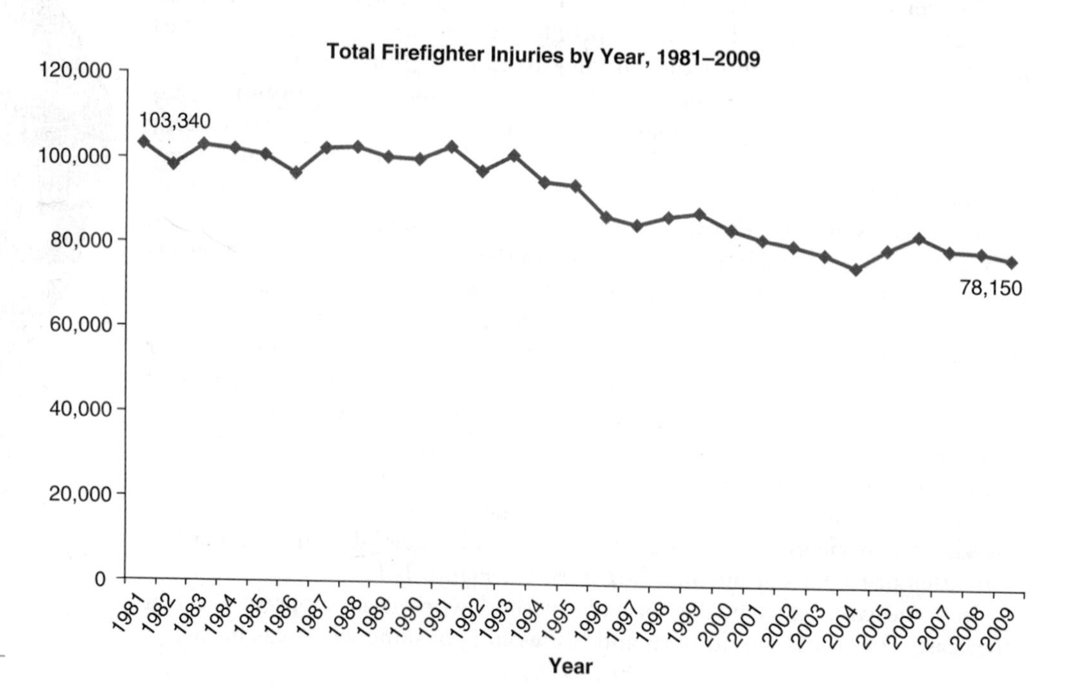

FIGURE 1–3 Firefighter injuries have declined over the years. *Reprinted with permission from "U.S. Firefighter Injuries—2009" by Michael J. Karter, Jr., and Joseph L. Molis, Copyright © 2010, National Fire Protection Association.*

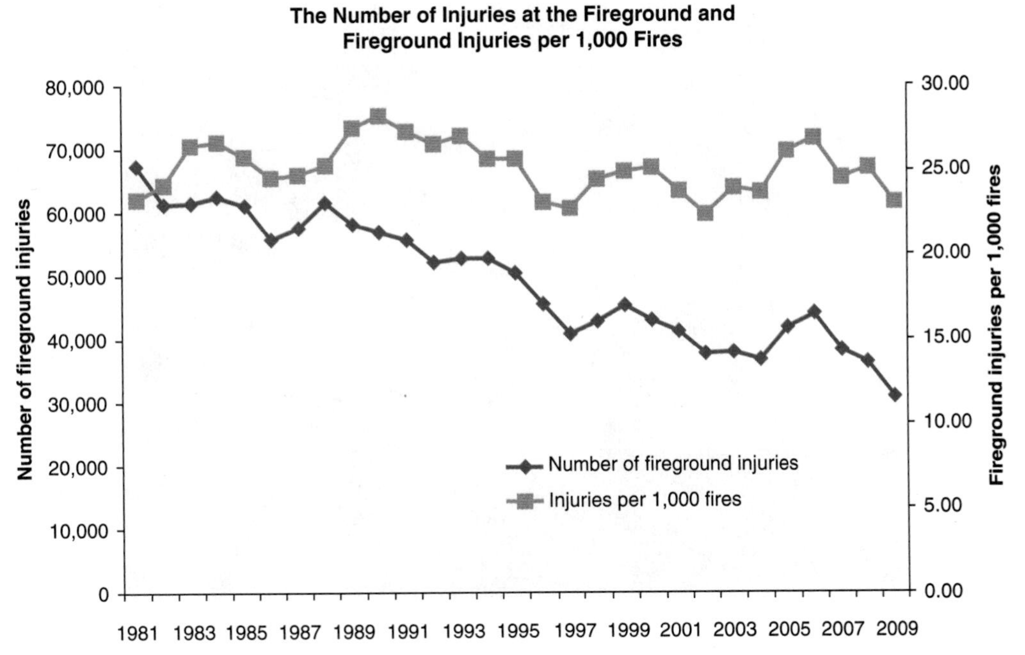

FIGURE 1–4 Injuries per 1,000 fires have remained the same. *Reprinted with permission from "U.S. Firefighter Injuries—2009" by Michael J. Karter, Jr., and Joseph L. Molis, Copyright © 2010, National Fire Protection Association.*

completely motorized fire department in America. Today, firefighting vehicles have evolved into highly specialized apparatus and equipment enabling fire service personnel to respond faster to varied emergencies.

Along with the common engine and ladder truck, other specialty vehicles include hazardous materials units, technical rescue apparatus, foam units, air and generator trucks, and water rescue units. The modern diesel-driven engine and pump can deliver as much as 2,000 gallons per minute. Ladder trucks can reach as high as 120 feet into the air. Fireboats can deliver as much as 10,000 gallons per minute. Airports are equipped with specialty crash-rescue apparatus. Today, most apparatus is diesel powered, and multiple variations of equipment allow emergency personnel to respond to a wide array of incidents. See Figure 1–5.

Early Protective Firefighter Equipment

Early firefighters had very little in terms of firefighting personal protective equipment. They wet their beards and shoved them into their mouths to act as a filter against the products of combustion. The fire service has come a long way since those days. Firefighting is a dangerous profession that requires specialized equipment to mitigate the various types of emergencies effectively and safely.

The most important part of a firefighter's protective gear is the apparatus to protect the respiratory system. Through the years the fire service has used several types of filter assemblies and masks to protect the wearer from the toxic effects of smoke and other hazards. The first successful self-contained breathing apparatus, known as the Gibbs, came into the fire service during the first decade of the 20th century.[1] The air pack introduced in late 1945 was developed by Scott Aviation to assist pilots during World War II in breathing at high altitudes. This early apparatus consisted of a steel oxygen cylinder and a pressure-regulating device that allowed firefighters to breathe oxygen from the cylinder in hazardous environments. These early breathing devices bear very little resemblance to the equipment used in the modern-day fire service.

FIGURE 1–5 Today's apparatus has evolved into specialized units for specific purposes. *Courtesy of Joe Bruni.*

Modern Respiratory Protection

Lightweight composite cylinders developed by the National Aeronautics and Space Administration (NASA) have replaced the heavy steel cylinders used earlier. Today's positive pressure systems provide constant pressure to the wearer's face piece and allow a "heads-up" display of the remaining air in the cylinder that can be viewed within the face piece. The combination of positive pressure and a heads-up display has increased firefighter safety. Positive pressure SCBA is intended to keep the products of combustion and other respiratory hazards out of the user's face piece, and the heads-up display helps to keep firefighters from going past the point of no return at incidents. There has also been an increase in the amount of air available to the individual firefighter: although the 15- to 20-minute cylinder is still available and commonly used by many departments, many others have changed to a cylinder that supplies breathable air for 45 minutes to an hour.

Personal Protective Equipment

Another important part of the firefighting gear is commonly referred to as the firefighter's personal protective equipment (PPE): coat, pants, hood, helmet, boots, gloves, breathing apparatus, and personal alert safety system (PASS) device. A drastic difference exists between the PPE of today and that of the early days of the fire service when firefighters wore everyday clothing and did the best they could. Many structures burned to the ground because adequate PPE was not available to firefighters, thus making interior operations impossible.

As firefighting evolved, so did the equipment firefighters wore. Jacobus Turck is credited with inventing the first fire helmet in the 1730s.[2] It was leather with a high crown and wide brim.

In the early 1800s, Henry T. Gratacap designed a helmet similar to the one used today, referred to as the "traditional" fire helmet. The advent of this helmet afforded firefighters protection from falling debris and water that would run off the back of the helmet. As the development of rubber progressed, it played a beneficial role in firefighter clothing. Firefighters began to wear a rubber slicker over their wool coats as an added layer of protection from heat and water. Rubber boots were also introduced.

As the firefighter's personal protective equipment continued to develop, the terms *bunker gear* and *turnout gear* become part of history. Bunker gear was derived in the mid-1800s; "bunking" was the practice of sleeping at New York City's volunteer firehouses, so the firefighter's bunking gear would be the clothing worn when responding or "turning out" to a fire during nighttime hours. Bunker pants were also derived from the padded pants worn by soldiers responsible for firing cannons from World War I bunkers. Those pants protected wearers from water, mud, shrapnel, and hot shell casings. Many of the war veterans became firefighters after the war and adapted those protective pants as part of the firefighting ensemble. It was common for firefighters to wear three-quarter rubber boots and long rubber trench coats. The traditional fire helmet was also a part of this protective gear.

After the wartime era, the National Fire Protection Association (NFPA) and other organizations developed testing and standards for firefighter protective gear. It became the standard for a firefighter's protective clothing to offer three layers of protection: a flame-resistant outer layer, a middle layer that prevented the wearer from becoming wet with water, and an inner layer that protected against the three heat transfer methods (convection, conduction, and radiation). A steel toe and shank were also required in firefighter protective footwear.

In the early 1980s, new protective gear took hold in the modern-day fire service. In 1982, the NFPA developed the **personal alert safety system (PASS)**. The PASS device sends out an audible alarm if the firefighter is immobilized or remains motionless for 30 seconds or more. Also in the 1980s, advanced fire-resistive materials such as Nomex® and Kevlar® were used to make the outer shell of the coats and pants. The protective equipment worn by firefighters today is a result of ongoing research and technology. Leather boots, gloves, and protective hoods now complete the firefighting protective ensemble. SCBA units are integrated with a personal alert safety system (PASS), which activates when the unit becomes charged with air.

personal alert safety system (PASS) ■ This device sounds a loud signal if the wearer becomes immobilized or is motionless for 30 seconds or more.

Personal escape ropes have also become an added measure of protection when operating on upper floors of a structure. As technology advances, the manufacturers of turnout clothing and SCBA units will continue to improve the protective gear for firefighting. Technology concerning PPE will directly impact the health and safety of personnel well into the future.

Along with the modifications in protective gear and equipment, the firehouse has also changed over the years. Early volunteer departments needed nothing more than a shed or garage to store apparatus and gear. As departments grew into full-time paid organizations, it was necessary to provide quarters for sleeping and cooking for firefighters who worked extended hours on shift. Modern firehouses provide individual personal spaces for members that are designed with privacy in mind.

#8

The History of Safety and Health in Emergency Services

Occupational safety and health is not a new concept in industry; however, the issue has advanced rapidly in emergency services. Over the course of time, many people and organizations have studied disease and injury related to specific occupations. The study of toxicology became an important area of research in both industry and emergency services. Many workers, including emergency responders, are faced with exposures to vapors, dusts, and the products of combustion. It is still common for workers in the fire service to suffer from some type of lung disease and certain types of cancers. The study of occupational safety and health has prompted the need for changes in the workplace. The fire and emergency services are no exception; however, it is common for emergency responders to resist changes in this and other important areas because of tradition. It must be kept in mind that tradition cannot take a lead role when it conflicts with a proactive approach in the field of health and safety.

NECESSARY CHANGES

In many cases, people and labor organizations have had to fight for necessary changes due to the cost of additional needed resources. Events and studies over the years had led to laws, regulations, and standards as many recognized the need for safe working conditions; however, the struggle to implement these conditions continues. In today's society, those laws, regulations, and standards have forced many organizations to provide safer working conditions. A number of organizations have safety and health divisions, and many institutions and universities offer education concerning safety and health at the collegiate level.

In certain occupations, medical surveillance, which offers the opportunity to find diseases early, is offered to employees and required by standards and regulations. The fire service continues to see advances in this important area. It is now recognized that certain occupational health and safety concepts that have long been in place in industry must also be applied to emergency services personnel.

The general public as well as fire service organizations has accepted the high injury and death rates for firefighters. The injury and death potential for firefighters has increased in recent times as firefighters have taken on the additional role of emergency medical responders. EMS has caused increased responses and a call volume that exposes personnel to diseases and the hazards related to the response to emergency incidents. There is a greater chance of vehicle accidents during response and return to quarters.

It is important to keep in mind that occupational safety and health has rapidly developed in the last 20 years in emergency services organizations. Unfortunately, it is often after a tragic event that regulations and standards in emergency services are enacted. In many cases, fire department and political leaders view improvements in health and safety as unfunded mandates. See Box 1-1.

HEALTH AND SAFETY AND THE NFPA STANDARDS

Over the years, many fire and EMS-related textbooks have clearly stated the injury and death rates for emergency services personnel, but made little to no reference to solving the ongoing health and safety problems. During the 1980s, an increased interest concerning the health and safety of emergency responders began to emerge and has continued into the 21st century. One of the most controversial standards published by the NFPA concerning the health and safety of emergency responders is the **NFPA 1500 standard**, *Standard on Fire Department Occupational Safety and Health Program.*

NFPA 1500 standard ▪ This standard shall contain the minimum requirements for a fire service–related occupational safety and health program.

The NFPA 1500 standard is actually a compilation of a number of NFPA standards regarding safety and health. Although many fire service organizations are still not in full compliance with this standard, others are striving toward this goal; and the standard has set the wheels in motion concerning the health and safety of emergency responders. Full compliance with this standard equates to adequate resources. The standard remains controversial due to operating costs and potential changes; however, it causes many volunteer and career departments to agree that the injury and death statistics in emergency services are unacceptable.

BOX 1-1: DATES IN THE DEVELOPMENT OF OCCUPATIONAL SAFETY AND HEALTH IN EMERGENCY SERVICES

1960 — International Association of Fire Fighters begins to publish a firefighter injury and death report.
1973 — *America Burning* is published by the Nixon administration.
1974 — NFPA begins to publish an annual firefighter injury report and annual death report.
1986 — NFPA 1403 standard is published.
1987 — NFPA 1500 standard is published.
1996 — OSHA publishes ruling on two-in/two-out.
1997 — Joint wellness-fitness program is developed by the International Association of Fire Chiefs and the International Association of Fire Fighters.
1998 — National Institute for Occupational Safety and Health begins the Fire Fighter Fatality Investigation and Prevention Program.
1998 — OSHA respiratory protection standard is applied to firefighters.
2001 — NFPA 1710 is adopted.

The **NFPA 1403 standard**, *Standard on Live Fire Training Evolutions*, was published and adopted in 1986. The premise behind this standard's development was the loss of life and injuries sustained during live fire training. The standard applies to both acquired structures and training center burn buildings. The two-in/two-out procedures developed and published by the NFPA in 1996 were followed by a respiratory protection standard in 1998.

NFPA 1403 standard ■ This standard shall contain the minimum requirements for training all fire suppression personnel engaged in firefighting operations under live fire conditions.

The **NFPA 1710 standard**, *Standard for the Organization and Deployment of Fire Suppression Operations, Emergency Medical Operations, and Special Operations to the Public by Career Fire Departments*, was approved in 2001. The NFPA 1710 standard was commonly viewed as a deployment and not a staffing standard; many of its components are concerned with the safety and health of responders with the aim of helping to achieve safe operations at the incident scene.

NFPA 1710 standard ■ This standard shall contain the minimum requirements relating to the organization and deployment of fire suppression operations, emergency medical operations, and special operations to the public by substantially all career fire departments.

TERRORISM

In the past decade, responders have also been concerned with the response to terrorist events. Since September 11, 2001, responders have had to shift their focus to encompass preparedness for weapons of mass destruction and the safety and health issues related to this important area. The threat of terrorism also uncovered a need for equipment and training. Along with terrorism came the development of the Department of Homeland Security and grant monies to be applied to equipment and training.

THE RESISTANCE TO CHANGE

Emergency services organizations and their responders have historically been resistant to change; however, in the last decade organizations have gradually accepted the laws, regulations, and standards because they have come to realize change is necessary to improve the safety and health of their members. Emergency services organizations have begun to comprehend that procedures, equipment, and technology do indeed change, and it is necessary to accept these changes to remain as proactive organizations. Unfortunately, many emergency services organizations are moving forward with change without considering the human element. Changes concerning standard operating procedures (SOPs) and policies do very little if personnel either do not follow the guidelines or do not have the necessary training and equipment to make the changes work properly.

It is also common for complacency to take hold in many emergency response organizations. Complacency often leads to personnel ignoring health and safety measures due to the common attitude of "it won't happen to me." It is an ongoing problem in emergency services as personnel continue to sustain approximately 100,000 injuries and 100 deaths per year. Fire department supervisors must continually guard against complacency and take necessary action when they recognize it. Without this action, personnel will continue to incur high injury and death rates.

HEALTH AND SAFETY PROGRAM MANAGER AND MEDICAL ISSUES

Since the adoption of NFPA 1500, many emergency services organizations have instituted the vital position of health and safety program manager. Health and safety should be a basic consideration in all emergency services organizations, and it is nearly impossible to improve health and safety without an effective health and safety program manager. It is the health and safety program manager's responsibility to work closely with the department's administration, physician, and members of the organization. It is also common to see heath and safety committees that include members from all levels of the organization as well as SOPs and policies in place that address the health and safety of members. The mind-set in emergency services has become one that accepts and encourages a safe working environment. Technology has developed better equipment and protective gear, and the use of the incident management system has become commonplace in both career and volunteer organizations. The improved safety and health of members come to the forefront of thinking in emergency services as we move further into the 21st century.

The acceptance and requirement of medical examinations at the time of appointment and then annually benefit both the employee and the organization greatly. The baseline medical examination can be used to compare against future medical examinations. Emergency services personnel can use the data from this annual medical evaluation to determine their own wellness and fitness.

Other strides concerning health and safety in emergency services include injury report statistics. National injury statistics allow organizations to identify trends and develop health and safety measures in an attempt to reduce various injuries and deaths. See Figure 1–6.

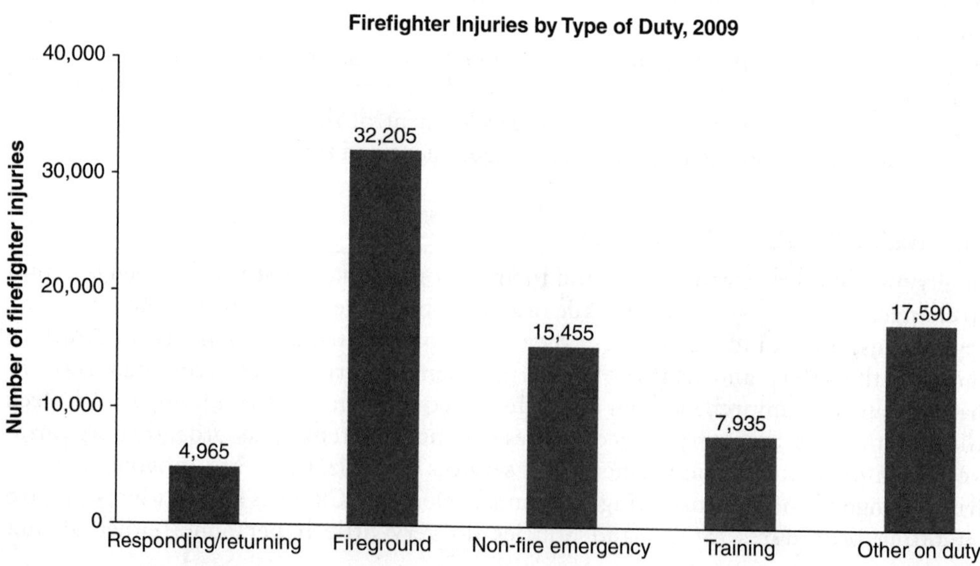

FIGURE 1–6 Injury statistics can be used to help reduce injuries. *Reprinted with permission from "U.S. Firefighter Injuries—2009" by Michael J. Karter, Jr., and Joseph L. Molis, Copyright © 2010, National Fire Protection Association.*

Safety Problem Identification

There are many resources available to the health and safety program manager in every emergency services organization. Some are at the national level; however, health and safety program managers must also remember to focus on local issues specific to their organization. For example, if a department is experiencing a high amount of back injuries related to its emergency medical transport service, there may be issues concerning patient movement and lifting. The health and safety manager needs to compare national, state, and local data for a clear assessment. The following organizations publish annual data related to the health and safety of emergency services responders.

#7

International Association of Fire Fighters (IAFF) ■ This is the labor union that represents most career firefighters.

INTERNATIONAL ASSOCIATIONS OF FIRE FIGHTERS AND FIRE CHIEFS

The **International Association of Fire Fighters (IAFF)** is the labor organization that represents most career firefighters in North America. The IAFF has published an annual injury and death report since 1960. The data from the IAFF concern various health and safety issues, provide some of the most in-depth information related to exposure to infectious disease, and are a valuable resource for the health and safety program manager. The health and safety program manager must keep in mind that the gathered data are only from career paid fire departments, which limits the breadth of the information.

The IAFF and the **International Association of Fire Chiefs (IAFC)** formed a joint labor agreement in 1997 concerning the wellness and fitness of firefighters called the Fire Service Joint Labor Management Wellness/Fitness Initiative. Now known as the Wellness-Fitness Initiative, the program is a valuable resource for health and safety program managers; however, it only contains data from departments willing to participate.

#6

International Association of Fire Chiefs (IAFC) ■ This organization provides leadership to career and volunteer chiefs, chief fire officers, company officers, and managers of emergency services organizations throughout the international community through vision, information, education, services, and representation to enhance their professionalism and capabilities.

NATIONAL FIRE PROTECTION ASSOCIATION

The **National Fire Protection Association (NFPA)** publishes annual reports concerning injury and deaths in the fire service with statistics concerning characteristics, frequency, and extent. The NFPA has been compiling data and publishing the injury and death report since 1974. The NFPA's death survey concerns all firefighter deaths. The annual NFPA injury reports and death surveys both provide an in-depth look at causes, nature, and type of duty, and permit a comparison between injuries and infectious disease exposure.

The injury and death surveys also document the causes of occupational injury and illness retirements. The NFPA injury survey can be used to project national firefighter injuries. The NFPA data provide a framework to help identify the nature, causes, and types of activities most responsible for injuries and fatalities.

National Fire Protection Association (NFPA) ■ This organization develops, publishes, and disseminates more than 300 consensus codes and standards intended to minimize the possibility and effects of fire and other risks.

UNITED STATES FIRE ADMINISTRATION

The **United States Fire Administration (USFA)** is an entity of the U.S. Department of Homeland Security's Federal Emergency Management Agency. The USFA oversees the **National Fire Incident Reporting System (NFIRS)**, which is the largest and most detailed fire incident database in the world. The purpose of the NFIRS is to gather information focused on the magnitude and scope of the fire problem in the United States to assist state and local governments in developing fire reporting and analysis for their own use, and to collect data that can be used accurately. Within the NFIRS reporting system is a section that captures fire reports and firefighter casualties. Local departments report their data through NFIRS to their individual states. The states then submit or release their data to the USFA. States voluntarily participate in the NFIRS system; thus, incomplete information creates a problem for compiling accurate data and research. NFIRS allows analysis at the local, state, and national levels. Data obtained at the national level can be compared to local data.

United States Fire Administration (USFA) ■ This is an entity of the U.S. Department of Homeland Security's Federal Emergency Management Agency. The mission of the USFA is to provide national leadership to local fire and emergency services.

National Fire Incident Reporting System (NFIRS) ▪ This is the standard national reporting system used by U.S. fire departments to report fires and other incidents to which they respond and to maintain records of these incidents in a uniform manner. It was established by the National Fire Data Center in the U.S. Fire Administration (USFA), which gathers and analyzes information on the magnitude of the nation's fire problem.

Occupational Safety and Health Administration (OSHA) ▪ The mission of this federal agency is to assure that every U.S. worker goes home whole and healthy every day. Toward that end, the agency sets and enforces workplace safety and health standards, encourages voluntary compliance through consultation and partnerships, and promotes safety training and education for workers and employers.

National Institute for Occupational Safety and Health (NIOSH) ▪ This federal agency is responsible for conducting research and making recommendations for the prevention of work-related injury and illness.

During the late 1970s and 1980s, the USFA provided funding, materials, and mainframe software for states to implement NFIRS. Due to the voluntary nature of NFIRS, injuries and deaths may go unreported because a local agency chooses not to participate in the program. The NFIRS system contains casualty information related to injuries and deaths that occur during actual incidents. Unless an organization generates an incident number and completes the casualty portion of the incident report, injuries that occur while on duty but not associated with an actual incident are not captured and reported. The information contained within NFIRS is used to formulate reports, which the USFA makes available. The USFA also provides case analysis studies related to actual incidents that have occurred as well as information concerning protective clothing. The USFA has an extensive Web site from which publications and documents can be downloaded, adding to the toolbox of any health and safety program manager.

OCCUPATIONAL SAFETY AND HEALTH ADMINISTRATION

The mission of the **Occupational Safety and Health Administration (OSHA)**, a division of the U.S. Department of Labor, is to prevent work-related injuries, occupational fatalities, and illnesses. OSHA is responsible for issuing and enforcing standards concerning safety and health. The Occupational Safety and Health Act that created OSHA also created the National Institute for Occupational Safety and Health (NIOSH), covered next. Many OSHA regulations are applicable to the fire service. The information concerning OSHA in Chapter 2 addresses specific regulations that apply to fire agencies. OSHA requires agencies to report occupation-related injuries and deaths, and then uses its reporting system to compile statistics and identify causes of injuries and deaths. This gathering of information allows prevention measures to be determined, established, and hopefully implemented. The data from OSHA are also incomplete for emergency services use because not all fire departments comply with OSHA's regulations. Some of the most important OSHA rules concerning the fire service are hazard communication (the right-to-know law concerning chemicals) and blood-borne pathogens.

NATIONAL INSTITUTE FOR OCCUPATIONAL SAFETY AND HEALTH

The **National Institute for Occupational Safety and Health (NIOSH)** is best known to emergency services members as the institution that investigates firefighter fatalities. NIOSH is a division of the Centers for Disease Control and Prevention. The Occupational Safety and Health Act of 1970 created both NIOSH and the Occupational Safety and Health Administration (OSHA). In 1998, NIOSH started the Fire Fighter Fatality Investigation and Prevention Program to investigate firefighter line-of-duty deaths and injuries. NIOSH also provides research, information, education, and training in the field of occupational safety and health. NIOSH is one of the leading agencies that conduct on-site investigations concerning safety and health evaluations. Through its investigative and research methods, NIOSH has become critical to the health and safety of firefighters.

Part of the NIOSH research process is recommendations for preventing future injuries and deaths. NIOSH works to help firefighters learn from fire department mistakes, although it does not attempt to place blame or find fault with fire departments or individual firefighters. NIOSH and OSHA often work together toward protecting worker safety and health. NIOSH offers the fire service investigative reports as well as firefighter safety resources. The Fire Fighter Fatality Investigation and Prevention Program's goal is to formulate recommendations for preventing future deaths and injuries, and to understand the characteristics of line-of-duty deaths. NIOSH also develops prevention strategies for fire service organizations related to injury and death. Its Web-based database serves as a valuable resource for all emergency services organizations to help identify trends and risk factors associated with firefighting and other emergency services.

Many emergency services organizations have utilized NIOSH's resources to develop injury prevention programs. Their goal is to prevent similar occurrences. NIOSH does not identify individual fire departments in its firefighter fatality investigation reports. Its research and programs are valuable not only to a department health and safety program manager but also to administrators and supervisors with the ability to implement necessary health and safety changes. Since the inception of the Fire Fighter Fatality Investigation and Prevention Program, many departments have implemented actions resulting in injury and death reduction.

FIRE DEPARTMENT SAFETY OFFICERS ASSOCIATION

The Fire Department Safety Officers Association (FDSOA) was established in 1989 as a nonprofit association in direct response to a demand for more education and networking among safety officers. The FDSOA represents over 12,000 safety officers, fire officers, and firefighters internationally. The FDSOA shares the concerns and represents the points of view of fire department safety officers. A volunteer board of directors leads the association while a small staff handles day-to-day operations.

The FDSOA is dedicated to the issues that affect the safety officer's critical role in protecting and promoting the safety and health responsibilities found in every fire service organization. The FDSOA certification for safety officers recognizes that individuals have demonstrated proficiency in and an ability to do the job in accordance with nationally recognized peer-developed standards. Individuals may seek certification in more than one discipline and as either an incident safety officer or a health and safety officer. The incident safety officer certification and testing process focuses heavily on specific areas such as building construction and smoke and fire conditions. The National Board on Fire Service Professional Qualifications (Pro Board) maintains a permanent registry containing the names and levels of certification of all nationally certified individuals. The purpose of the Pro Board was to develop peer-driven professional standards for the fire service. The NFPA has agreed to publish these documents and charged its technical committees with the task of creating clear and concise performance standards to determine whether any person measured truly possesses the required skill.

The Fire Department Safety Officers Association is an accredited certifying agency of the National Board on Fire Service Professional Qualifications. Individuals who wish to become certified must contact the FDSOA to obtain the available study materials. Then they study at their own pace and have the choice of testing individually or at national fire service conferences where testing is offered five times a year. The FDSOA offers a monthly publication titled *Health & Safety* and has become the networking resource of the safety officers throughout the nation. The FDSOA also presents programs related to firefighter and apparatus safety that can be downloaded from its Web site.

NATIONAL FIRE FIGHTER NEAR-MISS AND FIREFIGHTER CLOSE-CALLS PROGRAMS

Recently, the National Fire Fighter Near-Miss Reporting System has been established by and is funded by grants from the Assistance to Firefighters Grant Program of the U.S. Department of Homeland Security (DHS). Fireman's Fund Insurance Company has also awarded grant money to the program. The near-miss program is affiliated with the IAFF, IAFC, and FEMA. Fire service personnel submit reports of a near miss, and identification of the individual departments is removed to protect identity. The individual near-miss report is then posted on the National Fire Fighter Near-Miss Reporting System Web site as a resource and tool to be used by any fire service professional. The goals of the program are to promote safety, training, and awareness in the fire service. Collecting and analyzing information on near-miss events promotes improvements in incident command, education, operations, and training.

The goals allow firefighters to learn from other firefighters' real-life experiences; to help formulate strategies to reduce the frequency of firefighter injuries and fatalities; and to enhance the safety culture of the fire and emergency services. These reports can be used as educational tools for fire service professionals. Any member of the fire service community can submit a report on any incident he or she deems a near-miss occurrence. The Near-Miss Reporting System does not take the place of any other agency required by a department, local, state, or federal government; but its key is the lessons learned from others' experiences.

Chief Billy Goldfeder, a deputy fire chief of the Loveland-Symmes Fire Department in southwestern Ohio, developed the firefighter close-calls program, commonly referred to as the secret list. The secret list was derived from an independent newsletter produced since 1998 to highlight the issues involving injury and death to firefighters. In many cases, these issues had been ignored, forgotten, or not openly discussed. The close-calls program and secret list started as an e-mail group; the secret list is now received by thousands of firefighters across the nation through e-mail on a daily basis with reports of firefighters who have been killed or injured in the line of duty. Its philosophy states that, to survive the dangers of the job, it is vital to learn how other firefighters have had close calls or even been injured or killed.

#9

The purpose of the secret list is to provide factual information that fire service personnel can access and use. As a grassroots attempt at creating a heightened awareness while recognizing that firefighting is a dirty and dangerous job, the secret list acknowledges that there must be a strong focus on safety and health in the fire service. It also provides a section on weekly fire training and standard operating procedures/guidelines (SOPs/SOGs). The near-miss and close-calls programs exemplify the proactive approach of dedicated fire service professionals in an attempt to reduce injury and death.

National Firefighter Injury and Death Statistics

Several agencies compile a great deal of information and statistics annually on firefighter injury and death. The NFPA publishes annual injury and death reports during the following year studied. For example, the report for 2009 would be published late in 2010. Each report displays the injury and death statistics from various perspectives. The reports from various agencies are useful in analyzing and comparing problems at the local, state, and national levels.

One of the most useful NFPA reports lists injuries by type of duty. The injuries to firefighters are classified in five categories (see Table 1–1):

- Firefighting/fireground
- Responding/returning
- On scene at non-fire calls
- Training
- Other on duty

As in many other years, fireground injuries account for the highest number of injuries.

It should come as no surprise that a higher number of fires per department are directly related to the population it serves. The number of injuries is directly related to the number of fire responses and exposures to fires. The rate of injuries per 1,000 fires has remained constant over the years.[3] Review of annual reports available to emergency services organizations should become and remain a valuable tool for every health and safety program manager.

TABLE 1–1 Firefighter Injuries by Type of Duty

	TOTAL	FIREFIGHTING, FIREGROUND	RESPONDING, RETURNING	ON SCENE AT NON-FIRE CALLS	TRAINING	OTHER ON DUTY
1981	103,340	67,510 (65.3%)	4,945 (4.8%)	9,600 (9.3%)	7,090 (6.9%)	14,195 (13.7%)
1982	98,150	61,370 (62.5%)	5,320 (5.4%)	9,385 (9.6%)	6,125 (6.2%)	15,950 (16.3%)
1983	103,150	61,740 (59.9%)	5,865 (5.7%)	11,105 (10.8%)	6,755 (6.5%)	17,685 (17.1%)
1984	102,300	62,700 (61.3%)	5,845 (5.7%)	10,630 (10.4%)	6,840 (6.7%)	16,285 (15.9%)
1985	100,900	61,255 (60.7%)	5,280 (5.2%)	12,500 (12.4%)	6,050 (6.0%)	15,815 (15.7%)
1986	96,450	55,990 (58.1%)	4,665 (4.8%)	12,545 (13.0%)	6,395 (6.6%)	16,855 (17.5%)
1987	102,600	57,755 (56.3%)	5,075 (4.9%)	13,940 (13.6%)	6,075 (5.9%)	19,755 (19.3%)
1988	102,900	61,790 (60.0%)	5,080 (4.9%)	12,325 (12.0%)	5,840 (5.7%)	17,865 (17.4%)
1989	100,700	58,250 (57.8%)	6,000 (6.0%)	12,580 (12.5%)	6,010 (6.0%)	17,860 (17.7%)
1990	100,300	57,100 (56.9%)	6,115 (6.1%)	14,200 (14.2%)	6,630 (6.6%)	16,255 (16.2%)
1991	103,300	55,830 (54.0%)	5,355 (5.2%)	15,065 (14.6%)	6,600 (6.4%)	20,450 (19.8%)
1992	97,700	52,290 (53.5%)	5,580 (5.7%)	14,645 (15.0%)	7,045 (7.2%)	18,140 (18.6%)
1993	101,500	52,885 (52.1%)	5,595 (5.5%)	16,675 (16.4%)	6,545 (6.5%)	19,800 (19.5%)
1994	95,400	52,875 (55.4%)	5,930 (6.2%)	11,810 (12.4%)	6,780 (7.1%)	18,005 (18.9%)
1995	94,500	50,640 (53.6%)	5,230 (5.5%)	13,500 (14.3%)	7,275 (7.7%)	17,855 (18.9%)
1996	87,150	45,725 (52.5%)	6,315 (7.2%)	12,630 (14.5%)	6,200 (7.1%)	16,280 (18.7%)
1997	85,400	40,920 (47.9%)	5,410 (6.3%)	14,880 (17.4%)	6,510 (7.6%)	17,680 (20.7%)
1998	87,500	43,080 (49.2%)	7,070 (8.1%)	13,960 (16.0%)	7,055 (8.1%)	16,335 (18.7%)
1999	88,500	45,550 (51.5%)	5,890 (6.7%)	13,565 (15.5%)	7,705 (8.7%)	15,790 (17.8%)
2000	84,550	43,065 (51.0%)	4,700 (5.6%)	13,660 (16.2%)	7,400 (8.8%)	15,725 (18.6%)
2001	82,250	41,395 (50.3%)	4,640 (5.6%)	14,140 (17.2%)	6,915 (8.4%)	15,160 (18.4%)
2002	80,800	37,860 (46.9%)	5,805 (7.2%)	15,095 (18.7%)	7,600 (9.4%)	14,440 (17.9%)
2003	78,750	38,045 (48.3%)	5,200 (6.6%)	13,855 (17.6%)	7,100 (9.0%)	14,550 (18.5%)
2004	75,840	36,880 (48.6%)	4,840 (6.4%)	13,150 (17.3%)	6,720 (8.9%)	14,250 (18.8%)
2005	80,100	41,950 (52.4%)	5,455 (6.8%)	12,250 (15.3%)	7,120 (8.9%)	13,325 (16.6%)
2006	83,400	44,210 (53.0%)	4,745 (5.7%)	13,090 (15.7%)	7,665 (9.2%)	13,690 (16.4%)
2007	80,100	38,340 (47.9%)	4,925 (6.1%)	15,435 (19.3%)	7,735 (9.7%)	13,665 (17.1%)
2008	79,700	36,595 (46.0%)	4,965 (6.2%)	15,745 (19.8%)	8,145 (10.2%)	14,250 (17.9%)
2009	78,150	32,205 (41.2%)	4,965 (6.4%)	15,455 (19.8%)	7,935 (10.2%)	17,590 (22.5%)

Source: NFPA's "U.S. Firefighter Injuries—2009," Michael J. Karter, Jr., and Joseph L. Molis, October 2010, and previous reports in the series.

Much Has Been Accomplished and Much Is Still Needed

Much change has taken place since the implementation of the NFPA 1500 standard. Safety and health seem to be in the forefront of thinking and in the approach taken in many emergency services organizations. As in the private sector, risk management has taken a foothold in emergency services.

Often organizations now use an individual appointed or assigned as the department health and safety program manager to conduct study and research, and to become proactive in the area of worker health and safety. Many fire service organizations are striving to meet the requirements of NFPA 1500. Unfortunately, the changes in the area of health and safety in emergency services do not reflect a drastic reduction in injuries and deaths. The numbers of injuries, deaths, and illnesses remain constant.

The trend in cardiac-related events continues, as do injuries and death related to emergency response and returning from incidents. Injuries and death commonly occur while responders respond to incidents in privately owned vehicles. Current efforts are a necessary step in the right direction; however, efforts need to continue and possibly be increased to reduce injuries and deaths in the emergency services and promote education of members.

Safety and health in emergency services must become and remain a top priority. Research needs to continue, and new prevention measures must be persistently sought. Theories and philosophies are only part of the equation related to health and safety. Continual training and hands-on experience are necessary to prove whether a theory will work.

Many new programs have been and continue to be developed; however, emergency services organizations must not lose sight of the basics. Technology has led to great advancements in the field, but a large measure of common sense and mechanical aptitude are necessary for personnel to work safely in the field. Emergency services personnel require continual training, as the profession is dirty, dangerous, and aggressive. It may be time for emergency services organizations to shift their everyday focus to truly important issues and matters in order to make strides in the area of health and safety. A large dose of risk management is also needed early in basic firefighter training classes.

Summary

Although firefighting is considered to be one of the most dangerous occupations, emergency services organizations have only recently begun to deal with occupational health and safety.

Certain characteristics are common to all fire departments, whether career, paid-on-call, combination, or volunteer. Most firefighters have a strong set of values and an above-average work ethic and professional pride. Great strides continue to be made in personal protective equipment for firefighters and the technology concerning equipment. In today's society, laws, regulations, and standards have forced safer working conditions for emergency services responders.

Even though emergency services organizations have made great improvements in the health and safety of their members, it is unlikely that injury and death will be eliminated. Application of risk management procedures in emergency services organizations will help to reduce the number of injuries and deaths. Every emergency services organization will need a health and safety program manager in place to identify and understand the issues concerning health and safety. An analysis of local, state, and national data helps identify the problems and issues as well as implement prevention measures. Firefighters and supervisors must continually address compliancy issues for members to remain safer and to ensure better health for emergency response personnel.

Many organizations provide annual data related to the safety and health of emergency response personnel. The IAFF puts out some of the most in-depth information related to exposure to infectious disease. The NFPA publishes an annual report of firefighter injuries and death. The USFA compiles data related to injury and death through the National Fire Incident Reporting System. OSHA collects, analyzes, and publishes data related to occupational injuries. NIOSH investigates and publishes data concerning firefighter fatalities. The Fire Department Safety Officers Association (FDSOA) as well as the National Firefighter Near-Miss and Close-Calls Programs publish resources and tools to highlight issues involving near misses, close calls, injury, and death to firefighters.

Many issues have arisen and changes have been made regarding health and safety in emergency services in the last 20 years. For example, a proactive effort of the IAFF and the IAFC, called the Wellness-Fitness Initiative, between labor and management addresses firefighter wellness. This program is a valuable resource for health and safety program managers in all emergency services organizations. It is also now recognized that a higher level of education is needed in emergency services for all involved. It should come as no surprise that the number of firefighter injuries is directly related to the number of fire responses and exposures to fires. As in the private sector, risk management has taken a foothold in emergency services.

In spite of the changes that have been made, no significant reduction of death and injury in emergency services has occurred. Progress in this important area requires research and the development of new procedures. Technology will continue to advance as the emergency services continues to make strides in health and safety and serve the public it protects. Emergency services personnel still need better training methods because the number of fires is down and the number of EMS-related incidents continues to rise. Better programs, research methods, and data—along with organizational commitment for significant improvement—are necessary to improve the health and safety of members.

Review Questions

1. Describe the many areas contained within the field of study of fire science.
2. What is one of the most controversial standards published by the NFPA?
   a. NFPA 1410
   b. NFPA 1720
   c. NFPA 1500
   d. NFPA 1962

3. Describe why organizations that move forward with change without consideration of the human element are at a disadvantage.
4. The common mind-set in emergency services has become one that:
   a. Accepts injury and death are here to stay
   b. Accepts a safe working environment
   c. Believes change is unnecessary
   d. Believes SOPs and policies are made to be broken
5. Which of the following organizations has published an annual injury and death report since 1960?
   a. The NFPA
   b. The IAFC
   c. The IAFF
   d. NIOSH
6. Occupational health and safety progressed rapidly in the fire service during which decade?
   a. 1960s
   b. 1970s
   c. 1980s
   d. 1990s
7. Which is the major initiative worked on by the IAFF and the IAFC?
   a. NFPA 1710
   b. NIOSH fatality reports
   c. HAZWOPER
   d. The Wellness-Fitness Initiative
8. Which of the following organizations publishes annual data related to the health and safety of emergency services responders?
   a. The IAFF
   b. OSHA
   c. The NFPA
   d. All of the above
9. The National Fire Incident Reporting System (NFIRS) is overseen by:
   a. The USFA
   b. The NFPA
   c. OSHA
   d. NIOSH
10. Which organization publishes annual injury and death reports during the following year studied?
    a. NIOSH
    b. The NFPA
    c. The IAFF
    d. The IAFC

Case Study

On July 30, 2002, a 32-year-old male career lieutenant (Victim 1) and a 20-year-old male career firefighter (Victim 2) died while participating in a live-fire training evolution. A flashover occurred several minutes after the fire had been lit in the acquired vacant structure while both of the victims were performing a simulated search and rescue. The lieutenant and the firefighter were both transported by ambulance to a local hospital where they were pronounced dead.

Before the start of the training, the incident commander (IC) and the participants walked through the structure so that the IC could give them a pre-burn briefing. The IC pointed out the ingress and egress routes, and he told them that a mannequin dressed in firefighter bunker gear would serve as a simulated rescue victim in the training exercise. He did not tell the participants that the mannequin would be located in the kitchen area. The IC told the participants that the live fire would be built inside a closet on the northwest corner of the burn room. The participants helped put the fuel—wooden pallets and straw—inside and outside the closet.

At approximately 1010 hours, the ignition officer/interior safety used a road flare to ignite the items in the closet and radioed the IC that the fire had been lit. When the ignition officer/interior safety left the burn room, the live fire was producing some flames, and the smoke had diminished visibility in the room. To produce a larger fire, some of the firefighters retrieved a twin-size mattress from another bedroom and put it on the live fire in the burn room.

The ignition officer/interior safety and one of the participants who was acting as interior safety assumed their positions in the hallway outside the burn room while the other two interior safety firefighters staged in the living room. At approximately 1011 hours, the ignition officer/interior safety radioed the IC that they were ready to begin the first training evolution. The IC ordered the search and rescue team (Victim 1 and Victim 2) to enter the structure.

Victim 1 and Victim 2 crawled through the front door (A-side) and performed a right-hand search to look for the simulated victim. A very brief time later, after receiving orders from the IC, the crew on Attack Line 1 entered the structure through the front door (A-side) with a charged 1¾-inch hose line.

While Victim 1 and Victim 2 were conducting a search of the living room where two of the interior safety firefighters were positioned, Victim 1 was overheard giving instructions on searching techniques to Victim 2. After both of the victims performed their search in the living room, they crawled down the hallway to the burn room, followed by one of the interior safety firefighters from the living room. *Note: Conditions in the structure at this time were heavy smoke with very little visibility.*

As both victims were conducting their search, one of the victims collided with one of the interior safety firefighters in the hallway outside the burn room. The interior safety firefighter in the hallway identified himself to the victims as one of the interior safety personnel and instructed them to continue their search. The interior safety firefighter that had followed both of the victims from the living room into the hallway told one of the other interior safety firefighters in the hallway outside the burn room that he was going to look for the crew with the first attack line. Victim 1 was overheard in the burn room asking Victim 2 if the entire room had been searched and receiving an affirmative response.

As the interior safety firefighter went back down the hallway to look for the first attack line crew, he encountered them entering the hallway, and he told them to put some water on the fire. He then headed back toward the burn room followed by the crew from Attack Line 1. Once he reached the section of hallway outside the burn room, he asked one of the interior safety firefighters in that area for the location of both victims. Receiving a reply that they were out, he then asked a second time if both victims were out of the burn room and received an affirmative response. He left to search for both the victims in the bedrooms on the B-side and the kitchen on the C-side.

At approximately 1013 hours, the IC radioed Attack Line 1 that the window in the burn room was going to be vented, and the exterior ventilation person broke out the window. When the window (56 inches in height by 42½ inches in width and made of ¼-inch-thick plate glass) was vented, it emitted very heavy black smoke followed a few seconds later by intense flames. According to the Office of the State Fire Marshal, and the fire analysis performed by NIST, a flashover is believed to have occurred in the burn room after the window was broken. See Figure 1–7.

NIOSH had the following recommendations:

Recommendation #1: Fire departments should ensure that the fuels used in live-fire training have known burning characteristics. Fuels for training fires should have known burning characteristics, and the quantities used should be the minimum necessary that are controllable and able to create the desired fire conditions.

Recommendation #2: Fire departments should ensure that ventilation is closely coordinated with

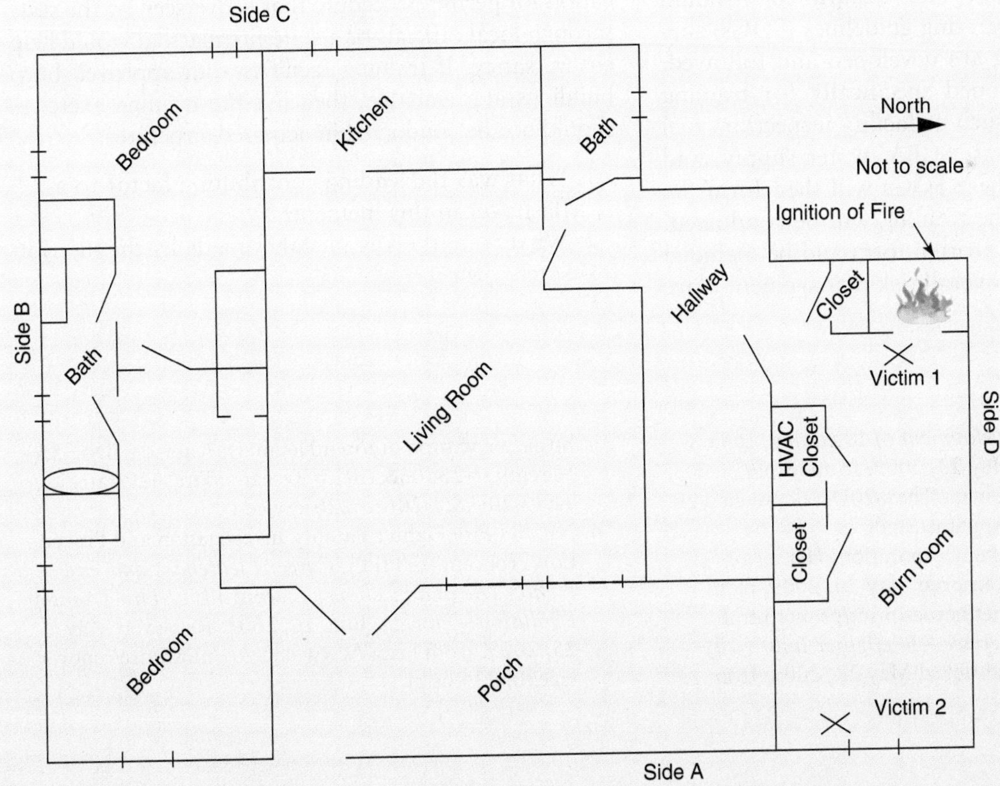

FIGURE 1–7 Overhead view of the floor plan. *Courtesy of National Institute for Occupational Safety and Health (NIOSH) Division of Safety Research.*

interior operations. Chapter 10 of *Essentials of Fire Fighting*, 4th edition, states that "ventilation must be closely coordinated with fire attack." Fire can quickly spread in a structure, causing problems such as flashover, a backdraft, or an explosion. Ventilation timing is extremely important and must be carefully coordinated between interior operating crews and ventilation crews.

Recommendation #3: Fire departments should ensure that fires are not located in designated exit paths. During a training exercise, every effort must be made to ensure the exit paths are free from obstructions. To provide a protected area of travel, fires should not be located in the vicinity of exit paths. Once the closet area in the burn room was ignited, the fire continued to increase in size, which produced fire, heat, and smoke in the exit path of the only doorway in the room.

Recommendation #4: Fire departments should ensure that a method of fireground communication is established to enable coordination between the incident commander and firefighters. It is imperative that communication shall be established between the incident commander and firefighters performing any interior operations, sector leaders, and the safety officer. Proper communication is a must at any incident site. Portable radios should be used to keep all personnel on the scene in communication with the IC. The use of a portable radio that is located in a radio coat or pants pocket impairs the performance of the unit.

Recommendation #5: Fire departments should ensure that standard operating guidelines (SOGs) specific to live-fire training are developed and followed. SOGs should be developed specifically for training fires and include areas such as facility inspection, fuel materials, RIT operations, SCBA, water supply, and hose-line operations. These SOGs will then form the foundation as to how the training will be conducted. The SOGs should be in written form and be included in the fire department's overall risk management plan. If these procedures are changed, appropriate training should be provided to all affected members.

Recommendation #6: Fire departments should consider using a thermal imaging camera during live-fire training situations. Thermal imaging cameras may assist firefighters by allowing them to see through blinding smoke and in zero visibility conditions. With the help of a thermal imaging camera, training instructors, interior safety officers, and firefighters may observe and critique participants, ensuring that they develop good foundational skills in areas including accountability, conducting effective search patterns, and handling a hose.

Recommendation #7: Develop a permitting procedure for live-fire training to be conducted at acquired structures. States should ensure that all the requirements of NFPA 1403 have been met before issuing the permit. NFPA 1403, *Standard on Live Fire Training Evolutions*, is the guideline for conducting live-fire training evolutions at approved training centers and, in this case, acquired structures. Approved training centers have burn buildings that are specifically designed for repeated live-fire training evolutions. The structures that are acquired for live-fire training are usually in disrepair and were never designed for live-fire training. Any building that is acquired for live-fire training must go through an inspection process to identify and eliminate any hazards or potential hazards that may be present to the participants, the public, and the environment. An application for permit procedure that is overseen by the state through local officials or a state fire marshal would help ensure safety. If training facilities with approved burn buildings are available, then live-fire training exercises should not be conducted in acquired structures.[4]

1. What was the leading contributing factor to a flashover in this incident?
2. What were the key mistakes made by the interior safety officer?

References

1. P. Hashagen,. *The Development of Breathing Apparatus*, 1997. Retrieved May 23, 2009, from http://lacountyfirefighters.org/items/The%20Development%20of%20Breathing%20Apparatus.pdf
2. Educational Broadcasting Corporation, *Heroes of Ground Zero*, 2002. Retrieved May 24, 2009, from http://www.pbs.org/wnet/heroes/print/primer.html
3. NFPA. *The U.S. Fire Service—Firefighter Injuries by Type of Duty*, 2009. Retrieved May 22, 2009, from http://www.nfpa.org/itemDetail.asp?categoryID=955&itemID=23466&URL=Research/Fire%20statistics/The%20U.S.%20fire%20service
4. NIOSH Fire Fighter Fatality Investigation and Prevention Program (FFFIPP), *Career Lieutenant and Fire Fighter Die in a Flashover During a Live-Fire Training Evolution—Florida*, June 16, 2003. Retrieved May 25, 2009, from http://www.cdc.gov/niosh/fire/reports/face200234.html

CHAPTER 2
Safety-Related Regulations and Standards Applied to Emergency Services

KEY TERMS

candidate physical ability test (CPAT), *p. 39*
Code of Federal Regulations (CFR), *p. 22*
consensus standards, *p. 22*
Environmental Protection Agency, *p. 23*
fire brigades, *p. 34*
Hazardous Waste Operations and Emergency Response Standard (HAZWOPER), *p. 27*
health and safety officer, *p. 35*
immediately dangerous to life or health (IDLH), *p. 29*
incident management system (IMS), *p. 36*
incident safety officer (ISO), *p. 35*
interior structural firefighting, *p. 29*
regulations, *p. 22*
self-contained breathing apparatus (SCBA), *p. 28*
standard of care, *p. 25*
standard operating guidelines (SOGs), *p. 29*
standard operating procedures (SOPs), *p. 29*
standards, *p. 22*

OBJECTIVES

After reading this chapter, you should be able to:

- Discuss the differences between regulations and standards.
- Discuss the concept of standard of care.
- List and discuss both the federal regulations and the National Fire Protection Association standards that have an impact on emergency responder occupational safety and health programs.
- Discuss how regulations and standards play a role in the safety and health of emergency responders.

Resource**Central**

For additional review and practice tests, visit **www.bradybooks.com** and click on Resource Central to access book-specific resources for this text! To access Resource Central, follow directions on the Student Access Card provided with this text. If there is no card, go to **www.bradybooks.com** and follow the Resource Central link to Buy Access from there.

Safety-related regulations and standards play a significant role in emergency response organizations. A thorough understanding and knowledge of regulations and standards is essential. Just because certain organizations take the stance that "we have not adopted that standard" or some states have not adopted a state plan does not mean others should view these as valid reasons to disregard the existence of regulations and standards.

regulations ■ These are laws or specific regulatory requirements developed by an agency of the federal, state, or local government.

standards ■ Written as a cooperative agreement by committees utilizing the consensus process, standards are not law or mandatory unless adopted by the governmental agency.

consensus standards ■ A consensus of subject area experts develop standards to be published that may or may not be adopted. These standards can be used as a gauge addressing the standard of care even if they are not adopted as law.

Code of Federal Regulations (CFR) ■ The codification of the general and permanent rules and regulations (sometimes called administrative law) is published in the *Federal Register* by the executive departments and agencies of the U.S. federal government.

The terms **regulations** and **standards** are not interchangeable; they have different meanings. Regulations are developed and issued by federal, state, and local governments to enforce a principle, rule, or laws designed to control or govern conduct. Regulations have the force of law. Standards do not carry the weight of the law unless they are adopted and enforced by the authority having jurisdiction. Many times, standards are referred to as **consensus standards**. Standards are different from laws in a number of aspects; whereas laws are written by elected officials to codify open or hidden agendas, standards are written as a cooperative effort by committees comprised of volunteers to set minimum requirements.

Regulations and standards can provide a road map for an agency's safety and health program, and an understanding of them is invaluable to any agency. The focus of this chapter is the regulations and standards that apply to emergency services safety and health, which will have the greatest impact on health and safety.

Regulations

Emergency agencies must understand that regulations are law. Without this understanding, an emergency services organization may be in direct violation of mandated laws and open itself to possible litigation and fines. Regulations are mandatory legislative acts based on federal, state, and sometimes local government law. A compilation of federal regulations is contained within the 50 titles of the **Code of Federal Regulations (CFR)**. Title 29 CFR, Part 1910, Occupational Safety and Health, contains the OSHA regulations that have substantial impact on emergency services health and safety programs.[1] These regulations, discussed further in this chapter, have the same authority and muscle as law.

Twenty-five states have adopted state plans that enforce OSHA safety regulations. The Obama administration is proposing that Congress amend the OSHA Act to make it applicable to all state and local governments. When a state adopts a federal regulation, state legislators or officials dictate the mandatory requirements and enforcement. For example, many states have adopted CFR 1910.120, Hazardous Waste Operations and Emergency Response, which covers the required cleanup operations. CFR section 1910.120 must be followed and enforced by federal, state, or local governments, including all emergency response agencies involved with hazardous substances, as well as ongoing operations conducted at uncontrolled hazardous waste sites. This regulation also requires those personnel operating at a hazardous materials incident to be trained and those in supervisory positions to have additional training specific to incident command.

Another commonly adopted regulation as applied to emergency services organizations is CFR 1910.1030, Bloodborne Pathogens, which applies to all occupational exposures to blood and other potentially infectious materials. It considers all bodily type fluids as potentially infectious materials. Certain states have adopted different standards applicable to this topic or may have different enforcement policies.

Various states have also adopted regulations or portions of OSHA regulations as they apply to firefighters and fire department employers. When there is no agreement between a state and the federal government, OSHA regulations do not apply to a state or its fire departments; however, there are various ways to bring fire departments into compliance with OSHA 1910.120 and OSHA 1910.1030. For example, some states have adopted state occupational safety and health regulations for state and local employees. Others are brought into compliance with the OSHA 1910.120 regulation through the Code of Federal Regulations 40 CFR 311 (known as EPA 311) of the **Environmental Protection Agency.**[2] Managing hazardous materials training at the state and local levels can be a complex task. The challenge is to conduct training with limited resources that meets the public sector response training requirements of OSHA 1910.120 and Environmental Protection Agency 40 CFR 311 (EPA 311).

Environmental Protection Agency ▪ This independent federal agency was established to coordinate programs aimed at reducing pollution and protecting the environment.

Occupational Safety and Health Administration Regulations

The Occupational Safety and Health Act of 1970 established OSHA as a branch of the U.S. Department of Labor, as mentioned in Chapter 1. Federal OSHA has no authority or direct enforcement at the state or local government level, including publicly owned fire and EMS departments; however, OSHA may very well apply to private fire and EMS companies. Any state can implement its own enforcement program or policies provided that federal OSHA approves that state's safety and health plan.

Section 18 of the Occupational Safety and Health Act lists the following requirements for an individual state to have an approved plan:

Designated state agency or agencies to administer the plan
Satisfactory assurances that such agency or agencies have/will have the legal authority and qualified personnel necessary to enforce such standards
Satisfactory assurances of finances available to administer and enforce such standards
Satisfactory assurances of an effective, comprehensive occupational safety and health program applicable to all employees of public agencies of the state and its political subdivisions
Employers in the state are to report to the secretary in the same manner and to the same extent as if the plan were not in effect

A number of OSHA standards and regulations affect emergency services safety and health: those covering hazardous waste operations, respiratory protection, protective clothing for emergency medical operations, and confined space operations. In addition, emergency services organizations must comply with certain regulations that fall outside the common standards. For example, activities other than those directly related to emergency services may be subject to the same OSHA regulations as any other private employer activity in their state. Emergency services organizations that have welders or oxyacetylene torches must provide OSHA training to their employees or members in that specific area. Volunteer organizations that hold fund-raisers such as pancake breakfast functions or bingo may also be subject to OSHA training in those specific areas. The sections later in the chapter describe and review the content of those standards as applicable to health and safety. See Figure 2-1.

#5

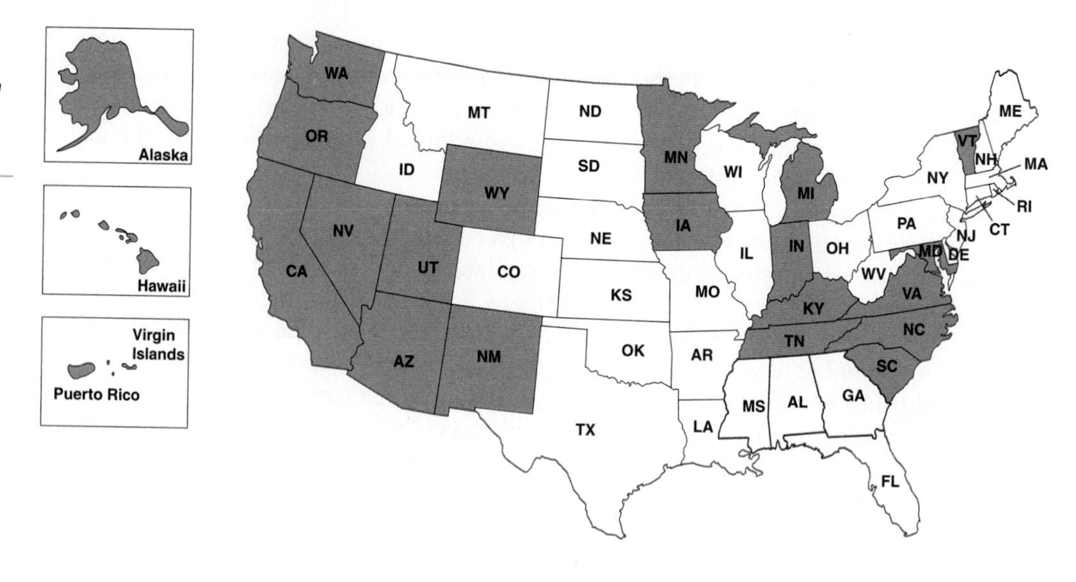

FIGURE 2-1 States and territories with OSHA plans. *Source: OSHA, from http://www.osha.gov/dcsp/osp/oshspa/oshspa_2007_report.html*

Standards

Documents that are published as standards do not have to be adopted as law, nor is there any requirement for compliance, unless state or local regulations of the authority having jurisdiction have adopted a standard. Individuals who each have a specific area of expertise, known as subject matter experts, agree on how to perform particular tasks and then develop standards, referred to as consensus standards. For example, the National Fire Protection Association (NFPA) is committed as a standards-making group that directly affects the safety and health of firefighters. The NFPA not only publishes standards, and recommended practices and codes, but also provides research, education, and training to agencies.

#11

NFPA's three hundred codes and standards influence every building, process, service, design, and installation in the United States as well as many of those in other countries. More than six thousand volunteers from diverse professional backgrounds serve on several hundred technical code and standard-development committees.

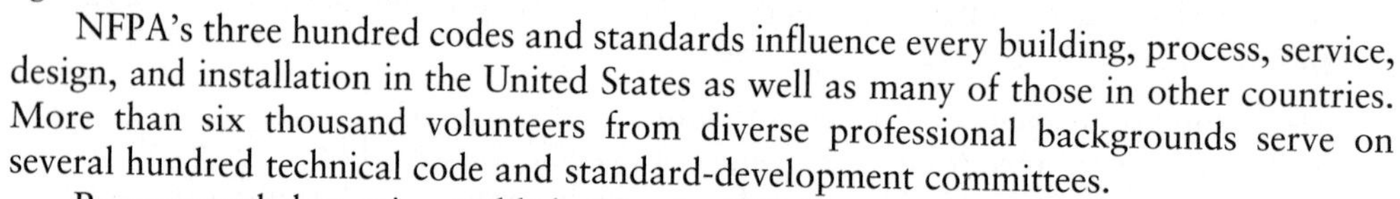

Recommended practices published by the NFPA are similar in content and format to other types of codes and standards; however, they are not mandatory when it comes to compliance. Codes that reference NFPA standards have the force of law and may be independently adopted by the authority having jurisdiction. For example, Ohio safety regulations require that all firefighter escape ropes comply with the NFPA 1983 standard: each escape rope shall have a product label that is permanently and conspicuously attached. At least the following information should be legibly printed on the label: "this rope meets the escape rope requirements of NFPA 1983, *Standard on Life Safety Rope and Equipment for Emergency Services*" (Ohio Administrative Code 4123:1–21).

The authority having jurisdiction can adopt any NFPA standard or code and then has the authority, legal privileges, and rights to enforce the adopted code or standard. The most common example would be the adoption of NFPA 101, *Life Safety Code*®. It has become common for many governmental agencies to adopt this code as applied to fire prevention and code enforcement even though NFPA 101 is not a legal code and has no statutory power of law unless it has been adopted by the authority having jurisdiction. The purpose of NFPA 101, *Life Safety Code*, is to minimize the hazards and dangers to life from the threat of fire; it applies to existing structures as well as new structures.

#15

The NFPA performs an advisory role to emergency response agencies. Although it has no enforcement power or authority, the NFPA has developed a vast amount of standards that impact the safety and health of firefighters. For example, NFPA 1403, *Standard on*

Live Fire Training Evolutions, was developed in response to the number of firefighter injuries and deaths that have occurred over the course of time. Due to those injuries and deaths, certain states have developed and implemented a testing process and certification for fire instructors who conduct live-fire training. In Florida, this certification has been adopted as law. There will be further discussion of the NFPA standards later in this chapter.

It is important to remember that NFPA standards are consensus standards developed by subject matter experts who come together to agree on the minimum level of performance. These standards usually become the standard of care in a particular area.

The Standard of Care

In emergency services, the **standard of care** is better known because it commonly applies to the emergency medical field more than to firefighting operations. Workers trained in the emergency medical field must perform to an acceptable level of other EMS workers with the same level of certification. In legal terms, this is the level at which the average prudent provider in a given community would practice. It is how similarly qualified practitioners would have managed the patient's care under the same or similar circumstances. In conceptual terms, the standard of care is simple—practitioners are expected to perform at a certain level. It is a diagnostic and treatment process that a clinician should follow for a certain type of patient, illness, or clinical circumstance. Court decisions have further defined the standard of care:

standard of care ■ This is the degree of prudence and caution required of an individual with similar training and equipment who is under a duty of care.

> Willful misconduct is intentionally doing that which is wrong or intentionally failing to do that which should be done. The circumstances must also disclose that the defendant knew or should have known that such conduct would probably cause injury to the plaintiff. It is a general rule that every person may be presumed to intend the natural and probable consequences of his acts. Willful misconduct implies an intentional disregard of a clear duty or of a definite rule of conduct, a purpose not to discharge such duty, or the performance of wrongful acts with knowledge of the likelihood of resulting injury. Knowledge of surrounding circumstances and existing conditions is essential; actual ill will or intent to injure need not be present.
>
> Wanton misconduct must be under such surrounding circumstances and existing conditions that the party doing the act or failing to act must be aware, from his knowledge of such circumstances and conditions, that his conduct will probably result in injury. Wanton misconduct implies a failure to use any care for the plaintiff and an indifference to the consequences, when the probability that harm would result from such failure is great, and such probability is known, or ought to have been known, to the defendant. (*Weber v. City Council, Huber Heights, Ohio*, 2001 OH 630 [OHCA, 2001]).

In 2000, a lawsuit was filed in Montgomery County, Ohio, against the Huber Heights City Council and two of the city's emergency medical technicians. It was based on these technicians' alleged willful and wanton misconduct causing injury to Rudolf Weber. Mr. Weber claimed the EMTs did not assist him to the ambulance stretcher while he was experiencing a stroke. When the EMTs told him he would have to walk to their stretcher, Mr. Weber asserted that he could not walk. The EMTs spoke with Mr. Weber's wife, claiming that his vital signs were fine, he was experiencing a panic attack, and he was not having a stroke. The EMTs claimed that Mr. Weber refused transport to a hospital. They then left his residence. The next morning his daughter took him to his doctor, who determined that Mr. Weber had experienced a stroke. An ambulance was summoned to the doctor's office to transport Mr. Weber to the hospital. The court determined that the EMTs may have been negligent, but did not commit willful and wanton misconduct because Mr. Weber's vital signs were normal during their examination, and he did not wish to be transported to a hospital. The court ruled that willful and wanton misconduct must be behavior demonstrating a deliberate or reckless disregard for the safety of others.

The same concept can be applied to health and safety issues as related to established standards. Many NFPA documents define the standard of care in the area of health and safety with emergency services. For example, NFPA 1403 requires that buildings used for live-fire training be prepared in such a way that the training evolution takes place in the safest way possible. Should you find yourself in court because a firefighter died in a live-fire training evolution of a building that was not prepared properly, you could be asked why the building was not prepared in accordance with NFPA 1403. It would not be a viable defense to answer that the NFPA standard is not law in your jurisdiction. Emergency services organizations must follow the established reasonable industry standards in this area because the NFPA document defines the standard of care.

The concept of the standard of care is dynamic in nature. Emergency services organizations must continually keep abreast of changing technology, regulations, standards, and procedures.

In most cases, standards of care were developed with the invention of new products, such as the standard to wear a self-contained breathing apparatus (SCBA) during interior firefighting. Today, the published NFPA standards require this device; therefore, it is the established standard of care. By law, those firefighters who fall under OSHA enforcement via "state approved plans" must also wear an SCBA as required by OSHA. In certain industries and professions, the standard of care is determined by the standard that would be exercised by the reasonably prudent manufacturer of a product or the reasonably prudent professional in that line of work. For example, in 1998 both OSHA and the NFPA published documents dealing with the minimum number of firefighters required to launch an interior fire attack during interior structural firefighting. Except for extenuating circumstances, the standards require a minimum of four trained and equipped personnel on scene before an interior attack can take place; this has become the standard of care. Well-advised local officials must have current information concerning all relevant federal, state, and provincial requirements, as well as applicable NFPA codes and standards, OSHA standards, and recommended practices that are issued from time to time.

Section 5(a)(1) General Duty Requirement

OSHA enforces many standards and regulations related to health and safety. When Congress developed and implemented the Occupational Safety and Health Act in 1970, it included the general duty clause to address situations that could not be addressed and covered by the OSHA regulations. The general duty clause of the OSHA regulation is Section 5(a)(1); it is not a specific standard, but has been applied by OSHA as a catchall regulation to require employers to ensure that workplaces are free of recognized hazards and to require employees to comply with standards, rules, regulations, and orders issued by OSHA. This one requirement covers all hazardous conditions. The general duty clause is an important weapon for workers because employers have the specific duty to provide a safe workplace.

Under the general duty clause, the employer has an obligation to protect workers from serious and recognized workplace hazards even where there is no standard. Employers must take whatever abatement actions are feasible to eliminate these hazards. The following elements would be necessary to prove a violation of the general duty clause:

- There must be a hazard—the employer failed to keep the workplace free of a hazard.
- The hazard must be recognized.
- The hazard causes or is likely to cause serious harm or death.
- The hazard must be correctable.

A relationship exists between this clause and the safety and health programs and national consensus standards such as those of the NFPA. When an OSHA regulation of a specific type and nature is not in place to address a particular hazard, a national consensus standard (such as the NFPA standards) can be used as a reference. In certain situations, the NFPA standard or standards can be used as a guide when it comes to enforcement of the general duty clause. The safety and health program manager of the emergency services department must keep this in mind when assessing the need to comply with any consensus standard.

OSHA 1910.120 (29 CFR 1910.120) Hazardous Waste Operations

The **Hazardous Waste Operations and Emergency Response Standard (HAZWOPER)** applies to five distinct groups of employers and their employees. These include the following situations:

Hazardous Waste Operations and Emergency Response Standard (HAZWOPER) ■ This refers to five types of hazardous waste operations conducted under OSHA Standard 1910.120, Hazardous Waste Operations and Emergency Response. The standard contains the safety requirements employers must meet in order to conduct these operations.

Any employees who are exposed, or potentially exposed, to hazardous substances—including hazardous waste—and who are engaged in cleanup operations required by a governmental body, whether federal, state, local, or others involving hazardous substances that are conducted at uncontrolled hazardous waste sites

Corrective actions involving cleanup operations at sites covered by the Resource Conservation and Recovery Act of 1976

Voluntary cleanup operations at sites recognized by federal, state, local, or other governmental bodies as uncontrolled hazardous waste sites

Operations involving hazardous waste that are conducted at treatment, storage, and disposal facilities

Emergency response operations for releases of, or substantial threats of releases of, hazardous substances without regard to the location of the hazard.

This OSHA standard has requirements related to the health and safety of employees working with hazardous waste operations.

The contents of the plan must be made available to all employees, and it is required to contain the following at a minimum:

- A way to identify, evaluate, and control safety and health hazards and provide for emergency response for hazardous waste operations
- An organizational structure
- A comprehensive work plan
- A site-specific safety and health plan, which need not repeat the employer's standard operating procedures
- A safety and health training program
- A medical surveillance program
- The employer's standard operating procedures for safety and health
- Any necessary interface between general program and site-specific activities.[3]

The regulation addresses the necessary training requirements, protective equipment, and personnel required to operate at an incident. The training requirements are broken down into specific areas, based on what the employees are expected to do at a hazardous materials incident. Employees are not permitted to supervise or participate in hazardous materials activities until they have received a level of training required of their job function and responsibility. The training must include the elements of the department's emergency response plan, standard operating procedures, the personal protective equipment to be worn, and procedures for handling hazardous material emergencies.

#5

Departments that respond to hazardous materials incidents should be familiar with this regulation as compliance with it must be a part of the department health and safety

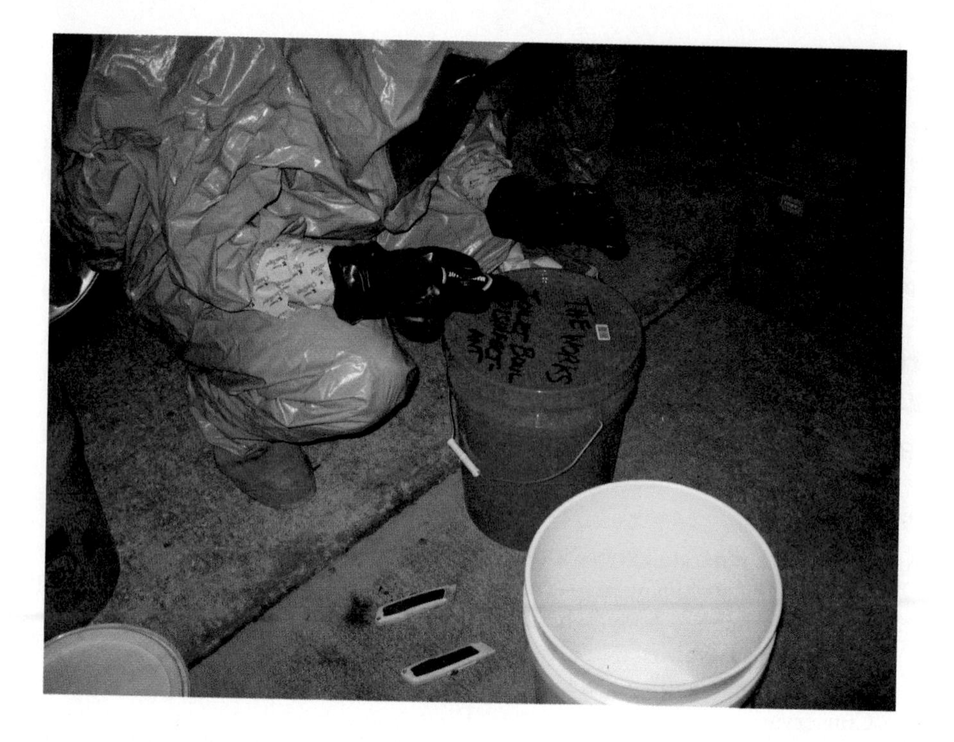

FIGURE 2-2 Regulations identify the minimum requirements for hazardous materials operations. *Courtesy of St. Petersburg Fire/Rescue.*

program. In January 1998, an OSHA interpretation addressed staffing utilizing the buddy system when responding to a hazardous materials incident. See Figure 2-2.

OSHA 1910.134 (29 CFR 1910.134) Personal Protective Equipment and Respiratory Protection

self-contained breathing apparatus (SCBA) ▪ This is an atmosphere-supplying respirator for which the breathing air source is designed to be carried by the user.

This regulation requires the employer to provide respiratory protection that is applicable and suitable for the situation to protect the health and safety of the employee up to and including **self-contained breathing apparatus (SCBA)**. A specific section of this 1910.134 standard applies to interior structural firefighting. See Figure 2-3.

An SCBA must be used by the first-entry team into an atmosphere that is immediately dangerous to life or health (IDLH). The firefighters who enter the IDLH atmosphere must be in visual or voice contact at all times. Two members are required to remain outside of the IDLH atmosphere at all times, commonly known as the two-in/two-out rule. These two members make up the initial rapid intervention team (IRIT). The regulation 1910.120(q)(3)(vi) also calls for the backup members to be equipped and prepared to provide rescue, which equates to having SCBA readily available. The individuals assigned as IRIT members may be permitted to function in another capacity or role, such as driver/operator or safety officer, provided they are able to provide rescue assistance when needed without jeopardizing the health and safety of firefighters working at the incident.

#5

This regulation requires that the department establish and maintain a respiratory protection program, and all employees must be trained to use the program. This comprehensive and understandable training must be provided annually or more often, if necessary. In addition, the employer must also provide qualitative and quantitative fit tests for the wearer's face piece or respirator and keep records of the fit testing. The 1910.134 regulation mandates the development and implementation of a written respiratory protection program stating the necessary work-site-specific procedures and elements of required respiratory use. A trained program administrator must deliver the program. The employer

FIGURE 2-3 The use of respirators and SCBA falls under the OSHA respiratory protection regulation. *Courtesy of Joe Bruni.*

standard operating procedures (SOPs) ■ This written procedure or set of procedures describes how to perform a given operation. SOPs should indicate the procedure, who has responsibility for carrying it out, and what actions should be taken if the procedures are not performed according to the written protocol or if there is not the expected outcome.

standard operating guidelines (SOGs) ■ This written statement guides the performance or behavior of departmental staff whether functioning alone or in groups. All major assignments are defined in general terms. The use of words such as *shall* and *will* leave no room for modification or flexibility.

immediately dangerous to life or health (IDLH) ■ This is a limit for personal exposure to a substance defined by the U.S. National Institute for Occupational Safety and Health (NIOSH), normally expressed in parts per million (ppm). The OSHA regulation 1910.134(b) defines IDLH as an atmosphere that poses an immediate threat to life, would cause irreversible adverse health effects, or would impair an individual's ability to escape from a dangerous atmosphere.

also needs to periodically consult employees are required to use respirators to assess their views of the respiratory protection program.

Most if not all of these requirements should be integrated into a department's **standard operating procedures (SOPs)** or **standard operating guidelines (SOGs)**. Although these are the minimum requirements, it is important to include them in the department's health and safety program.

Included in the regulation are the procedures for operating in an environment that is **immediately dangerous to life or health (IDLH)**. The requirements for operating in an IDLH environment can be found in section 1910.134(g)(3) of the regulation. The requirements for **interior structural firefighting** are found in 1910.134(g)(4) of the regulation. See Box 2-1.

OSHA 1910.1030 (29 CFR 1910.1030) Occupational Exposure to Blood-Borne Pathogens

In 1992, the blood-borne pathogen regulation became effective with the intent to minimize or eliminate occupational exposures to the hepatitis B virus (HBV), human immunodeficiency virus (HIV), and other blood-borne pathogens. The regulation specifies that universal precautions must be observed to prevent contact with blood or other potentially infectious materials, and that all bodily fluids are considered as potentially infectious materials. The regulation is based on the premise that exposures can be reduced or eliminated through planning and practices in the workplace that include control, personal protective clothing and equipment, training, hepatitis B vaccination, medical surveillance, signs and labels, and other provisions. It applies to all employees who may have an occupational exposure to blood or other infectious material. See Figure 2-4.

Employers must have an infectious control plan that they review and update annually, or as needed, to mirror new procedures or equipment. Changes and updates in technology that eliminates or reduces exposure to blood-borne pathogens must be considered during

BOX 2-1: REQUIREMENTS OF OSHA 1910.134(g)(3) AND 1910.134(g)(4)

For IDLH atmospheres, 1910.134(g)(3) requires that the employer shall ensure the following:

- One employee or, when needed, more than one employee is located outside the IDLH atmosphere.
- Visual, voice, or signal line communication is maintained between the employee(s) in the IDLH atmosphere and the employee(s) located outside the IDLH atmosphere.
- The employee(s) located outside the IDLH atmosphere are trained and equipped to provide effective emergency rescue.
- The employer or designee is notified before the employee(s) located outside the IDLH atmosphere enter the IDLH atmosphere to provide emergency rescue.
- The employer or designee authorized to do so by the employer, once notified, provides necessary assistance appropriate to the situation.
- Employee(s) located outside the IDLH atmospheres are equipped with:
    - Pressure demand or other positive pressure SCBAs, or a pressure demand or other positive pressure supplied-air respirator with auxiliary SCBA; and either
    - Appropriate retrieval equipment for removing the employee(s) who enter(s) these hazardous atmospheres where retrieval equipment would contribute to the rescue of the employee(s) and would not increase the overall risk resulting from entry; or
    - Equivalent means for rescue where retrieval equipment is not required under paragraph (g)(3)(vi)(B).

According to 1910.134(g)(4), in addition to the requirements set forth under paragraph (g)(3), in interior structural fires, the employer shall ensure that:

- At least two employees enter the IDLH atmosphere and remain in visual or voice contact with one another at all times.
- At least two employees are located outside the IDLH atmosphere.
- All employees engaged in interior structural firefighting use SCBAs.

Note 1 to paragraph (g): One of the two individuals located outside the IDLH atmosphere may be assigned to an additional role, such as incident commander in charge of the emergency or safety officer, so long as this individual is able to perform assistance or rescue activities without jeopardizing the safety or health of any firefighter working at the incident.

Note 2 to paragraph (g): Nothing in this section is meant to preclude firefighters from performing emergency rescue activities before an entire team has assembled.

Source: U.S. Department of Labor, Occupational Safety and Health Administration, 1910.134 Respiratory Protection. Retrieved May 7, 2010, from http://www.osha.gov/pls/oshaweb/owadisp.show_document?p_table=STANDARDS&p_id=12716

interior structural firefighting ■ This activity of fire suppression, rescue activities, or both takes place inside structures or enclosed buildings that are involved in fire beyond the beginning stages.

the annual review of the plan with a focus on effective and safer medical devices that are commercially available. The plan must include how a department will provide communication of hazards and training to employees and protect their health and safety. Every employer must establish that a copy of its exposure control plan is available and accessible to employees in accordance with this OSHA regulation. Additional requirements of the regulation include exposure determination and assessment of circumstances surrounding exposure, engineering and work practice controls, labeling of waste, housekeeping, personal protective equipment, record keeping, and decontamination procedures. The department must consult the 1910.1030 regulation when establishing and writing its health and safety plan and program and creating response procedures. Almost all employees of the emergency services response agency commonly fall under the guidelines of this regulation as it relates to occupational exposure.

The regulation also requires that the employer provide hepatitis B vaccinations to employees who have an occupational risk of exposure. The employer must provide

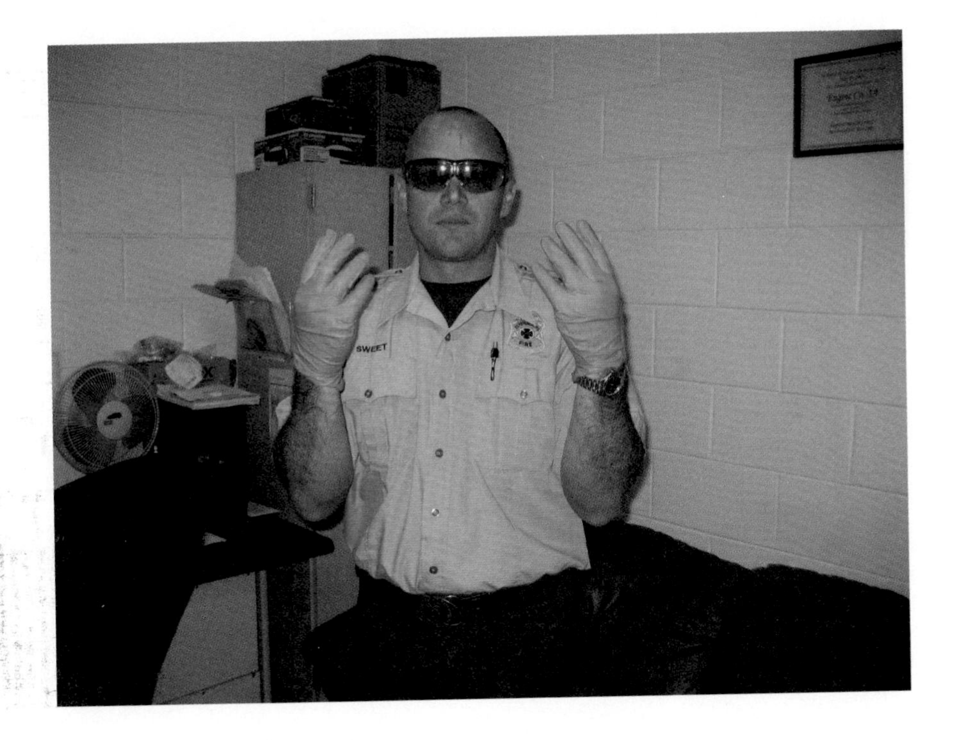

FIGURE 2-4 PPE is required to limit the exposure to blood and other infectious materials.
Courtesy of Joe Bruni.

postexposure evaluation and ensure follow-up of all employees suffering from an exposure. Medical evaluations, the hepatitis B vaccine, postexposure evaluation and follow-up, and prophylaxis administration must be done according to strict guidelines.

OSHA 1910.146 (29 CFR 1910.146) Permit-Required Confined Spaces

This regulation addresses practices and procedures to protect employees in general industry from the hazards of entry into permit-required confined spaces. Several sections apply to emergency response agencies' practices and procedures. *Confined space*, according to the OSHA regulation, means a space that (see Figure 2-5):

- Is large enough and so configured that an employee can bodily enter and perform assigned work
- Has limited or restricted means for entry or exit (for example, tanks, vessels, silos, storage bins, hoppers, vaults, and pits are spaces that may have limited means of entry)
- Is not designed for continuous employee occupancy

Companies or workplaces that have a permit-required confined space in which employees may work fall under certain requirements according to the OSHA regulation. A permit space is a confined space that has one or more of the following characteristics:

- Contains or has a potential to contain a hazardous atmosphere;
- Contains a material that has the potential for engulfing an entrant;
- Has an internal configuration such that an entrant could be trapped or asphyxiated by inwardly converging walls or by a floor that slopes downward and tapers to a smaller cross section; or
- Contains any other recognized serious safety or health hazard

FIGURE 2-5 Entry into a confined space is regulated by OSHA. *Courtesy of Joe Bruni.*

Hence, these companies or workplaces must adhere to the following requirements:

- Evaluation
- Training
- Equipment
- Rescue services

The company representative is responsible for evaluating the workplace to determine whether any permit-required confined spaces exist there. The goal is to identify all permit-required confined spaces and develop the procedures to eliminate or control hazards associated with permit-required confined space operations. If the workplace contains permit spaces, the employer needs to inform exposed employees. The employer can accomplish this by posting danger signs or by any other equally effective means to make employees aware of the existence of, location of, and danger posed by the permit spaces. If the employer decides that its employees will enter permit spaces, the employer then develops and implements a written permit space program that complies with the OSHA 1910.146 regulation and makes it available for employees and their authorized representatives to inspect. Before an employee enters the space, the internal atmosphere must be tested with a calibrated direct-reading instrument for oxygen content, for flammable gases and vapors, and for potential toxic air contaminants, in that order.

The company representative is responsible for ensuring that all affected personnel are properly trained and that refresher training is given. The employer must establish the training and prove it has been provided. Personnel who may be trained are any authorized entrants, attendants, entry supervisors, on-site rescue team members, and employees who may potentially enter the space. The employer is also in charge of ensuring employees acquire the understanding, knowledge, and skills necessary for the safe performance of the duties assigned. The training provided must establish employee proficiency in the duties required for a confined space entry and needs to introduce new or revised procedures, as necessary.

The company representative also ensures that all equipment needed for safe entry into any permit spaces and non-permit spaces is available and in proper working order. This equipment might include the following: testing and monitoring equipment, ventilating equipment, communications equipment, personal protective equipment, lighting equipment, ladders, barriers and shields, and rescue and emergency equipment.

One of the most important issues is for the employer to evaluate a prospective rescue company's ability to respond to a rescue request in a timely manner, considering the hazard(s) identified. The employer must evaluate a prospective rescue company's ability, in terms of proficiency with rescue-related tasks and equipment, to function appropriately while rescuing entrants from the particular confined space or types of permit spaces identified. The rescue company must be able to reach the victim(s) within an appropriate time frame as well as be equipped and proficient within that time frame. The employer must make the rescue company aware of the hazards it may encounter when called upon to perform a rescue. It must also provide the rescue company with access to all permit spaces from which rescue may be necessary so that the rescue company can develop appropriate rescue plans and train on rescue operations.

Additional OSHA Information

It is vital for emergency response organizations and health and safety program managers to understand the specifics of the OSHA Act, which covers employers and employees either directly through federal OSHA or through OSHA-approved state programs. States that choose to adopt certain OSHA regulations must meet or exceed the federal OSHA standards as they relate to workplace health and safety. All emergency services organizations must understand there are additional OSHA regulations, which are not mentioned in this book, that apply to emergency responders as well as general industry. Therefore, it would be in the best interest of emergency response organizations to familiarize themselves with the OSHA regulations that presently apply to their organization and to those regulations that will apply in the future.

The administration and managers in both the private and the public sector must embrace the concept that it is the employer's responsibility to provide a safe and healthy place of employment to preserve its most valuable resource: its employees. This is also OSHA's goal, which it accomplishes by establishing and enforcing standards. OSHA strives to form partnerships with employers through education, training, and outreach to encourage continued health and safety improvement in the workplace. Identifying the OSHA requirements as they pertain to all emergency response agencies will pay off in dividends that equate to safer operations and healthy employees.

NFPA Standards

Over the course of time, the NFPA has become very involved in the area of fire service safety, including NFPA 1500, *Standard on Fire Department Occupational Safety and Health Program*. The initial NFPA 1501, *Standard on Fire Department Safety Officer*, was developed during the 1970s. A new technical committee formed in the early 1980s developed the components of a fire service safety and health program. During the 1980s, a greater concern for safety and health surfaced in the fire and emergency services fields. NFPA 1500 was developed and produced in 1987 as a standard as well as a minimum performance criteria standard. Unfortunately, few fire departments have the available resources and finances to be in full compliance with NFPA 1500, although many departments have made great strides in doing so.

NFPA 1500, STANDARD ON FIRE DEPARTMENT OCCUPATIONAL SAFETY AND HEALTH PROGRAM

The NFPA 1500 standard became the blanket standard for fire service occupational health and safety and is on a five-year revision cycle to develop and introduce additional safety and health-related standards. NFPA 1500 includes a focus on the safety officer, safety and health program, medical requirements, respiratory protection, incident management, risk management, and training and education, to name just a few. Each of these has an impact on the management of a health and safety program.

#1

The NFPA 1500 standard was the first significant standard to deal directly with the health and safety of firefighters and other emergency response workers at an in-depth level. The 1997 edition of the NFPA 1500 standard included a major revision in risk management at emergency scenes and it was also the first consensus standard to deal with occupational health and safety.

The development of NFPA 1500 initially caused a great deal of unrest in the American fire service due to the anticipated cost associated with compliance with this standard. Many fire service organizations felt that this standard would put their agencies, or the fire service in general, in a position financially in which they could no longer afford to operate as a fire department. Obviously, this did not occur. The standard also generated a great deal of animosity due to its section dealing with safe staffing levels during an "initial operation" of an atmosphere deemed to be potentially IDLH. The NFPA 1500 standard contained at least 45 NFPA standards. Understanding the NFPA 1500 standard and implementation of its provisions has helped limit injuries on the fireground and reduce firefighter fatalities.

#3

NFPA 1500 provides a framework for fire service–related health and safety programs, including EMS response–related programs. It contains the minimum requirements related directly to any emergency response agency that responds to fire suppression, medical emergencies, and other emergency response functions. In the modern emergency response agency, NFPA 1500 is considered the acceptable standard of care. It applies to any full-time or part-time agency, as well as military, public, private, governmental, or industrial fire departments. The standard does not apply to industrial **fire brigades**, which are covered in NFPA 600, *Standard on Industrial Fire Brigades*.

fire brigades ■ These private or temporary organizations of individuals are equipped to fight fires.

The NFPA 1500 standard was originally developed to specify the minimum fire department criteria in such areas as management, emergency operations, facility safety, apparatus safety, fitness and wellness, critical incident stress, PPE use, and medical and physical requirements. It focuses heavily on improving the health and safety of members of fire departments and other emergency response organizations. NFPA ensures the implementation of certain practices concerning health and safety. Compliance with the provisions set forth in NFPA 1500 ensures and maintains high levels of health and safety for firefighters.

The NFPA 1500 standard specifies the requirements for use of the different kinds of firefighter PPE by addressing the various types of work they perform. A specific chapter in the standard covers protective equipment and clothing. NFPA 1500 contains references to other NFPA standards as well as general criteria developed to address specific types of PPE. See Figure 2-6.

The topics covered by the NFPA 1500 standard can be considered the basis for safety and health program development. Although NFPA 1500 is a minimum standard, nothing prevents a safety and health program manager from going beyond this level. The appendixes of this standard contain valuable information and ample direction that a safety and health program manager can use as a basis for program development.

- The employer must provide firefighters with the proper PPE appropriate with the encountered hazards and exposures.
- The employer and firefighters must select PPE and equipment based on an assessment of the hazards in operating areas in which firefighters are likely to be exposed.
- The wearing of compliant NFPA 1975 work and station uniforms ensures that firefighters do not use or wear unsafe clothing as a function of emergency operations.

FIGURE 2-6 General criteria for PPE selection.

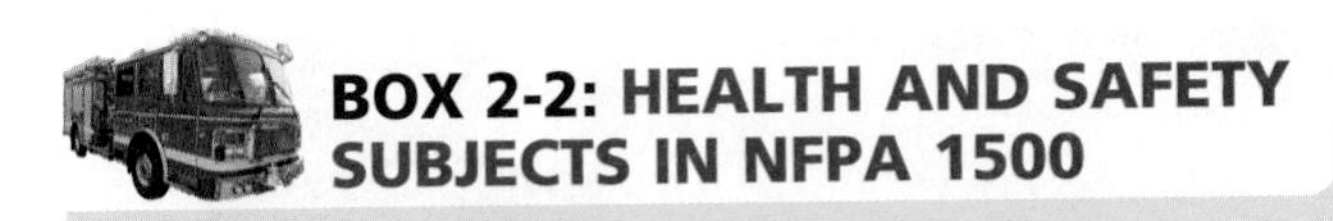

BOX 2-2: HEALTH AND SAFETY SUBJECTS IN NFPA 1500

Fire Department Administration
Training, Education, and Professional Development
Fire Apparatus, Equipment, and Drivers/Operators
Protective Clothing and Protective Equipment
Emergency Operations
Facility Safety
Medical and Physical Requirements
Member Assistance and Wellness Programs
Critical Incident Stress Program

The NFPA 1500 standard requires an emergency agency to utilize a risk management matrix that defines what risks the agency will respond to within its jurisdiction. The plan should outline what risks are within the community, the level of these risks, and whether the response agency is trained and equipped to handle them. Its focus is for an agency to operate in a safe and efficient manner to reduce or eliminate the potential for injury and death. The NFPA 1500 standard is one of the most important references for fire departments and other emergency response organizations to use in the development of an effective health and safety program. See Box 2-2.

NFPA 1521, STANDARD FOR FIRE DEPARTMENT SAFETY OFFICER

NFPA 1521 became the first fire service safety document in a series to deal with firefighter health and safety. The first two areas of NFPA 1521 dealt specifically with the roles and responsibilities of the fire department safety officer. These roles were interpreted differently from user to user, and the responsibilities varied from department to department. The NFPA committee struggled with how to make the document a useful tool and have it play a larger role in firefighter safety. A movement began not only to enlarge the roles and responsibilities of the safety officer but also to separate the roles and responsibilities between the **incident safety officer (ISO)** and the **health and safety officer**.

Along with the NFPA committee, the curriculum developers at the National Fire Academy developed courses on the tasks of the incident safety officer (ISO) and the health and safety officer. During this period the roles and responsibilities of both officers became defined. The developed courses emphasized the decision-making process, how to read the NFPA standards, federal laws, and rules and regulations. It is now accepted that both roles should be included in a department's health and safety program. Unfortunately, in some departments, one person may assume both roles, causing either the individual to be overworked and frustrated, or a lack of program implementation in one or both areas.

In the current fire service organization, the fire chief is responsible for the appointment of the department heath and safety officer. The incident safety officer is a qualified individual of the command staff at the emergency scene who works directly for the incident commander. See Figure 2-7.

The health and safety officer focuses on the development of the health and safety program for his or her department, comparing the areas of responsibility for both the health and safety officer and the incident safety officer. The health and safety officer should meet the NFPA 1521 standard and also ensure that the incident safety officer(s) does so. See Table 2-1.

incident safety officer (ISO) ■ This officer must monitor and assess the incident for existing or potential hazards, and develop measures for ensuring personnel safety.

health and safety officer ■ This appointed individual within an organization is familiar with policy and procedure issues that affect the health and safety of firefighters. Risk analysis, wellness, program management, and other occupational safety issues are the main emphasis of the health and safety officer.

FIGURE 2-7 An incident safety officer should be a priority at every incident.
Courtesy of Lt. Joel Granata.

TABLE 2-1 The Responsibilities and Functions Required by NFPA 1521

HEALTH AND SAFETY OFFICER	INCIDENT SAFETY OFFICER
Risk management	Incident management system
Training and education	Incident scene safety
Laws, codes, and standards	ISO for fire operations incidents
Accident prevention	ISO for hazardous materials incidents
Accident investigation, procedures, and review	ISO for emergency medical operations
Records management and data analysis	ISO for special operations
Apparatus and equipment compliance with standards	Accident investigation and review
Facility inspection	Post-incident analysis
Health maintenance	
Liaison for the health and safety committee	
Occupational safety and health committee	
Post-incident analysis	
Critical incident stress management	
Infection control	

incident management system (IMS) ▪ This system defines the roles and responsibilities to be assumed by personnel and the operating procedures to be used in the management and direction of emergency operations; it is also referred to as an incident command system (ICS).

NFPA 1561, STANDARD ON EMERGENCY SERVICES INCIDENT MANAGEMENT SYSTEM

NFPA 1500 requires that an **incident management system (IMS)** be used during fire scene operations. The NFPA 1561 standard encompasses the minimum requirements for an IMS system. Its purpose and goals are to manage personnel and other resources while balancing the health and safety of members with effective service delivery. An effective IMS plays a critical role in the reduction of firefighter fatalities. The IMS is designed not only to be applicable to incidents of all sizes and types but also with clear communications and effective operations in mind. It is important for fire departments and other emergency response organizations not only to continually assess their daily operations with a focus on the IMS but also to note the importance of the incident safety officer's role in and knowledge of the IMS. The NFPA 1561 standard describes the following: the structure of the IMS, the implementation of the system, communications, multiagency response, command structure, resource accountability, personnel accountability, training and qualifications, and incident scene rehabilitation.

The system components of the IMS contain a description of the incident commander, command staff, public information officer, incident safety officer, liaison officer, general staff, staging, operations functions, planning functions, logistics functions, finance and administration, and supervisory personnel.

Annex C to the NFPA 1561 standard, which addresses managing responder safety, is not part of the standard's requirements, although it provides valuable information about the functions of the incident officer, incident safety during fire suppression, and risk management during scene operations. Annex C's information comes from the Firefighting Resources of California Organized for Potential Emergencies (FIRESCOPE), which was developed for fire departments to establish firefighter safety and accountability guidelines.

According to NFPA 1561 incident commanders appoint an incident safety officer to advise and support the incident commander at scene operations. This assignment should take place early in the incident. On large incidents, assistant safety officers should also be assigned and utilized. There should be a strong interpersonal relationship between the incident safety officer and other division and branch officers in the IMS. Their interaction at the incident scene must be professional and respectful at all times.

NFPA 1581, STANDARD ON FIRE DEPARTMENT INFECTION CONTROL PROGRAM

Many fire department employees perform some role in patient care and emergency medical response. Members place themselves at risk during these incidents, and some of the same risks are also found at department facilities. NFPA 1581 was developed with the guidelines from the Centers for Disease Control and Prevention to reduce exposures during emergency and nonemergency situations. This standard defines the minimum requirements for a department's infection control program. The standard's many chapters describe the requirements and components for an infection control program, which include education and training, risk management, the infection control officer, exposure incidents, and immunization and testing. NFPA 1581 also explains the requirements for cleaning and disinfecting areas and storage areas at a fire department's facilities.

Certain requirements for emergency medical operations directly affect personnel, such as those about equipment, protective clothing, and the handling of sharp objects. Chapter 6 discusses cleaning, washing, and disposal issues. The NFPA 1581 standard is of great value as it relates to compliance with the OSHA 1910.1030 regulation. See Figure 2-8.

NFPA 1582, STANDARD ON COMPREHENSIVE OCCUPATIONAL MEDICAL PROGRAM FOR FIRE DEPARTMENTS

Originally, the medical requirements for firefighters were included in NFPA 1001, *Standard for Fire Fighter Professional Qualifications*. In 1988, a subcommittee was formed to write the medical requirements standard for both candidate and incumbent firefighters.

The NFPA 1582 standard applies to both full-time and part-time employees; however, its first edition caused difficulty for fire department physicians or those hired to perform medical evaluations on firefighters. The original standard's medical requirements

FIGURE 2-8 Department facilities should include an infection control area to accomplish cleaning and disinfecting.

were categorized into A and B conditions: A conditions precluded employees either from becoming a firefighter or from continuing employment as a firefighter who conducted fire suppression and training operations. The fire department physician was to evaluate B conditions on an individual and case-by-case basis. In many instances, no firefighter job description or task analysis existed on which to base the medical evaluation. It became difficult to determine the medical condition necessary to be a firefighter. Many departments struggled with terminology of incumbent and candidate firefighters when applying this standard. Some departments made this practice a part of the labor agreement, whereas others considered it a routine evaluation of employees. The technical committee physicians then worked to identify areas that needed greater clarification and emphasis, including cardiovascular, neurological, vision, hearing, and metabolic processes. Such information was included in subsequent editions of the standard.

NFPA 1582 covers roles and responsibilities, medical evaluations of candidates, essential job tasks, occupational medical evaluation of current members, annual occupational fitness evaluation of existing members, and the essential job tasks in a specific evaluation of medical conditions. This standard provides direction to the department physician as to what conditions may prohibit an individual from performing firefighter functions. The standard is also helpful in determining both the frequency of medical evaluation and its criteria. In light of today's statistics concerning firefighter injuries and fatalities, the fire department physician has a vital role in the occupational safety and health program.

NFPA 1583, STANDARD ON HEALTH-RELATED FITNESS PROGRAMS FOR FIRE DEPARTMENT MEMBERS

NFPA 1500 requires a physical performance program as well as a fitness program as a part of a fire department's occupational safety and health program. Many fire departments and labor organizations found it difficult to assess the validity and content of physical performance testing and evaluation. Validation of physical test skills by a task analysis became a very difficult process, and the NFPA technical committee struggled with reaching a consensus with this NFPA 1583 standard. Many departments expressed a need for a fitness and wellness program for use in hiring new members, and the International Association of Fire Fighters (IAFF) and the International Association of Fire Chiefs (IAFC) jointly developed the **candidate physical ability test (CPAT)** to work on a joint fire service fitness and wellness program. Their Fire Service Joint Labor Management Wellness/Fitness Initiative benefits all of the fire service. The NFPA 1583 standard can be used as a companion standard with NFPA 1582 and the IAFF/IAFC Wellness-Fitness Initiative.

candidate physical ability test (CPAT) ■ This is an indoor facility test that measures the capabilities of firefighting recruits in eight job-specific areas.

Other standardized physical ability tests that are in place and used by various organizations are the Biddle Physical Ability Test and the Firefighter Combat Challenge; the latter was created by Dr. Paul Davis, a past firefighter/paramedic who conducted research in the areas of physical fitness and employment standards. Another alternative is when individual fire departments, such as the City of Phoenix, create their own type of physical testing.

The minimum requirements of the NFPA 1583 standard include program development, implementation, and management of a health-related fitness program. These are applicable to public, governmental, military, private, and industrial fire department organizations providing rescue, fire suppression, emergency medical services, hazardous materials mitigation, special operations, and other emergency services. This standard is not applicable to industrial fire brigades or to industrial fire departments meeting the requirements of NFPA 600, *Standard on Industrial Fire Brigades*. The subjects of NFPA 1583 include the assignment of a qualified health and fitness coordinator, a periodic fitness assessment for all members, an exercise training program available to all members, education and counseling regarding health promotion for all members, and data collection. Many departments have added a wellness and fitness program coordinator or an occupational health specialist from the medical community to their staff in an attempt to meet the intent of this standard. Other organizations have gotten assistance from local colleges and universities to promote employee wellness and fitness.

Numerous states now require heart and lung screenings for firefighters prior to employment or membership. Heart and lung screening also takes place during the annual physical in some fire service organizations. Some states have also combined community resources and partnerships to promote and achieve health, wellness, and safety initiatives. Many organizations can tap into these resources to directly benefit their members. The National Volunteer Fire Council (NVFC, see http://www.nvfc.org) has resources and tools available concerning wellness and fitness for firefighters.

The insurance providers for emergency services have the potential to assist emergency services organizations in the wellness and fitness areas by providing additional benefits for members. Leading insurance companies such as the Volunteer Firemen's Insurance Service (VFIS) provide incentives such as the length of service awards program (LOSAP). The purpose of the LOSAP program is to recruit, retain, and reward emergency services volunteers; its benefits include life insurance, a disability benefit, and a supplemental retirement income. The insurance industry could extend this program into the areas of wellness and fitness for the emergency services workers and could also put it in place for members of volunteer ambulance services and rescue squads. Such programs would benefit both the individual members and the organization as a whole.

NFPA 1584, STANDARD ON THE REHABILITATION PROCESS FOR MEMBERS DURING EMERGENCY OPERATIONS AND TRAINING EXERCISES

This standard, which is an integral component of both an occupational safety and health program and incident scene management, provides an organized approach for fire department members' rehabilitation at training functions and incident scene operations. Its recommendations illustrate practices for medical evaluation and treatment, food and fluid replacement, relief from environmental conditions, rest and recovery, crew rotation, and accountability. NFPA 1584 discusses pre-incident considerations as well as SOPs and training that should be in place concerning rehabilitation. One of the new additions to the NFPA 1584 standard involves the requirement of oximeters for monitoring vital signs during training sessions and rapid assessment of carbon monoxide levels in firefighters. It has recently become clear that monitoring of pulse rate and blood oxygen saturation levels is critical during training evolutions. EMS personnel must evaluate and confirm that members are safely able to return to full duty before being permitted to exit the rehab area. Vital signs listed in the NFPA 1584 annex that should be monitored include temperature, pulse, respirations, blood pressure, pulse oximetry, and carbon monoxide assessment.

The standard also covers rehabilitation site characteristics, rehabilitation facilities, and rehabilitation resources that should be considered both at incident scenes and during training functions. Compliance with the NFPA 1584 standard involves an organization's commitment to a safer emergency incident environment. Emergency services organizations are concerned about the dangers of heat and cold at training events as well as real incidents, and must be prepared at all events to intervene and prevent a tragedy. The standard helps with the development of rehabilitation procedures, as there is also a component that addresses post-incident recovery. Post-incident measures include patient care reports for all personnel who have received medical care during the rehab process, and a recommendation to command and company officers to continue rehab measures prior to personnel returning to full duty.

The NFPA 1584 standard is a move in the right direction to promote the health of emergency services workers.

NFPA 1710, STANDARD FOR THE ORGANIZATION AND DEVELOPMENT OF FIRE SUPPRESSION OPERATIONS, EMERGENCY MEDICAL OPERATIONS, AND SPECIAL OPERATIONS TO THE PUBLIC BY CAREER FIRE DEPARTMENTS

#11

The NFPA 1710 standard deals with the organization and deployment of fire suppression and EMS operations in career fire departments. A component of the standard covers the time frame when responding to fire and medical emergencies for the first-due unit; the entire first-due alarm assignment; and the roles of the basic and advanced emergency medical responders.

The standard covers the minimum number and roles of firefighters necessary to conduct initial interior operations. The NFPA 1710 standard also includes special operations regarding the occupational health and safety of fire department members. It addresses the requirements for fire department services, airport rescue and firefighting, marine rescue and firefighting, as well as wildfire incidents.

Chapter 6 of NFPA 1710 covers the requirements for fire department systems as well as health and safety, communications, training, pre-incident planning, and incident management. The standard not only addresses health and safety issues but also fire department organization and deployment.

NFPA 1720, STANDARD FOR THE ORGANIZATION AND DEPLOYMENT OF FIRE SUPPRESSION OPERATIONS, EMERGENCY MEDICAL OPERATIONS, AND SPECIAL OPERATIONS TO THE PUBLIC BY VOLUNTEER FIRE DEPARTMENTS

The NFPA 1720 standard deals with the organization, operation, and deployment of fire suppression and EMS operations that are substantially volunteer in design and makeup.

The standard is not as specific as the NFPA 1710 standard in the areas of required responders or the maximum time frame; however, it does address prompt initiation of suppression attack and the presence of sufficient personnel to develop a sustained fire attack. NFPA 1720 covers the uniqueness of volunteer fire departments, how they respond and deploy personnel, and the services they provide. Due to the distinctiveness of volunteer fire departments, specific requirements of this standard are left up to the individual authority having jurisdiction to determine needs.

Chapter 4 of NFPA 1720 covers the organization of the volunteer fire department, operations and deployment, fire suppression, intercommunity organization, emergency medical services, and special operations. Similar to NFPA 1710, this standard addresses fire department systems in Chapter 5.

The health and safety officer in a volunteer fire department should refer to this standard when developing a health and safety program. This officer should also be familiar with the minimum requirements of NFPA 1720 as it relates to deployment and operations.

ADDITIONAL NFPA STANDARDS RELATED TO HEALTH AND SAFETY

A number of NFPA standards that directly affect the health and safety of firefighters were united and merged into the 2007 edition of NFPA 1500. It is important to recognize the standards incorporated into NFPA 1500 as they are related to occupational health and safety. NFPA technical committees can continue to develop and revise standards; however, a behavioral change must occur with firefighters and fire departments' views as they practice occupational health and safety.

Fire departments must continually educate their members and the citizens in their communities to the hazards of firefighting and emergency medical response. It is up to the administration of each specific emergency response organization to take the lead in the areas of marketing and education by keeping abreast of the NFPA standards and regulations in order to inform the public in their individual communities. The public expects the best service that is available and affordable; without public education efforts and outreach, citizens will be less willing to pay higher taxes for what they perceive to be a waste of taxpayer money. Fire departments and firefighters must be given the training and resources to practice occupational health and safety in the fire service, or the fire service will regress in this important area.

Related Standards and Regulations Beyond NFPA and OSHA

NATIONAL INSTITUTE FOR OCCUPATIONAL SAFETY AND HEALTH (NIOSH)

The same act that created OSHA also formed the National Institute for Occupational Safety and Health (NIOSH), which is part of the Centers for Disease Control and Prevention (CDC). NIOSH is the only federal institution responsible for conducting research in and making recommendations for the prevention of work-related illnesses and injuries. NIOSH plays an important role investigating firefighter fatalities through its Fire Fighter Fatality Investigation and Prevention Program.

NIOSH publishes many documents relevant to emergency responders in the areas of chemical hazards, firefighter cardiovascular disease, safety management in disaster and terrorism response, and traffic hazards to firefighters while working on roadways. NIOSH continues to expand outreach and partnership activities to fire service organizations, and it pursues activities that complement and support other organizations' prevention efforts.

NIOSH also gathers data from occupational injury and illness databases such as the Bureau of Labor Statistics' Census of Fatal Occupational Injuries. One of the goals of

NIOSH is to increase coordination among its divisions conducting research on firefighter safety and health, including the NIOSH National Personal Protective Technology Laboratory. NIOSH provides technical assistance to fire departments that have experienced a line-of-duty death. The NIOSH Fire Fighter Fatality Investigation and Prevention Program directly benefits departments that are seeking information on incidents that personnel can view as a post-incident analysis from which to learn. Its valuable material helps the fire service avoid repeating its tragic and bloody history.

AMERICAN NATIONAL STANDARDS INSTITUTE (ANSI)

The American National Standards Institute (ANSI) is a voluntary standards-making organization at the national level. ANSI helps to ensure the safety and health of consumers and the protection of the environment. ANSI is comprised of government agencies, organizations, companies, academic and international bodies, and individuals. The American National Standards Institute represents the interests of more than 125,000 companies and 3.5 million professionals as it develops standards in response to the requirements of other standards and regulations. Many of the ANSI standards are related and directly affect firefighters and emergency medical personnel in the area of personal protective equipment. General consensus standards may become ANSI standards. The ANSI Safety and Health Standards Board has adopted and approved standards developed by the NFPA. It is important for other organizations to pay particular attention to the ANSI standards as they relate to the health and safety of their members.

AMERICAN SOCIETY FOR TESTING AND MATERIALS (ASTM)

The American Society for Testing and Materials (ASTM) develops standards addressing several areas. For example, it has become a leader in the development of international voluntary consensus standards. ASTM is known for improving product quality and enhancing safety. The specific areas addressed by ASTM include not only products and materials but also systems and services. As a consensus organization, ASTM uses the expertise of individuals and organizations from various disciplines of science and engineering. Although much of the focus of ASTM is directly related to consumer protection, it also impacts the safety of emergency services and concentrates on the training area of emergency medical systems.

CENTERS FOR DISEASE CONTROL AND PREVENTION (CDC)

The Centers for Disease Control and Prevention (CDC) addresses many health care–related topics concerning firefighters and emergency medical response providers. These topics include blood-borne pathogens, the H1N1 influenza, hepatitis, tuberculosis, chemical hazards, and HIV. In 1989 the CDC released the *Guidelines for Prevention of Transmission of Human Immunodeficiency Virus and Hepatitis B to Health-Care and Public-Safety Workers*. The most important component of this document concerns the medical management of persons who have been exposed in the workplace. The infection control officer can make valuable use of this document as it relates to an organization's exposure control plan, as well as other CDC guidelines for infection control. The one most applicable to emergency response personnel is the *Guideline for Infection Control in Health Care Personnel, 1998*, which replaces the 1983 *Guideline for Infection Control in Hospital Personnel*. The 1998 guideline provides prevention strategies, objectives, and elements to reduce the transmission of infectious disease from patients to health care personnel and from personnel to patients. See Figure 2-9.

The infection control strategies, objectives, and elements should be an integral part of the emergency response organization's infection control plan. The objectives of the infection control plan cannot be met without the support of the organization's administration, medical staff and personnel, and the other health care personnel. See Figure 2-10.

- Immunizations for vaccine-preventable diseases
- Isolation precautions to prevent exposures to infectious agents
- Management of health care personnel exposure to infected persons, including postexposure prophylaxis
- Work restrictions for exposed or infected health care personnel
- Issues related to latex hypersensitivity and provides recommendations to prevent sensitization and reactions among health care personnel

FIGURE 2-9 Prevention strategies for infection control in health care personnel. *Courtesy of Centers for Disease Control and Prevention.*

Elements of Infection Control Prevention

- Coordination with other departments
- Medical evaluations
- Health and safety education
- Immunization programs
- Management of job-related illnesses and exposures to infectious diseases
- Policies for work restrictions for infected or exposed personnel
- Counseling services for personnel on infection
- Risks related to employment or special conditions
- Maintenance and confidentiality of personnel
- Health records

Objectives of Infection Control Prevention

- Educating personnel about the principles of infection control and stressing individual responsibility for infection control
- Collaborating with the infection control department in monitoring and investigating potentially harmful infectious exposures and outbreaks among personnel
- Providing care to personnel for work-related illnesses or exposure
- Identifying work-related infection risks and instituting appropriate preventive measures
- Containing costs by preventing infectious diseases that result in absenteeism and disability

FIGURE 2-10 The elements and objectives of infection control and prevention. *Courtesy of Centers for Disease Control and Prevention.*

Recently, the CDC has published information concerning bioterrorism, mass casualties, radiation emergencies, and natural disasters and severe weather emergencies. This essential information enables emergency response agencies to become better prepared for certain emergencies.

The CDC has many documents available to emergency response agencies. The department safety and health officer needs to remember that CDC offers a wealth of information dealing directly with responder safety, predominantly in the area of emergency medical and terrorism response. The CDC also provides information concerning disaster site management and the potential associated hazards. One area of particular interest concerns the personal protective equipment guidelines for structural collapse events.

THE FIREFIGHTER AUTOPSY PROTOCOL

The Firefighter Autopsy Protocol, originally issued by the U.S. Fire Administration (USFA) in 1994, was extensively revised in 2008. Its purpose remains the same: to gain a better understanding of firefighter deaths and contribute to better health and safety standards, operational procedures, and technology for the fire service. It is widely accepted that an improved understanding of firefighter fatalities will aid the fire service in prevention issues in the future. The protocol provides a general background and emerging issues concerning the conduct of firefighter autopsies, as well as information about current autopsy protocol and practices. This information may be used to determine mechanisms and causes of firefighter fatalities. The protocol also addresses personnel protective clothing and toxicology. The Firefighter Autopsy Protocol manual recommends that an autopsy be performed not only for every firefighter line-of-duty death but also for any non-line-of-duty death that is connected to a line-of-duty exposure. The protocol gives specific supplemental information with additional evaluations and considerations in certain areas in which current autopsy practices are used. See Figure 2-11.

The intent of the autopsy protocol is to improve firefighter health and safety, and it is critical in helping to determine eligibility for death benefits under the federal government's Public Safety Officers' Benefits (PSOB) Program. The protocol also helps with the study of deaths of both active and retired firefighters related to occupational illnesses by addressing those particular attributes of firefighter casualties that distinguish them from casualties of the general population.

FIGURE 2-11 Firefighter autopsy protocol supplemental factors and considerations. *Source: U.S. Fire Administration, USFA, 2008, from http://www.usfa.dhs.gov/downloads/pdf/publications/firefighter_autopsy_protocol.pdf*

- Evaluation of victim work history with specific attention to prior exposures
- Examination of personal protective equipment (PPE) for relating effects of clothing and equipment on individual parts of the body, particularly in cases of trauma and burn injury
- Details in the physical examination for identifying signs of smoke asphyxiation and burn injury as contributing causes of firefighter fatality
- Implementation of appropriate carbon monoxide and cyanide evaluation protocols as part of the toxicological evaluation
- Detailed toxicological evaluations where hazardous atmospheres have been encountered

Summary

A great deal of standards and regulations directly relate to emergency responder occupational safety and health. Regulations are developed and prescribed at the government level to enforce a principle, rule, or law designed to control or govern conduct. Standards are developed by a consensus of experts and do not carry the weight of the law unless they are adopted and enforced by the authority having jurisdiction. It is up to the authority having jurisdiction to determine its needs and adopt the necessary codes and standards to suit the emergency response organization and the community at large. Documents that are published as standards do not have to be adopted as law, nor is there any requirement for compliance; however, standards may be used to establish a standard of care. Standards may also be referenced by a regulation and used during litigation.

A number of organizations provide standards and regulations that apply to emergency response agencies. For example, OSHA publishes many regulations related to emergency response organizations, and the NFPA puts out a number of consensus standards directly related to firefighter safety and health. Other sources are the American National Standards Institute, the Centers for Disease Control and Prevention, the National Institute for Occupational Safety and Health, and the Environmental Protection Agency.

The health and safety program officer within an emergency response agency must consult the various standards and regulations to develop a comprehensive health and safety program. The health and safety officer needs to keep in mind that regulations are law and compliance must be in place, and that standards can provide a standard of care and a structure for the health and safety program.

Review Questions

1. A regulation is law, and compliance must occur in an emergency response organization.
   a. True
   b. False
2. Standards are written by:
   a. Consensus of experts
   b. Politicians
   c. Scientists
   d. Engineers
3. NFPA standards are not mandatory when it comes to compliance.
   a. True
   b. False
4. Discuss the standard of care.
5. Which NFPA document deals with the minimum number of firefighters in a career department required to launch an interior fire attack during interior structural firefighting?
   a. NFPA 1583
   b. NFPA 1403
   c. NFPA 1710
   d. NFPA 1582
6. What is one of the requirements of the OSHA 1910.120 (29 CFR 1910.120) Hazardous Waste Operations and Emergency Response standard related to the health and safety of employees working with hazardous waste operations?
   a. An organizational structure
   b. A comprehensive work plan
   c. A safety and health training program
   d. All of the above
7. The NFPA 1500 standard requires an emergency agency to operate utilizing:
   a. A risk management matrix
   b. A general duty clause
   c. A task analysis
   d. None of the above
8. The OSHA 1910.156 regulation that applies to fire brigades, industrial fire departments, and private or contracted fire departments purposely excludes wildland firefighting and airport crash fire rescue operations.
   a. True
   b. False

9. The Centers for Disease Control and Prevention has published information concerning:
   a. Bioterrorism
   b. Mass casualties
   c. Radiation emergencies
   d. All of the above

10. The NFPA 1720 standard deals with the organization, operation, and deployment of fire suppression and EMS operations that are substantially ____________ in design and makeup.
   a. Industrial
   b. Volunteer
   c. Fire brigade
   d. None of the above

Case Study

On March 18, 1996, two male firefighters (38 and 32 years old) died while fighting a fire in an auto parts store. At 1129 hours, a call came into the fire/police dispatcher from an auto parts store in a strip shopping mall, reporting sparking and popping from an inside "fuse box." Engine 3, Engine 1, Ladder 2, and Battalion 2 were ordered to respond. Engine 3 was the first on the scene (1135 hours) and assumed command. When Engine 3 pulled up in front of the auto parts store, no smoke or fire was visible.

The firefighters arriving on the scene did not know that the reported sparking in the fuse box was caused when the boom of a power company truck had accidentally broken the neutral line on the 208/120 volt three-phase service drop to the auto parts store. An investigation conducted by the power company revealed the panel box in the auto parts store had been improperly grounded; therefore, when the neutral was broken, the power surge did not go to ground at the panel box but traveled throughout the electrical circuitry, causing electrical fires at each circuit connection. For example, the electric hot water tank caught fire, and the wiring in electrical junction boxes of the HVAC units on the roof of the store were fused together from the extreme heat created by the short circuit.

The acting lieutenant and a firefighter specialist entered the front door of the store to investigate, while the driver of Engine 3 went to the side door. Although the lights were off in the store, the large plate glass windows in the front provided enough light to see there was not any smoke inside the store, and that "it looked clear."

At 1137 hours, the driver of Engine 3 heard the acting lieutenant calling on the radio (Portable 3), so he went back to the engine and received instructions from Portable 3 to reposition Engine 3 to the rear of the building. While driving the engine to the rear of the building, the driver noticed a little smoke coming from the edge of the roof and also heard the transmission from Portable 3 to Battalion 2 (1138 hours) that Engine 3 and Ladder 2 could handle the situation. When the driver of Engine 3 arrived at the rear of the auto parts store, the acting lieutenant and the firefighter specialist were coming out the rear door. Battalion 2 now ordered Engine 1 back into service. The Engine 3 driver asked whether the others had noticed the smoke, which was now more intense and noticeable, coming from the roof, and they stated yes. At this point, the firefighters from Engine 3 pulled off the first 1 3/4-inch hose. The acting lieutenant took the charged line and went back inside the store, returning shortly to pull a second line. While the firefighter specialist was donning his self-contained breathing apparatus (SCBA), the acting lieutenant was using the second line to knock down the fire that was coming through the edge of the roof. When the firefighter specialist had donned his SCBA, he and the acting lieutenant entered the back door with the second charged line.

At 1142 hours, Engine 3 (Portable 3—inside the auto parts store) requested pike poles and assistance in removing the ceiling. Meanwhile, Engine 3 at the rear of the store was calling for an engine to lay a supply line because it would be out of water shortly.

At 1145 hours, two firefighters from Ladder 2 positioned their unit facing the auto parts store. They walked up to the front door and observed a brisk wind (approximately 30 miles per hour) blowing through the thick black smoke in the store. Although they could not see any fire, the blowing wind and the heavy smoke made it apparent that there was a heavy fire somewhere, so they decided not to enter the building.

At 1149 hours, Portable 3, inside the auto parts store, radioed that it was in trouble and could not get out. However, due to the heavy radio traffic, Battalion 2 (positioned in front of the store) did not understand the transmission. Battalion 1, en route to the fire scene, had picked up the radio transmission and radioed Battalion 2 that the transmission sounded as though someone were trapped inside the building.

At 1150 hours, without warning, the fire accelerated rapidly, and the entire roof collapsed into

Auto Parts Store Fire Concerns and Issues	Yes	No
Was a proper risk management assessment completed at this incident by the interior crew?		
Was it appropriate for the first-due engine to initiate an interior attack with limited personnel on scene?		
Did the lack of an accountability system play a role in these line-of-duty deaths?		
Did the dispatch channel also used as the tactical channel play a role in inadequate communications with the interior crew?		
Could the exterior pump operator and battalion chief have made a rescue of interior members when they recognized the interior crew was in trouble?		
Would it have made a difference if exterior crew members had provided a warning to interior crews regarding fire conditions?		
Did the lack of recognition of lightweight construction play a role at this incident?		
Would a use of a preplan at this incident have made a difference in the outcome?		
Did the initial response to this incident provide enough resources to mount an interior attack?		
Did an early and adequate water supply play a role in this incident?		

FIGURE 2-12 Auto parts store fire concerns and issues.

the auto parts store. The building was now totally engulfed in fire and conditions were changing rapidly: a firefighter from Engine 1 noticed that the hose line leading into the rear door of the building had burned through, allowing water to flow freely; numerous explosions were heard inside the store (overheating pressurized cans); and Engine 3 had to be moved for fear of losing the engine due to the extreme heat as the fire was being whipped over the engine.

At 1208 hours, Battalion 2 stated it might have two firefighters down inside the burning building. Fire suppression operations continued, using multiple streams to contain and extinguish the fire. The two firefighters (acting lieutenant and firefighter specialist) inside the building were unable to escape as the roof collapsed, and they died in the fire.

NIOSH had the following recommendations:

Recommendation #1: Fire departments should ensure that fire command always maintains close accountability for all personnel at the fire scene.

Recommendation #2: Fire departments should ensure that at least four firefighters be on the scene before initiating interior firefighting operations at a working structural fire.

Recommendation #3: Fire departments should ensure that standard operating procedures and equipment are adequate and sufficient to support the volume of radio traffic at multiple-responder fire scenes.

Recommendation #4: Fire departments should ensure that prefire planning and inspections cover all structural building materials/components.

Recommendation #5: Municipalities should ensure that all electrical circuits are installed in accordance with the National Electrical Code, and fire departments should include electrical inspection on prefire planning and inspection.[4] See Figure 2-12.

1. Explain how a lack of situational awareness played a role in this incident.
2. Describe the most important factor that would have prevented these firefighter fatalities.

References

1. U.S. Department of Labor, Occupational Safety and Health Administration, *Standard Number 1910*, 2009. Retrieved May 23, 2010, from http://www.osha.gov/pls/oshaweb/owastand.display_standard_group?p_toc_level=1&p_part_number=1910
2. U.S. Department of Labor, Occupational Safety and Health Administration, *1910.120 Hazardous Waste Operations and Emergency Response*, 2009. Retrieved May 3, 2010, from http://www.osha.gov/pls/oshaweb/owadisp.show_document?p_table=standards&p_id=9765
3. U.S. Department of Labor, Occupational Safety and Health Administration, *Standard Number 1910*, 2009. Retrieved May 23, 2010, from http://www.osha.gov/pls/oshaweb/owastand.display_standard_group?p_toc_level=1&p_part_number=1910
4. NIOSH Fire Fighter Fatality Investigation and Prevention Program (FFFIPP), *Sudden Roof Collapse of a Burning Auto Parts Store Claims the Lives of Two Fire Fighters—Virginia*, 1996. Retrieved May 17, 2009, from http://www.cdc.gov/niosh/fire/reports/face9617.html

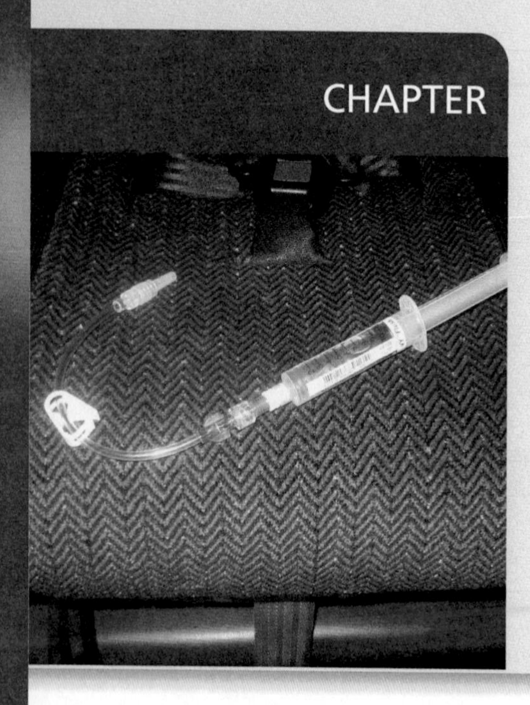

CHAPTER 3 Risk Management and Administration

KEY TERMS

frequency, *p. 56*
job task analysis, *p. 53*
risk control, *p. 57*
risk transfer, *p. 64*
risk, *p. 49*
safety audit, *p. 52*
severity, *p. 56*

OBJECTIVES

After reading this chapter, you should be able to:

- Discuss the meaning of and the process of risk identification.
- Discuss the meaning of and the process for risk evaluation.
- Discuss the meaning of and the process for risk control.
- Apply risk control strategies to an injury problem.

For additional review and practice tests, visit **www.bradybooks.com** and click on Resource Central to access book-specific resources for this text! To access Resource Central, follow directions on the Student Access Card provided with this text. If there is no card, go to **www.bradybooks.com** and follow the Resource Central link to Buy Access from there!

Risk management is a structured approach to managing uncertainty related to a threat. With regard to occupational health and safety, risk management means protecting the safety, health, and welfare of people on the job, which includes those involved in emergency scene operations and other fire department activities. Its strategies include avoiding the risk, reducing the negative effect of the risk, transferring the risk to another party, and accepting some or all of the consequences of a particular risk. Many incorrectly believe risk management to be the responsibility of the administration; instead, it is everyone's responsibility, from the firefighters to the fire chief. Risk management is an ongoing process directed toward the assessing, mitigating (to an acceptable level), and monitoring of risks. **Risk** can be defined as anything negative that poses a threat to an individual or an organization. Ineffective risk management can mean unnecessary injuries and loss of life.

risk ■ Anything negative that poses a threat to an individual or an organization.

Mitigation of risks involves the use of available technological, human, organizational, and financial resources. In today's culture of fiscal responsibility, it takes a great deal of marketing and skill to appropriate the necessary resources to properly conduct risk management. The NFPA requires four components of risk management as covered in NFPA 1500:

- Risk identification
- Risk evaluation
- Risk control
- Risk management and monitoring

Risk Assessment and Identification

Risk identification, the first NFPA component of risk management, is at times difficult and complicated because it is a function of hazard assessment and identification. It is the responsibility of the health and safety program manager to conduct an identification of risk in a selected domain of interest or a location. Risk identification draws on past experiences, research in the areas of historical and current data, local and national resources, and the experiences of emergency services responders on the job. See Figure 3-1.

Risk identification involves recognition of potential problems. It can be difficult to predict the future using risk management because the fire service and other emergency response agencies are involved in continual and constant change, which can be unpredictable at times, such as in the area of preparedness and response to terrorism. However, any emergency response organization would benefit from using risk management as a tool to complete a strategic plan. As the strategic plan forces the organization to take an in-depth look at the future, the organization becomes better prepared for upcoming changes.

In the fire service, identification of particular risks that firefighters are expected to encounter should become part of risk identification. A list should be developed to identify potential problems for every aspect of fire department operations. The following sources of information are useful in the process:

- A list of the risks to which members are exposed
- Records of accidents, illness, and injuries
- Reports on inspections and surveys of facilities and apparatus

Emergency responders face hazards and risks at the emergency scene, during training evolutions, and while operating apparatus. Risk management for an emergency response organization consists of two types: nonemergency risk identification and emergency risk

FIGURE 3-1 New types of lightweight construction present a risk to firefighters and should be evaluated as part of risk identification. *Courtesy of Chief Mike Zamparelli.*

identification. The following sources of information should be useful in the process of nonemergency risk identification:

- Physical fitness
- Training
- Nonemergency apparatus movement (routine driving)
- Station activities
- Vehicle and station maintenance

Firefighters no longer face risk only at fire scene emergencies. Because the fire service is now an "all risk and all hazards" service, new threats and additional responsibilities are involved, as the firefighting tasks increase in complexity. In present-day society, firefighters have been trained and take responsibility in several specialized areas such as emergency medicine, hazardous materials response and mitigation, technical rescue, terrorism, and water rescue, to name a few. Changes have been identified in building construction methods, passenger vehicles, and diseases. The following sources of information should be useful in the process of emergency risk identification:

- Fireground and EMS operations
- Hazmat incidents
- Special operations
- Emergency response driving

A particular risk that has been identified recently is a lack of situational awareness at fire scene emergencies, commonly during structural firefighting. Situational awareness, the perception of environmental elements and one's surroundings, is critical to decision makers and those directly involved with the incident. Because of the complex and dynamic nature of these incidents, the information flow can be dynamic and rapid. The ensuing lack of situational awareness causes near misses, injuries, and line-of-duty deaths in emergency services organizations.

Dynamic risk assessment must be carried out in this evolving environment in which the situation being assessed is developing as the process of mitigation is undertaken. It can become a particular problem for the incident commander when many incident priorities conflict or compete with one another. Demands and/or distractions may create a problem for the incident commander or individual firefighters prior to gathering critical or essential information and data concerning the incident. Situational awareness, coupled with dynamic risk assessment, is about informed and decisive decision making. Incident commanders, company officers, and firefighters must not only have an understanding of the fire incident and the building but also possess the ability to communicate and take the necessary actions to support the incident action plan as safely as possible. Situational awareness also concerns the proper attitude and knowledge that has been previously learned, and the ability to process new information as it rapidly becomes available during the incident in order to make effective and safer decisions to keep firefighters out of harm's way. Maintaining observation and a continual keen attentiveness are the keys to effective situational awareness. Everyone at the incident must continually stay aware of changing conditions. A thorough understanding and knowledge of building construction, fire behavior, and fire growth also play a major role in situational awareness.

Numerous sources have identified a drop in structural fires across America, which has had a direct impact on firefighting experience and exposure. Subsequently, many fire service organizations have increased their training efforts to compensate for the lack of fireground experience, whereas others have not, thereby placing their employees at risk. As a reflection of the society it protects, the fire service adapts many of society's tools, which bring risk into its organization. Traditionally firefighters have accepted the risk of injury and death associated with the job of responding to fires, accidents, and hazardous materials, technical rescue, and water-related incidents. Fire department officers must accept responsibility for the health and safety of the personnel they supervise, which is different from the responsibility to the community they protect; the risk assessment for each is also different.

TRENDS IN EMERGENCY SERVICES

Maintaining an awareness of emerging trends is a useful tool in risk identification. Certain information has been known for years, such as that the leading cause of firefighter deaths is directly related to cardiovascular disease and heart attacks. Also identified are several categories of injuries that occur during emergency operations. Today, firefighters and fire service organizations also have a vast amount of resources available: fire service magazines, periodicals, trade journals, fire service Internet sites, conferences, and seminars.

One of the important areas directly related to fire scene injury and death is the lack of flashover identification and recognition. The U.S. fire service has also come up with several sources as a way to bring recognition to firefighter injury and death and hopefully spread the word nationally. The Firefighter Close-Calls and the Firefighter Near-Miss Programs are valuable tools that any firefighter can access online to gain awareness. Such media tools can become part of the risk management process at the grassroots level of any fire service organization.

Risk management in emergency services organizations can be applied to every activity and function. As a multifaceted approach, it will involve:

- Safety
- Health
- Financial impact
- Loss control

Risk management should be an ongoing process that focuses on continual improvement. Each organization must work together to form a solid foundation of risk management principles and practices. To study how to manage data and information that will aid in organizational and operational risk management, the risk manager and health and safety program manager of emergency services organizations must have a thorough understanding of issues that constitute risk. Risk can be viewed as the probability that a harmful consequence will occur if risk management methods do not prevent it. Its anticipated or probable outcome should be rated on a scale of negative consequences. Risk management should be evaluated on three levels:

- The community at large
- The emergency response organization as a whole
- The operations of the emergency response organization

Emergency services organizations play a critical role in community risk management as they defend the community against fire, respond to medical emergencies, and handle other emergency situations. As an example, fire departments exist to reduce the risk associated with probable loss when fire occurs. If the fire departments cannot perform the mission of fire control due to inadequate resources, the risk to the community increases.

Although the overriding concern of emergency response organizations is to manage community risk, they must also be concerned with risk to the emergency response organization at large. The nature of risk and loss varies from organization to organization. Therefore, the goal of the risk manager or health and safety program manager in an emergency response organization is to protect its assets—personnel, equipment, and facilities—to ensure the organization can fulfill its mission. These managers must be aware of potential disruptions to service because they are responsible to ensure that their organizations are always willing and able to respond to emergencies.

The administrations of emergency response organizations are responsible for the public funds and assets budgeted to the organization. It is the responsibility of the risk manager or the health and safety program manager to prevent avoidable losses such as large costs related to vehicle accidents, paying medical bills for injured members, legal expenses to defend against claims, and paying overtime related to losses.

As it relates to organizational risk management, the risk management approach should focus on the risk of personal injury or death related to emergency response operations. In the hazardous work of the emergency services and the activities encountered by emergency response personnel, certain risks are unavoidable. Because of the high amount of inherent risk, many times the result is injury and death of workers. The risk manager, health and safety program manager, and personnel all must learn to identify which risks are too great to take. Sometimes it is appropriate not to take any action. Although danger cannot always be eliminated from the work environment, risk management serves to reduce the inherent risk of the work itself. For example, training prepares responders to recognize situations that pose a danger and enables their working safely in the encountered environment.

SAFETY AUDIT AND JOB TASK ANALYSIS

The use of a safety audit and job task analysis is a way for a fire service organization to complete risk identification by reducing high-risk behaviors and increasing safe behaviors. **Safety audits** can be performed in the station or by the administration of the organization. A checklist should be developed by the health and safety program manager as a way to identify inherent risk to the organization. NFPA 1500, *Standard on Fire Department Occupational Safety and Health Program*, can be used as a point of reference and a yardstick in the area of risk identification and to develop a safety audit, a job task analysis, and a checklist.

safety audit ■ This study of an organization's operations and its real and personal property is to both discover existing and potential hazards and determine the actions needed to render these hazards harmless.

The department should use the safety audit as a safety survey tool. It can use the scores from a safety audit to track upward and downward trends as well as to identify current and future trends in the fire service and the individual organization as a whole. For example, NFPA 1403, *Standard on Live Fire Training Evolutions*, specifies procedures to follow when conducting live-fire training exercises. The safety audit might identify such training procedures that are not in compliance with the current NFPA 1403 standard. Safety audits can be performed internally by the organization or contracted by an independent consultant.

There are two types of safety audits that can be performed by an emergency services organization: general and specific. Assessing its basic hazards and controls are the primary goal and focus of the general safety audit, which would include a broad review of a building's interior and exterior. This type of audit would focus on the entire facility.

Assessing the safety hazards of an organization's operations or a single piece of equipment is the purpose of the specific safety audit. This type of safety audit is particularly useful with high-hazard operations or in areas that have a high frequency and probability of accidents. The specific safety audit is very detailed and is usually very time consuming; examples would be an evaluation of firefighters' personal protective gear, an evaluation of hose and nozzles, and an evaluation of hydraulic rescue and extrication tools.

Both types of safety audits must be specific to the organization's day-to-day operations. Measurable and meaningful data must be gathered for both types of audits to be effective concerning the health and safety programs. In order for a safety audit to be effective, it must be ongoing, and an organization should alternate between the general and specific safety audits. See Figure 3-2.

Job task analysis is widely recognized as the foundation of successful training. As a part of risk identification, job task analysis is the process of determining which tasks each employee needs to perform and the standards at which he or she must perform them. The job analysis process produces three important tools:

job task analysis ■ The process of determining which tasks each employee needs to perform and the standards at which he or she must perform them.

- Task lists
- Job breakdowns
- Job performance standards

A task list helps in planning employee training by listing all the tasks that must be performed by an employee in a given position. It should be prepared for each category of employee to be trained (for example, firefighters, paramedics, hazardous materials technician, rope rescue technician, fire rescue divers).

Job breakdowns incorporate standard operating procedures (SOPs) and specify *how* job duties must be performed to meet the organizational standards. The amount of detail involved in a job breakdown depends upon the intricacies and complexities of the task.

Job performance standards describe what levels of employee performance are acceptable to the organization. They may be expressed as minimum performance levels or as desired performance levels and must also include measurements that correspond with each performance level. It is crucial for an organization's employees to be able to do their jobs at a level that meets basic quality and quantity standards. The same performance standards employed for training can also be used on an ongoing basis to evaluate and improve employee performance on the job throughout the employee's career. See Figure 3-3.

PAST INJURY IDENTIFICATION

A review of past injuries and deaths is part of the risk identification process. By reviewing national and individual department statistics, not only can the specific risks in an individual organization be identified, but also the risks associated with emergency response organizations worldwide. Within any organization, a review of injury reports and workers' compensation history help in this. Outside an organization, a review can be made by researching NIOSH and OSHA illness/injury reports. After a review, it is important to compare local statistics with those that have occurred nationally to understand why they may differ. Without researching these important areas, it will be difficult to predict future injuries. After a review, injury and risk should be categorized to help the health and safety program manager perform his or her job. The categories might include emergency response, in-station hazards, disasters, emergency medical hazards, training, hazardous materials hazards, technical rescue hazards, and water rescue hazards.

FIGURE 3-2 The specific safety audit identifies areas for safe operation.
Courtesy of Joe Bruni.

Specific Risk Management and Safety Audit Checklist	Yes	No	Compliance Needed	N/A
Training and Certifications				
Exposure Plan Training				
Blood-borne Pathogen/Universal Precaution Training				
Hazardous Materials Awareness Training				
Paramedic and EMT License Current				
Driver Certification Training				
Procedures and Policies Reviewed Annually				
New Equipment Training Prior to Use				
Hazardous Materials Technician Certification Current				
Water Rescue/Dive Certifications Current				
Water Craft Training and Certification Current				
Technical Rescue Training and Certification Current				
Sexual Harassment/Diversity Training Current				
Incident/Injury Reporting Training Completed				
Proper Work Practices Training Completed				
Personnel Wellness/Fitness Program Training Completed				
Proper Communications for Emergency Response Training Completed				
Safe Fireground Operations				
Fire Suppression Policies and Procedures Followed				
Safe Operations Followed (Lifting, Tool Use, Ergonomics)				
EMS Team Standing by During Operations				
PPE Worn and Used Safely				
Traffic Control Measures in Place During Operations				
Risk Management Procedures in Place				
Incident Command in Place				
Personnel Trained in Hazardous Materials Response				
Personnel Held Accountable for Unsafe Actions				
All Personnel Radio Equipped				
Safety Issues Reviewed by Safety Committee				
Proper Use of SCBA				
Proper Use of Wildland PPE				
Workplace Wellness and Fitness				
Physical Fitness Program in Place				
Annual Employee Physicals in Place				
Relevant Physical Abilities Testing in Place				
Employee Assistance Program for Employees				
Injury Rehabilitation Program in Place				
Preemployment Physicals and Mental Testing in Place				
Peer Fitness Trainers for Employee Assistance				

FIRE FIGHTER I
NFPA 1001, 2008

5.1 General Requirements

Candidate: ____________________ **Date:** ____________

STANDARD: 5.1.2 NFPA 1001, 2008 Edition	**TASK:** Don personal protective clothing within one (1) minute: doff personal protective clothing and prepare for reuse.				
PERFORMANCE OUTCOME: The candidate shall be able to properly don personal protective clothing in one minute and to prepare the personal protective clothing for reuse.					
EQUIPMENT REQUIRED: Bunker pants, coat, hood, gloves, and helmet.					
CONDITIONS: Given personal protective clothing, the candidate shall demonstrate the ability to:					
NO.	**TASK STEPS**	**FIRST TEST**		**RETEST**	
		PASS	**FAIL**	**PASS**	**FAIL**
	Dohhing				
1.	Don pants and boots properly—including suspenders in place.				
2.	Don hood.				
3.	Don coat—including storm flap closed and collar up and secured.				
4.	Don helmet.				
5.	Don gloves.				
6.	Complete all the above correctly and within one minute.				
	Doffing				
7.	Place all equipment in a ready state for reuse.				

Evaluator Comments: ______________________________

____

____

Evaluator (print & sign) ________ **Date** ________ **Candidate** ________ **Date** ________

Re-test Evaluator (print & sign) ________ **Date** ________ **Retest Candidate** ________ **Date** ________

FIGURE 3-3 Example of a job performance standard. *Courtesy of Joe Bruni.*

Risk Evaluation and Assessment

The second NFPA component of risk management is risk evaluation. Once risks have been identified, they must then be evaluated as to both their potential severity of loss and their probability of occurrence. In the assessment process, it is critical to make the best-educated guesses possible in order to prioritize the implementation of the risk management plan properly. After the risk identification process has been completed, the associated risks should be recorded for future reference. There are two risk management evaluation areas: frequency and severity.

frequency ■ How often a risk occurs or can be expected to occur.

Frequency of a hazard or risk addresses the likelihood of occurrence. For this reason, it is not possible to assign numerical measures to such data. The risk manager must assess how often the risk will occur in an effort to assign a priority to it. Unfortunately, no hard-and-fast rule can determine what an acceptable level of risk is. Using historical data that include local and nationwide statistics of similar size departments and demographics will help in compiling this data. Risks with high probability of occurrence and serious consequences must be immediately addressed; those nonserious risks with a low likelihood of occurrence are a lower priority.

severity ■ How bad a risk will be if it does occur.

Severity of a hazard or risk addresses the degree of seriousness of the incident. During the risk evaluation and assessment, the risk manager must determine how great the loss is. For example, suppose our department seems to have a significant number of steam burn injuries to firefighters every year. Using historical data we determine there are 650 working structure fires per year. The research indicates there have been nine steam burn injuries to firefighters over the past year; the risk may sound high. Is this frequency high, medium, or low? In a medium-size department, this may be a high number of steam burn injuries. What about a department with 95 working structure fires per year? In a small department, nine steam burn injuries may indeed be high; however, in a large department this injury may be a medium or high amount of risk.

The way to determine whether a risk has a high or low frequency is to compare its history locally to its history nationwide in departments of similar size. Severity is concerned with the greatness and consequences of the loss. Again, no numerical rating can be assigned to the severity category. Just like frequency, a factor is assigned as to whether the risk is low or high, and some of the judgment or opinion will be left up to the individual who conducts the measurement and evaluation. The organization must consider two additional areas: associated cost and organizational impact.

ASSOCIATED COSTS

The costs associated with risk evaluation and assessment are measurable: for example, the cost of time away from the workplace, damages, and disruption of service; legal costs; and the cost of repairs, time, or replacement. Using the information gathered in the evaluation and assessment step, the risks can then be classified on the basis of severity.

Certain other costs are more difficult to measure; for example, the cost of low morale, low productivity, employee stress, infection prevention, or the cost to replace an employee. In many cases, these types of costs will have to be estimated in emergency services organizations.

THE RISK TO THE ORGANIZATION

It is important to evaluate how the risk plays a role within the organization. For example, how would several employees being ill from a pandemic affect the organization? In a smaller department numerous illnesses could affect the ability to continue emergency response. In a larger department, the organization might be able to absorb the impact from these illnesses.

Another area that should be factored into organizational impact is the time it takes to recover from an event. This may include the replacement of personnel, equipment, or apparatus. Certain types of events can be devastating to an emergency services organization.

COMPLETION OF THE RISK ASSESSMENT AND EVALUATION

After the health and safety program manager or risk manager completes a risk evaluation and assessment, the unique combination of frequency and severity will determine the organization's priorities and risk control measures. Risks of a high probability of frequency are a higher priority than those nonserious incidents with a low likelihood of frequency; the latter can be placed near the bottom of the "action-required" and priority list. During

TABLE 3-1 Risk of Frequency and Severity of Identified Risks

RISK	FREQUENCY	SEVERITY
Caught and burned in a flashover	L	H
Eye injury related to overhaul	M	M
Involvement in a vehicle accident while responding	L	H
Back injury while pulling hose lines	M	H
Heat-related injury while working in hot weather	M	H
Muscle soreness after a fire event	H	L

the prioritizing process, even though determining frequency and severity is a subjective process, it will prove useful. Tables can be developed to evaluate risk, and an analysis can be made. Ratings will be made after a significant study, with the consequence being a move toward focusing on prevention. See Table 3-1.

THE NATIONAL FIRE FIGHTER NEAR-MISS REPORTING PROGRAM

The fire service and its members have always been proactive in their approach to risk management. An example of this is the National Fire Fighter Near-Miss Reporting System. The goal of this free, voluntary, confidential, nonpunitive, and secure reporting system is improving firefighter safety. Funded by the U.S. Department of Homeland Security's Assistance to Firefighters Grant Program and managed by the International Association of Fire Chiefs, the firefighter near-miss program has heightened awareness of incidents that have resulted in a "near miss." Firefighters and fire service organizations can now access this important program through use of the Internet, and many states and Canadian provinces are also using it. The near-miss program addresses both frequency and severity as it relates to risk management. Almost half of all near-miss reports involve fire emergency events; about 20% occur during non-fire emergency events, such as emergency medical calls and technical rescues; and another 20% are categorized as vehicle events, such as responding to or returning from a call. The remainder of the reports are divided between on-duty events and training events.

Early analysis of reports submitted in the category of fire emergency events points to several human factors being the leading contributing factors to near misses. This information is providing a foundation for formulating new program development strategies. These new initiatives seek to improve firefighter performance in the hazard zone, promote sound risk-versus-reward thinking, and move the fire service toward an intentional actions mind-set in lieu of reactive aggressive action. Near-miss statistics help departments make consequential sense of their reports and thereby help them save lives. A "resources" component of the system's Web site provides videos, photos, training presentations, statistics, and sample newsletter articles.

Risk Control

The third NFPA component of risk management is risk control. The effective management of risk is a process, and so ways must be developed to control risk once the identification, evaluation, and assessment process has been completed. **Risk control** is divided into three categories: risk avoidance; risk reduction, which includes implementation of control measures; and risk transfer. According to NFPA 1500, there must be solutions for elimination or mitigation of potential problems and implementation of a best solution.

risk control ■ This risk management process is designed to control the frequency and severity of losses to an organization.

Risk Management and Monitoring

After risk control measures have been put in place, risk management and monitoring measures must follow to evaluate the effectiveness of risk control techniques. Risk control also entails safety program development, adoption, and enforcement, as well as standard operating procedures (SOPs) development, dissemination, and enforcement. Training and inspections and evaluation for compliance will be needed.

Risk Avoidance

The best choice in any situation is risk avoidance and prevention. Obviously, in emergency services organizations, these tactics will not be possible in all situations because the public calls us when things have become an emergency and the situation is dangerous and unpredictable. However, it may be possible to reduce or control the risk. Risk could be eliminated through control measures (for example, not going into burning buildings eliminates injury from flashover); however, such actions would reduce the potential to save lives and property and would not justify an existence of the fire service. It has become more common in present-day firefighting for firefighters to get caught in a flashover situation, partly due to better protective gear enabling them to go deeper into a fire situation than had been previously possible. Better training in the area of fire behavior and flashover recognition is needed to help in avoiding the risk of burn injuries and flashover.

An example in which risk avoidance has worked well in emergency medical services is the "needleless" system, which does not eliminate the risk of being stuck with a needle, but reduces and controls the risk. Only one needle is used to establish an intravenous site; no other needles are used to administer drugs and medications after establishment of the intravenous site. See Figure 3-4.

The elimination of infectious disease could also be established by not responding to EMS calls, but again this would not justify the existence of an emergency medical response system. A better option is risk avoidance, risk reduction, or risk transfer. Although

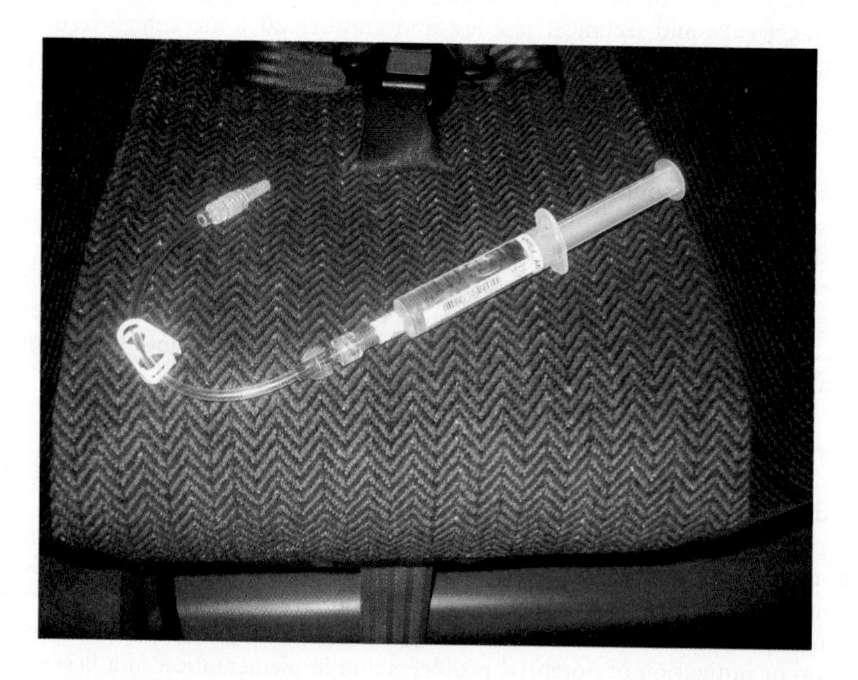

FIGURE 3-4 The "needleless" system is used as a way to reduce the possibility of a needlestick. *Courtesy of Joe Bruni.*

control measures will not eliminate the risk, they can reduce the likelihood of occurrence or mitigate severity.

RISK REDUCTION

If risk cannot be avoided, the next possible option is risk reduction to minimize the risk. The development of safety programs, ongoing training programs, and well-defined standard operating procedures are all effective control measures; however, it is important to recognize the value that input from employees will be in developing these measures. Some typical control and reduction measures instituted to manage fireground injuries include accountability, full protective clothing, a mandatory respiratory protection program, training and education, and well-constructed SOPs. See Box 3-1.

The Heinrich Theory

One method to accomplish risk reduction is to view the injury process as a chain of interruptible events. According to his theory, industrial accident specialist H. W. Heinrich viewed the accident sequence as a combination of five factors: the ancestry and social environment, the fault of the person or human factors, unsafe acts including mechanical and physical hazards, the accident, and the injury itself. Heinrich believed that any interruption in the chain of events of these factors would prevent the injury or risk from occurring. Heinrich[1] uses the analogy comparing the chain of events to dominoes standing in a line. The standing dominoes all will fall if one is pushed; however, if one is removed, the chain reaction stops. Events can be classified according to the five factors in the sequence and control measures developed. The process of using Heinrich's theory is simple and can be applied to more complex problems.

Let us take a look at eye injuries to firefighters that occur from overhaul procedures on the fireground. The risk or program manager breaks down the events that will lead to an injury and then classifies them. See Table 3-2.

The table of events displays the results of this process. Once the events have been viewed, procedures can be written to reduce the occurrence of eye injuries during overhaul. Training could be developed to enforce proper techniques; eye protection could be issued and required during overhaul events; and proper supervision could be put in place to prevent further eye injury.

Heinrich's theory also states that for every accident that causes a major injury, there are 29 accidents that cause minor injuries and 300 accidents that cause no injuries. Many accidents share common root causes, so addressing common accidents that cause no injuries will help to prevent accidents that cause injuries. Heinrich developed the theory of behavior-based safety. Behavior-based safety refers to a group of programs whose focus is based almost entirely on changing or shifting the behavior of workers to prevent injuries and illnesses on the job. Heinrich claimed that up to 88% of industrial accidents could be blamed on workers.[2]

Risk Assessment Matrices

Developing a risk matrix should be the first step toward risk reduction. The risk matrix is a methodology to aid an organization in risk identification and prioritization by looking at the environment, human factors, and measures. The measures will help to determine the costs associated with risk reduction. This process identifies safety risks and helps to develop the most cost-effective measures to reduce those risks. A ranking system should be the methodology used for making risk-based decisions. This ranking encompasses a prioritization and ranking of the determined risks. Regardless of the system or matrix used, the ranking process has two advantages: it differentiates the relative risks to help facilitate decision making, and it improves the consistency of the decisions toward the risk reduction efforts. Any matrix used should include a consequence, severity, and frequency

BOX 3.1 EXAMPLE OF STANDARD OPERATING PROCEDURE

St. Petersburg Fire & Rescue
Standard Operating Procedures **9.3**

Title: RESPONSE TO POWER LINES/ ENERGIZED ELECTRICAL EQUIPMENT

Date Issued: 3/12/2001R
Deletes:
Total Pages: 2

PURPOSE

This procedure will establish a standard approach and response to the report of power lines down. Power lines can come in contact with the ground as a result of storm related activity, fire, or vehicles striking power poles. In all cases the potential for electrical shock/electrocution and secondary fire must be considered.

Key Points

- **Down lines must always be considered energized with potentially lethal current.** Lines can reset and become "hot" or "energized" again by ***manual operation of a switch***, by ***automatic re-closing methods*** (either method from short or long distances away), by ***induction*** (where, a de-energized line can become hot if it's near an energized line), or through ***back feed conditions.***
- Power line tends to have "Reel Memory" and may curl back or roll on itself when down.
- Use caution when spraying water on or around energized electrical equipment. Hose streams conduct current! Never spray directly into the power lines. Use a fog spray at the base of the pole. Your primary responsibility is to protect the surrounding area.
- PCB hazards: Smoke potentially fatal—avoid and contain pools of oil around transformers.

RESPONSE TO POWER LINES DOWN

- Request utility company to respond.
- Consider all down wires as "energized."
- Place apparatus away from "down lines and power poles."
- Locate both ends of downed wires.
- Secure the area/deny entry.
- Periods of high activity: company officer may choose to leave one (1) crew member on-scene with a radio to wait for utility company.
- In the event of multiple lines/poles down over a large area, call additional resources.

DOWN POWER LINES AND VEHICLES

- Request utility company to respond.
- Do not touch vehicle.
- Have occupants remain inside the vehicle.
- Place apparatus a safe distance away from down lines.
- If occupants must leave the vehicle (fire or other threat to life) instruct them to open the door, not step out! They should jump free of the vehicle without touching vehicle and ground at the same time. **This guideline pertains to fire apparatus and emergency procedures if vehicle is contacted by a power line.**

SUB-STATION, TRANSFORMER, ELECTRICAL VAULT AND MANHOLE FIRE

- Request utility company to respond.
- Clear the area.
- Be aware of explosion potential.
- Place apparatus in a safe location away from overhead power lines.
- Protect exposures.
- ***Do not make entry until above electrical equipment has been de-energized.***

Courtesy of St. Petersburg Fire/Rescue

TABLE 3-2 Events Leading to an Eye Injury

EVENT	SOCIAL ENVIRONMENT	HUMAN FACTOR	UNSAFE ACT OR CONDITION	ACCIDENT	INJURY
Operator error		X	X	X	X
Faulty design of tool				X	X
Mechanical failure of tool				X	X
Procedures	X	X	X	X	X
Inadequate training	X	X	X	X	X
Environment (including management organization)	X	X	X	X	X
Failure to use eye protection	X	X	X	X	X

axis, and be easily read and understood by all who use it. A risk matrix should also provide clear guidance on the action(s) necessary to mitigate risks identified as intolerable. It must be understood that emergency services organizations will have both tolerable and intolerable risks associated with their work.

Although risk reduction measures can be generated using various matrices, not all control measures are equally effective. Brainstorming, a popular method to use in risk reduction, helps in selecting the priority risk by assessing control measures to fit into the blocks of a matrix. Brainstorming involves eliminating those impractical measures and conducting further research into those practical measures. An effective method of brainstorming involves an organization's health and safety committee focusing its efforts, goals, and objectives on risk reduction. Such efforts require a deeper look into the risks after the selection of the practical measures, which are divided into whether they are passive or active and voluntary or mandatory. Passive measures do not require any action for control measures to occur; active measures require physical action.

Individuals in the fire service continually receive reinforcement concerning safety. Training, policies, and procedures assist them in voluntary choices. Mandatory measures are required by laws, rules, or regulations and frequently training. The most effective measures are mandatory yet passive. For example, the individual choice to use an SCBA mask–mounted voice amplifier is an active and voluntary measure. See Table 3-3.

SOP/SOG Procedure Development

When writing standard operating procedures, developers can choose a number of different ways to organize and format them. Their goal is to create an easy-to-understand document to help in the type of work to be accomplished. Two factors determine what type of SOP to use: First, how many decisions will the worker need to make during the procedure? Second, how many steps and sub-steps are in the procedure? Short routine procedures that require few decisions can be written using the simple steps format. Long procedures consisting of more than 10 steps, with few decisions, should be written using the hierarchical steps format or graphic format. Procedures that require many decisions should be written in the form of a flowchart. See Table 3-4.

TABLE 3-3 Example of a Voluntary/Mandatory Matrix

	VOLUNTARY	MANDATORY
Passive	EMS at the scene to provide rapid care and transport	Wear a personal alert safety system (PASS)
Active	Have an escape route	Wear full PPE
	Provide a rapid intervention team	Isolate personnel and public from the hazard area

TABLE 3-4 Standard Operating Procedure Format Choices and Criteria

MANY DECISIONS?	MORE THAN 10 STEPS	BEST SOP FORMAT
No	No	Simple steps
No	Yes	Hierarchical or graphic
Yes	No	Flowchart
Yes	Yes	Flowchart

SIMPLE STEPS

Many routine procedures that require repetition and very few decisions can be classified as requiring simple steps. Such an SOP will not normally need much detail. A training program may still be needed to ensure all workers understand the procedure and the steps involved. Unfortunately, a simple steps format may provide few details and may leave a lot of leeway for workers to interpret it. SOPs should be written with sufficient detail to ensure that someone with basic understanding and limited experience or knowledge can successfully carry out the procedure in a safe manner when unsupervised. They should be written in a concise, logical, step-by-step, easy-to-read format.

HIERARCHICAL STEPS

Procedures that require very consistent types of work should make use of a detailed, precise format. The hierarchical steps format permits the use of easy-to-read steps for workers who have obtained experience in the field of study while including detailed sub-steps as well. Experienced users may refer to the sub-steps only when they need to, whereas beginners will use those sub-steps to learn the procedure. A staff member who has good knowledge of the task and has performed it should write the hierarchical SOP, consulting with others as required. Each task step should be recorded in the order of normal sequence, describing what is done, not how it is done. As a working guide, the task description should contain 10 broad steps, depending on the complexity and the hazardous nature of the job. For each potential hazard or risk, identify and list how to complete the work task, including what the worker(s) should or should not do to manage the level of risk. Specifically describe the SOPs and necessary precautions for each step.

GRAPHIC PROCEDURE

When writing procedures for very long activities, managers should consider using a graphic format that breaks long processes into shorter subprocesses of only a few steps. It is easier to learn several short subprocesses than one long procedure. Another possibility for the graphic format is to use photographs and diagrams to illustrate the procedure.

Many emergency services organizations have access to computer systems with powerful graphic systems that will allow development of the graphic type of SOPs. Digital cameras, with their ease to use, ready availability, and low cost, can also be of benefit here. Pictures and graphics are very helpful in SOPs regardless of the literacy level of the worker.

LEVEL OF DETAIL

Generally speaking, SOPs should provide only broad procedural guidelines, not specific details. SOPs are not training manuals, but broad organizational guidelines for performing tasks that members have already been trained to accomplish safely and effectively. SOPs should be clear, concise, and in plain English. It is easy to include regulatory language in SOPs; however such language can be difficult to understand and apply to operational situations. Clear, simple statements are best. An "outline" or "bulleted" style, instead of a continuous narrative, simplifies the presentation of information and helps clarify relationships among different components of the SOP. To be effective, organizational guidelines must be unambiguous, yet at the same time, provide enough flexibility for the on-scene commander to make decisions based on the situation at hand. Department SOPs should be precise but inherently flexible, permitting an acceptable level of discretion that reflects the nature of the situation and the judgment of the incident commander.

SOP DEVELOPMENT PROCESS

The necessary experience and qualifications of personnel should be a component of the SOP, as should notation of any additional training or qualifications necessary.

SOPs should be written by individuals who have experience in and are knowledgeable about the activity and the organization as a whole. They should be subject matter experts in the specific area who actually perform the work or use the process. In fact, a team approach is best, because the experiences of a group of members contribute to both development and use of SOPs. This approach will help promote buy-in from organization members that are going to perform the process.

SOP REVIEW AND APPROVAL

SOPs should be reviewed and validated by individuals with training in and experience with this process. It is helpful if these individuals were not involved in the development process prior to implementation. SOPs should also include a quality assurance component. For example, an SOP involving the health and safety of members should be reviewed and approved by the organization's health and safety program manager, whereas those concerning other areas should be reviewed and approved by a supervisor or chief officer at the administrative level.

SOP IMPLEMENTATION

The development or revision of SOPs must include a plan for implementation that allows time to think through the related tasks, schedules, assignments, and resource needs. The process can be either formal or informal, and one of the first steps is to decide on the scope and task(s) to be accomplished. The approach to the implementation phase may be as simple as a notification to members or employees, to training for selected members. It might also include formal classroom training for the entire organization. For example, a fire department that is developing and implementing its first set of SOPs will take a far different approach than will an established department that is completing minor updates to a few SOPs affecting its operations.

The implementation phase often requires personnel training in the new procedure, which may be formal or informal and conducted in the classroom or on the job. A needs assessment is necessary to determine training requirements; the first step would be to identify how many need the training and the categories of training needs. Of course, this depends upon the nature of the SOP, the scope of recent changes, individual responsibilities within the organization, and other factors. Every individual in the organization who needs the training should receive it.

REVISION AND FREQUENCY OF REVIEW

SOPs need to remain current to be useful. Whenever a procedure changes, the SOP should be updated—often only the pertinent section—reflecting such changes. The changes should be reflected by revision and date change, to maintain quality assurance, and then reapproved. Scheduling periodic review of SOPs ensures that procedures remain current and appropriate. It also helps determine whether the SOP is still needed. In order to encourage a timely review, the SOP review process should not be cumbersome for any one individual or group; so the frequency of review, as well as the individuals responsible for it, should be part of the organization's quality management plan or strategic plan.

Risk Transfer

risk transfer ■ The process of transferring risk to another party, such as buying insurance that will transfer the cost of an accident to the insurance company.

Although **risk transfer** in the emergency response service is commonly accomplished through insurance, it can also be achieved by transferring the risk to someone else. Risk transfer does not eliminate risk; it only reduces or limits the costs to the organization. For a fire or EMS organization, it may be difficult to transfer risk, but the risks associated with some activities—such as operating and maintaining responder helicopters—can be transferred to private contractors.

Purchasing insurance transfers financial risk only. Risk transfer does not affect the likelihood of occurrence, does nothing with certain associated personnel costs, and does little to nothing for the costs associated to replace an employee who is injured or killed in the line of duty. The amount paid for insurance is related to the department history or previous costs associated with accidents and injuries. For this reason, risk transfer should be associated with the health and safety program; and the health and safety program manager should continually look into previous injuries, accidents, and various other risks. The purchase of insurance is no substitute for effective control measures, as insurance companies will look for an organization to have policies and procedures in place that focus on safety-related issues and concerns.

Risk transfer through insurance coverage is an in-depth topic; however, liability, vehicle, and workers' compensation insurance are the most common types of insurance available to an emergency response organization. It would serve an organization well to research its areas of liability and insurance, and to check state and local laws pertaining to insurance coverage.

Training as a Risk Management Tool

Some of the many components of the risk management toolbox are standard operating procedures, pre-incident planning, protective clothing and equipment, incident management system, and personnel accountability. Although the training department is the most important program a department can develop, implement, and manage to maintain effective and efficient members, it is often underfunded and understaffed. Training is very crucial to the successful outcome of any incident and should be based on pre-incident planning, post-incident analysis, and the forecasting of future events. The fire service must develop short- and long-term strategies for training based on each position/function, accident and injury data, and the individual department's current needs.

In the fire service, practical work without preparation is a setup for continued injury and death. Unfortunately, the fire service has a long, bloody history; however, many proactive individuals recognize the need to change this history. Training must also be outlined in a matrix as a way to accomplish successful risk management. The risk management program is a continuous, proactive approach that must be a part of all department operations. Training is the best avenue for developing, implementing, and managing the risk management process. A direct correlation exists between the operations conducted during an emergency incident and the manner in which departments train. Department health and safety program managers need to recognize that training and safety go hand in hand. Unfortunately, due to its culture and resistance to change, the fire service has had a major problem changing its attitude toward safety and risk management.

An excellent example of risk management involved in training came about as a result of serious problems during live-fire training. As a result, the NFPA passed NFPA 1403, *Standard on Live Fire Training Evolutions*. Before NFPA 1403, the safety of members during live-fire training depended on the staff conducting the training. Unfortunately, many instructors and training staff were not experienced enough to conduct live-fire training, which resulted in injury and death. The accident and injury data for a department highlights weak areas that might require training or retraining programs. Another source of data is on injuries associated with failure to wear, or wear properly, protective clothing and equipment.

An individual department's risk management plan is an excellent source for forecasting training needs as the fire service moves forward in utilizing a proactive approach. Another forecasting tool for training is the use of incident response statistics.

CHAPTER REVIEW

Summary

The process of risk management involves four components: risk identification, risk evaluation, risk control, and risk management monitoring. Risk identification is a function of hazard assessment and identification. The health and safety program manager must utilize a historical approach identify events that occurred in the past and be able to recognize what may occur in the future. Risk identification must analyze the local experience of the department as well as trends in emergency services, while continuing to perform a safety audit and reviewing previous injuries.

Risk evaluation and assessment is a way to evaluate the risks that have been identified as to potential severity of loss and probability of occurrence. The two risk management evaluation areas are frequency and severity. Frequency is concerned with how often a risk can occur. Severity is concerned with how bad the risk will be if it does occur. The costs associated with risk evaluation and assessment are measurable. It is important to evaluate how the risks impact the organization. The unique combination of frequency and severity will determine the organization's priorities and risk control measures.

Risk control, as the third component of the risk management process, is divided into three categories: risk avoidance; risk reduction, which includes implementation of control measures; and risk transfer. Risk control can be accomplished by avoiding, reducing, or transferring the risk. In emergency services organizations, risk avoidance is commonly impractical; however, it does have useful application at times. Risk reduction measures are commonly used in emergency services as a way to reduce risk. The development of safety programs, ongoing training programs, and well-defined standard operating procedures are all effective control measures. The Heinrich theory and several additional theories can be used to develop a matrix to reduce the exposure to risk. Risk transfer in the emergency response service is commonly accomplished through the purchase of insurance. Risk transfer does not eliminate risk; it only reduces or limits the costs to the organization.

Although the training department is the most important program a department can develop, implement, and manage to maintain effective and efficient members, it is often underfunded and understaffed. The fire service must develop short and long-term strategies for training based upon pre-incident planning, post-incident analysis, and the forecasting of future events. Training is the best avenue for developing, implementing, and managing the risk management process. Having a risk management plan and compiling incident response statistics are two of the best tools to forecast the training needs of an emergency services organization.

Review Questions

1. Discuss the meaning of and the process of risk identification.
2. The transfer of risk is commonly associated with:
   a. The purchase of insurance
   b. A safety audit
   c. Training
   d. None of the above
3. The most common way to avoid risk in the fire service is accomplished through:
   a. Risk identification
   b. Risk evaluation
   c. Risk reduction
   d. Risk transfer
4. Explain the differences between risk frequency and risk severity.
5. Explain how the Heinrich theory can be applied to the fire service.
6. Explain the types of insurance that are common in the fire service.
7. The job analysis process is comprised of:
   a. Task lists
   b. Job breakdowns
   c. Job performance standards
   d. All of the above
8. The risk matrix is commonly used to accomplish:
   a. Risk identification
   b. Risk assessment
   c. Risk avoidance
   d. Risk reduction

9. Risk management plays an important role in:
   a. Emergency scene operations
   b. Other fire department activities
   c. Training
   d. All of the above

10. A forecasting tool for training is the use of:
   a. Incident response statistics
   b. The Heinrich and Haddon theories
   c. Trends
   d. Risk control measures

Case Study

On June 18, 2007, nine career firefighters in South Carolina died when they became disoriented and ran out of air in rapidly deteriorating conditions inside a burning commercial furniture showroom and warehouse facility. The first arriving engine company found a rapidly growing fire at the enclosed loading dock connecting the showroom to the warehouse. At least seven other municipal firefighters and two mutual-aid firefighters barely escaped serious injury. At the time of the incident, the career fire department was an ISO Class I rated department with 19 fire companies located throughout the city. The department operated 16 engine companies and 3 ladder truck companies at 14 stations in the city. Each apparatus was staffed with four firefighters but routinely operated with three firefighters per apparatus (a captain, engineer, and firefighter), depending on the staffing available each shift. Each shift was supervised by an assistant chief. On the day of the incident, the department had 61 firefighters, 4 battalion chiefs, and an assistant chief working on duty.

The structure involved in this incident was a one-story, commercial furniture showroom and warehouse facility totaling over 51,500 square feet that incorporated mixed-construction types. The structure was non-sprinklered. The facility had been renovated and expanded a number of times over the past 15 years. The original structure was constructed in the 1960s as a 17,500-square-foot grocery store with concrete block walls and lightweight metal bar joists (metal roof trusses) supporting the roof to create an open floor plan. After being converted to a furniture retail store, the original structure was expanded by adding a 6,970-square-foot addition to the right side (D-side) in 1994 and a 7,020-square-foot addition to the left (B-side) in 1995. Both additions were attached to the original exterior walls and consisted of steel beams supporting the walls and roof. To provide access between the original structure and the two additions, the exterior walls on the B and D sides of the original structure were each penetrated in three locations to form six 8′ × 8′ openings that were equipped with metal roll-up fire doors. These fire doors were equipped with fusible links designed to automatically close the doors in the event of a fire. In 1996, a 15,600-square-foot warehouse was added to the rear of the main showroom. The main showroom and the warehouse were connected by an enclosed wood-framed loading dock of approximately 2,250 square feet. Double metal doors connected the rear of the right-side addition to the loading dock area. These metal doors swung outward (opened into the loading dock). Additional access to the loading dock area was available from the rear of the original structure.

The furniture store fire on June 18, 2007, was originally dispatched as a possible fire behind a commercial retail furniture store. The initial incident commander radioed dispatch that the fire was a "bunch of trash free-burning against the side of the structure." The fire very rapidly grew into an incident of major proportions. At approximately 1907 hours, the fire department was dispatched to a possible fire behind a large commercial retail furniture store. Two engines (Engine 11 and Engine 10), one ladder truck (Ladder 5), and the Battalion Chief (BC-4) were dispatched per department procedures. The on-duty assistant chief (AC) was at Station 11 and responded to the scene. While en route, BC-4 observed heavy dark smoke rising into the air and radioed dispatch that smoke was coming from the direction of the store. Per department procedures, this initiated the response of the third-due engine (Engine 16) to the scene. BC-4 arrived on scene driving east to west, pulled past the store, and drove down the alley to the loading dock located on the D-side of the structure. BC-4 observed fire burning from ground level to over the roofline outside of the covered loading dock. BC-4 radioed dispatch that the fire was a "bunch of trash free-burning against the side of the structure."

When the assistant chief arrived on scene, he parked in the parking lot in front of the main showroom right addition. The assistant chief and BC-4 briefly discussed their observations and directed Engine 10 to back down the alley to the loading dock area. The assistant chief entered the store through the main entrance located in the center of the front of the structure (A-side). The assistant chief walked down the center of the showroom to the rear (in the original structure) and then went back outside. He did not observe any smoke or fire in the main showroom. BC-4 drove his car to the front of the showroom and observed the assistant chief coming out of the showroom's main entrance. The assistant chief remained at the front of the store while BC-4 returned to the D-side. Note: Departmental policy was that the highest-ranking officer on-scene was the incident commander. Incident command (IC) was never formally announced at this incident. The assistant chief detected fire when he opened a door connecting the rear of the right showroom addition to the loading dock area. The E-11 acting captain radioed that he needed a 1½″ hand line inside the building. When E-15 radioed that they had relocated to the west side, the assistant chief instructed E-15 to come to the scene. The assistant chief also instructed E-15 to bring a 1½″ hand line inside to the rear right-side of the structure. The assistant chief radioed that the fire was inside the rear of the structure and was moving toward the showroom.

The burning furniture quickly generated large volumes of smoke, toxic gases, and soot that added to the fuel load. At approximately 1926 hours, a store employee called the city's 911 dispatch center and reported that he was trapped inside the back of the building. The city assistant fire chief and a battalion chief (BC-5) quickly instructed a crew of four firefighters from the mutual-aid department to initiate the rescue attempt on the B-side of the warehouse. This crew quickly located the point where the trapped civilian was banging on the exterior wall. They were able to cut through the exterior wall (metal siding) using a Haligan bar and axe. The firefighters were able to safely extricate the civilian at approximately 1933 hours. The civilian employee rescue was announced over the radio. As the civilian was being rescued, the fire was extending into the main showroom. The fire quickly outgrew the available suppression water supply. The interior fire attack crews could not contain the spread of the fire. At this point, three hose lines were inside the main showroom—the initial 1½-inch hose line, a 2½-inch hose line, and a 1-inch booster line. All three hose lines were pulled off Engine 11, which was being supplied by Engine 16 through a single 2½-inch supply line approximately 1,850 feet long. The interior crews from Engine 11, Ladder 5, Engine 16, Engine 15, Engine 19, and Engine 6 became disoriented as the heat rapidly intensified and visibility dropped to zero as the thick black smoke filled the showroom from ceiling to floor. The interior firefighters realized they were in trouble and began to radio for assistance. At least one Mayday was called. Another firefighter radioed that he had lost contact with the hose line and needed help. One firefighter activated the emergency button on his radio. The Engine 6 crew and three firefighters from E-15 were able to find the front door and exit the showroom. The front showroom windows were knocked out to improve visibility. Firefighters, including two firefighters from the mutual-aid crew who extricated the trapped civilian, were sent inside to search for the missing firefighters at approximately 1936 hours. The two mutual-aid firefighters made brief contact with two disoriented firefighters just as the flammable mixture of gases and combustion by-products in the showroom ignited, filling the showroom with flames. The two mutual-aid firefighters lost contact with the two disoriented firefighters and were driven outside by the intense heat and flames. While firefighters were known to be trapped inside, the number and their identities were not known. Interior firefighters were caught in the rapid fire progression and nine firefighters from the first responding fire department were killed. See Figure 3-5.

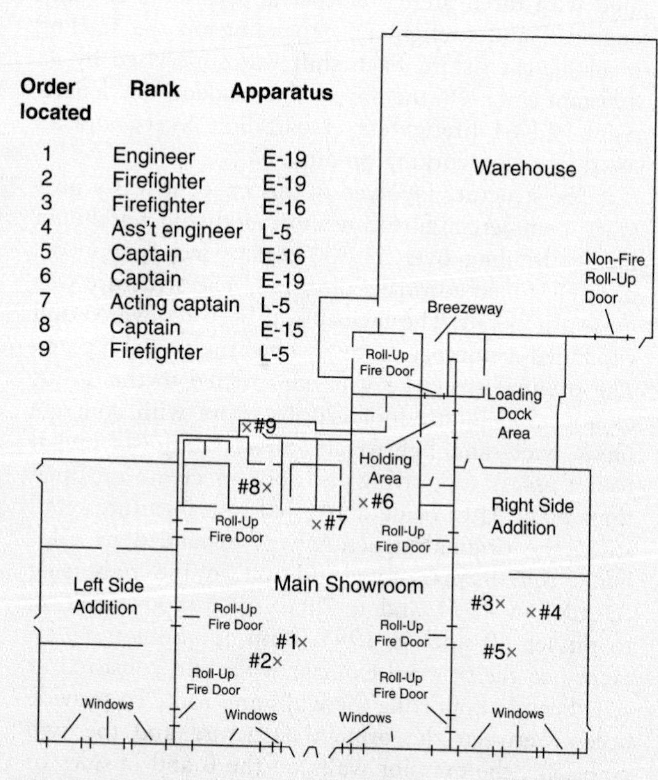

FIGURE 3-5 Floor plan of the Sofa Super Store with firefighter fatality locations. *Source: Firefighter Fatality Investigative Report, June 18, 2007. http://downloads.pennnet.com/fe/misc/20080515charlestonreport.pdf*

NIOSH had the following recommendations:

Recommendation #1: Fire departments should develop, implement, and enforce written standard operating procedures (SOPs) for an occupational safety and health program in accordance with NFPA 1500.

Recommendation #2: Fire departments should develop, implement, and enforce a written incident management system to be followed at all emergency incident operations.

Recommendation #3: Fire departments should develop, implement and enforce written SOPs that identify incident management training standards and requirements for members expected to serve in command roles.

Recommendation #4: Fire departments should ensure that the incident commander is clearly identified as the only individual with overall authority and responsibility for management of all activities at an incident.

Recommendation #5: Fire departments should ensure that the incident commander conducts an initial size-up and risk assessment of the incident scene before beginning interior firefighting operations.

Recommendation #6: Fire departments should train firefighters to communicate interior conditions to the incident commander as soon as possible and to provide regular updates.

Recommendation #7: Fire departments should ensure that the incident commander establishes a stationary command post, maintains the role of director of fireground operations, and does not become involved in firefighting efforts.

Recommendation #8: Fire departments should ensure the early implementation of division and group command into the incident command system.

Recommendation #9: Fire departments should ensure that the incident commander continuously evaluates the risk versus gain when determining whether the fire suppression operation will be offensive or defensive.

Recommendation #10: Fire departments should ensure that the incident commander maintains close accountability for all personnel operating on the fireground.

Recommendation #11: Fire departments should ensure that a separate incident safety officer, independent from the incident commander, is appointed at each structure fire.

Recommendation #12: Fire departments should ensure that crew integrity is maintained during fire suppression operations.

Recommendation #13: Fire departments should ensure that a rapid intervention crew (RIC)/rapid intervention team (RIT) is established and available to immediately respond to emergency rescue incidents.

Recommendation #14: Fire departments should ensure that adequate numbers of staff are available to immediately respond to emergency incidents.

Recommendation #15: Fire departments should ensure that ventilation to release heat and smoke is closely coordinated with interior fire suppression operations.

Recommendation #16: Fire departments should conduct pre-incident planning inspections of buildings within their jurisdictions to facilitate development of safe fireground strategies and tactics.

Recommendation #17: Fire departments should consider establishing and enforcing standardized resource deployment approaches and utilize dispatch entities to move resources to fill service gaps.

Recommendation #18: Fire departments should develop and coordinate pre-incident planning protocols with mutual-aid departments.

Recommendation #19: Fire departments should ensure that any offensive attack is conducted using adequate fire streams based on characteristics of the structure and fuel load present.

Recommendation #20: Fire departments should ensure that an adequate water supply is established and maintained.

Recommendation #21: Fire departments should consider using exit locators such as high intensity floodlights, flashing strobe lights, hose markings, or safety ropes to guide lost or disoriented firefighters to the exit.

Recommendation #22: Fire departments should ensure that Mayday transmissions are received and prioritized by the incident commander.

Recommendation #23: Fire departments should train firefighters on actions to take if they become trapped or disoriented inside a burning structure.

Recommendation #24: Fire departments should ensure that all firefighters and line officers receive fundamental and annual refresher training according to NFPA 1001 and NFPA 1021.

Recommendation #25: Fire departments should implement joint training on response protocols with mutual-aid departments.

Recommendation #26: Fire departments should ensure apparatus operators are properly trained and familiar with their apparatus.

Recommendation #27: Fire departments should protect stretched hose lines from vehicular traffic and work with law enforcement or other appropriate agencies to provide traffic control.

FIGURE 3-6 Concerns and issues of the Sofa Super Store. *Courtesy of Joe Bruni*

Sofa Super Store Fire Concerns and Issues	Yes	No
Did interior crews perform a proper risk versus benefit survey?		
Were hose line and fire stream selection adequate for this type of event?		
Was there an adequate incident management system in place?		
Did the interior crew provide adequate and recent update of conditions to the incident commander?		
Did the incident commander continuously evaluate changing conditions?		
Was crew integrity maintained throughout this event?		
Did the lack of recognition of the void space and fire conditions play a role at this event?		
Was the established water supply adequate for this size event?		
Would adequate firefighter survival training have played a role at this fire?		
Was personal protective gear worn properly at this event?		

Recommendation #28: Fire departments should ensure that firefighters wear a full array of turnout clothing and personal protective equipment appropriate for the assigned task while participating in fire suppression and overhaul activities.

Recommendation #29: Fire departments should ensure that firefighters are trained in air management techniques to ensure they receive the maximum benefit from their self-contained breathing apparatus (SCBA).

Recommendation #30: Fire departments should develop, implement, and enforce written SOPs to ensure that SCBA cylinders are fully charged and ready for use.

Recommendation #31: Fire departments should use thermal imaging cameras (TICs) during the initial size-up and search phases of a fire.

Recommendation #32: Fire departments should develop, implement, and enforce written SOPs and provide firefighters with training on the hazards of truss construction.

Recommendation #33: Fire departments should establish a system to facilitate the reporting of unsafe conditions or code violations to the appropriate authorities.

Recommendation #34: Fire departments should ensure that firefighters and emergency responders are provided with effective incident rehabilitation.

Recommendation #35: Fire departments should provide firefighters with station/work uniforms (e.g., pants and shirts) that are compliant with NFPA 1975 and ensure the use and proper care of these garments.

Recommendation #36: Federal and state occupational safety and health administrations should consider developing additional regulations to improve the safety of firefighters, including adopting National Fire Protection Association (NFPA) consensus standards.

Recommendation #37: Manufacturers, equipment designers, and researchers should continue to develop and refine durable, easy-to-use radio systems to enhance verbal and radio communication in conjunction with properly worn SCBA.

Recommendation #38: Manufacturers, equipment designers, and researchers should conduct research into refining existing and developing new technology to track the movement of firefighters inside structures.

Recommendation #39: Code setting organizations and municipalities should require the use of sprinkler systems in commercial structures, especially ones having high fuel loads and other unique life safety hazards, and establish retroactive requirements for the installation of fire sprinkler systems when additions to commercial buildings increase the fire and life safety hazards.

Recommendation #40: Code setting organizations and municipalities should require the use of automatic ventilation systems in commercial structures, especially ones having high fuel loads and other unique life safety hazards.

Recommendation #41: Municipalities and local authorities having jurisdiction should coordinate the collection of building information and the sharing of information between building authorities and fire departments.

Recommendation #42: Municipalities and local authorities having jurisdiction should consider establishing one central dispatch center to coordinate and communicate activities involving units from multiple jurisdictions.

Recommendation #43: Municipalities and local authorities having jurisdiction should ensure that fire departments responding to mutual-aid incidents are equipped with mobile and portable communications equipment that are capable of handling the volume of radio traffic and allow communications among all responding companies within their jurisdiction.[3] See Figure 3-6.

1. Explain how a proper risk verses benefit analysis would have changed the incident strategy and tactics.
2. Describe how building information contained within a preplan would have been beneficial at this incident.

References

1. H. W. Heinrich, *Industrial Accident Prevention: A Scientific Approach* (4th ed.), New York: McGraw-Hill, 1959.
2. *Who is H. W. Heinrich*? Retrieved May 19, 2011, from http://www.insurancefreefaq.com/insurance/34381-insurancefreefaq.html
3. NIOSH Fire Fighter Fatality Investigation and Prevention Program (FFFIPP), *Nine Career Fire Fighters Die in Rapid Fire Progression at Commercial Furniture Showroom—South Carolina*, February 11, 2009. Retrieved May 15, 2009, from http://www.cdc.gov/niosh/fire/reports/face200718.html

CHAPTER 4

The Development and Management of the Safety Program

KEY TERMS

goals, *p. 75* **objectives,** *p. 76* **policies,** *p. 82*

OBJECTIVES

After reading this chapter, you should be able to:

- Describe the essential elements of a safety and health program.
- List the tools available concerning program development.
- Describe the process to establish goals and objectives.
- Explain the difference between goals and objectives.
- Describe the process for developing an action plan.
- Explain why knowledge of the laws, regulations, and standards is an important aspect concerning health and safety program development.
- Establish and perform a cost–benefit analysis.
- Explain the relationship between training and the health and safety program.
- Describe the process for developing standard operating procedures.
- Explain why the evaluation of SOPs is an important function.

A comprehensive health and safety program in today's emergency services organizations requires a great deal of groundwork. The health and safety program manager, as well as the organization as a whole, is inundated with a vast amount of information from many areas that must be sorted through. This chapter identifies the essential elements necessary to formulate a successful program. The tools concerning program development will be presented in a specific order; however, the order of use of these tools in each organization differs depending upon its specific needs. The goal is to enable fire departments to design effective safety programs based on community hazards and service commitment, enhance firefighter safety, and provide tools for continual evaluation of emergency response systems.

Historically, the fire service has been based solely on those activities related to fire prevention and suppression. As has been mentioned, times have changed; many emergency services organizations now provide emergency medical services, terrorism response, hazardous materials response and mitigation, natural disaster response, specialized rescue, and response to other community needs. The concern for health and safety is no longer concerned with just firefighting. The additional roles and job responsibilities of emergency responders has caused significant challenges in this important area.

Health and safety program development involving emergency services requires both managers and employees to have knowledge of not only the pertinent regulations and standards but also the written policies and procedures. The latter help to ensure compliance with mandated regulations and suggested standards. Emergency services organizations must be able to determine and design an acceptable level of service and resource deployment. Fire service organizations must also focus on fatality and injury reports from reputable sources at the local, state, and federal levels to ensure the proper focus of the health and safety program. OSHA's electronic assistance tools (eTools) help in developing a comprehensive health and safety program by recommending good industry practices.

Several organizations focus their efforts on data collection so all emergency services organizations can analyze the same data across the nation. These organizations include the National Fire Protection Association, International Association of Firefighters, and United States Fire Administration, just to name a few. Standardized data collection establishes a knowledge base concerning employee injury and death as a way to further risk management, as well as injury and death prevention.

Safety Program Management

Safety program management is important to the success of day-to-day operations in emergency services organizations. Ineffective management of the safety program, even of a good program, will eventually lead to a lack of support. Effective planning, resources, communication, training, and evaluation tools must be in place to ensure support for a good program. Sound management techniques engender the success and effectiveness of the safety program.

Effective safety program management is not only about having control over the tasks associated with management of the program but also about gathering data associated with health and safety in order to express and evaluate mathematical information. Charts, graphs, computers, and cost analysis will be necessary to determine ongoing program needs. The safety program may contain "projects" that will produce tangible end products or results; these projects require organizing, planning, coordinating, staffing, budgeting, and reporting. The three basic concepts of safety program development are (1) determination of and setting goals and objectives; (2) development and implementation of an action plan; (3) and evaluation of the results and methods.

It is imperative that the safety program manager consider the political environment that will be associated with safety program development and be able to navigate effectively in it. In most cases, the emergency services organization is just one of many departments or organizations competing for limited resources and funds. Necessary resources may not be readily available in the political environment. In its simplest form, politics is nothing more than relationships with people and groups of people. By the organization's establishing good relationships with people both within it and external to it, the development of the safety program will move forward in a positive and proactive way. The political forces within the organization and the community's political leaders must empower the safety program manager to successfully develop the safety program. He or she needs to do a very good job selling the program to the department administration and the political leaders. The program manager must understand and know the groups, individuals, and people that will favorably influence safety program development.

Safety Program Issues

One of the first steps in program development is identification and review of the basic program issues. Program issues will vary from department to department based on several factors including whether the department is volunteer or career and the typical responses to which the department responds. Many emergency services organizations have cross-trained employees who perform more than one function. Emergency services organizations can access the information contained in NFPA 1500, *Standard on Fire Department Occupational Safety and Health Program*, to establish the following program issues:

- An administration that is positive and committed to the safety program
- A fire department organizational statement
- A safety and health program manager
- A risk management assessment and plan
- Policies related to health and safety
- A computerized data collection system for data collection and analysis
- An occupational safety and health committee
- Incident safety officer(s) and a program developer
- Standard operating procedures, guidelines, and policies
- A training program including training officer(s)
- Accident and injury investigation, procedures, and review process
- Apparatus and protective equipment that meets requirements and current standards
- Facility inspection that includes safety audits
- A department physician with programs including medical surveillance, wellness programs, physical fitness, nutrition, and injury and illness rehabilitation
- An infection control program and plan
- A critical incident stress management program and employee assistance programs
- A post-incident analysis process
- Local, state, and national injury and death statistical review
- A process to analyze policies and procedures in order to comply with local, state, and national regulations and standards

The difficult aspect of health and safety program development and management is the cost of the necessary resources. Local government leaders and communities must continually be made aware of how important it is to dedicate resources toward this type of program. Injury and death rates for workers can be reduced only if the department administration realizes the benefits of this reduction and makes the local government leaders aware of this critical component.

#2

Emergency services organizations must learn to market themselves better than they ever have. Many health and safety programs must be developed gradually as money and resources become available. It certainly does not cost anything to obtain the commitment of local government leaders and emergency administrations. Then, once a commitment from administration is obtained, a safety program manager, data analysis system, safety committee, and access to local, state, and national data are all that is needed to bring a health and safety program out of the starting block.

Strategic Planning

Once the safety program issues have been evaluated, it is time to start setting goals and objectives for the program. Strategic planning is an important step during the goals and objectives process. Health and safety training needs must be developed for all employees, as the goal for them is to learn and understand their jobs.

GOALS

Setting **goals** is a key beginning step in the process. Goal setting ideally involves establishing specific, measurable, and time-targeted objectives. Although it is the health and safety program manager's responsibility to set goals, the input of a health and safety committee or strategic planning group should be encouraged. Goal setting helps employees and the organization develop a clear awareness of what must be done to achieve an objective.

goals ▪ Broad statements of concerns and issues that need to be accomplished.

A main focus of realistic and attainable goals should be effective operations of the organization and the internal and external support services. Brainstorming and the process of involving more than one to a few individuals will help to produce a better result, as well as a greater chance of acceptance by members of the organization. An organizational goal-management solution ensures that individual employee goals and objectives align with the entire organization's vision and strategic goals in order to steer the health and safety program.

Goals may never be reached if there is too much to do with limited resources. The health and safety program manager, as well as the health and safety committee, must recognize that members and employees might lose interest in the goals of the program, and the goals might never be accomplished if realistic and adequate resources have not been assigned to the task at hand. With goal management, every involved employee understands how his or her efforts contribute to an organization's success. When developing goals, employees should know exactly what is expected:

- To do in performance of their tasks
- To do as a way to improve their performance
- To do as a way to improve their behavior

Goals should be measurable and have a specific time frame. Once a goal has been developed, specific statements, objectives, and a course of action should be developed to meet the goal. These provide the plan or course of action to reach the goal. During program development, it must be kept in mind that goals are broad based, whereas objectives are specific.

OBJECTIVES

objectives ■ Clearly defined desired results that are specific and measurable within a given time frame.

In many situations the terms *goals* and *objectives* are used interchangeably. **Objectives** are also goals, but further down in the hierarchy. Goals can be viewed as "where we want to be," with objectives being the steps needed to get there. Specific measures are used to determine whether or not the goals have been met.

As with goals, objectives should be measurable and obtainable within a given time frame, as well as clear and to the point. The goal may be to get all employees to wear their seat belt during emergency response and upon returning to quarters. The objective would be to develop a standard operating procedure or policy and have employees sign a document within 60 days requiring their compliance. Objectives should be easily analyzed, so that near the end of the time frame it can be determined whether they have been completed. It is common in emergency services organizations to write objectives to give specific direction to training events.

Virtually every public and private institution sets goals or objectives to be used as motivational and management tools. Objectives can be set for both short and long time periods. In the training realm, effective instructional objectives describe a measurable outcome at a specific level of accuracy that can be observed and performed by the student or employee. Because objectives are a planning tool, they are concrete, tangible, narrow, and precise. At the end of the process, objectives should be validated. Should an objective or activity not work to achieve organizational goals, change or replace that objective so that it does.

Achieving goals and objectives involves persistence and flexibility, and change does not come easily in most organizations, particularly in emergency services. If the process involved in achieving objectives is not working, change the means not the end. Keep trying until a method is discovered that will work. By continually evaluating goals and objectives, problems and possible solutions can be identified and solutions sought. The health and safety program manager should work together with the safety committee on the goals and objectives for a safety program. After goals and objectives have been established, they should be published so that all department members recognize and work toward them. It is important to know the difference between goals and objectives: objectives are performance oriented. Once objectives have been established, it is time to move forward and begin to manage those established objectives as a method to produce results.

Management by Objectives

The focus of management by objectives is to develop a functional operating system that operates efficiently beyond the current system that is in place. Specific steps must be identified to achieve necessary results. In order to establish a more efficient and effective system, it must be determined whether the current system is capable of accomplishing the desired objectives. It is more than likely that changes in the present system will be necessary. An effective process would involve (1) identification of the core values and services offered by the organization and (2) identification of the supporting services that help the organization offer its core values and services. The organization then recognizes any service gaps, which will lead to establishing goals and objectives for the organization's strategic plan.

The next step is writing enabling objectives concerning the changes and improvements that need to be made. The first consideration with enabling objectives is how to accomplish these changes and improvements in the system currently in place. These objectives must be performance based. Expected and actual results should be monitored at timed intervals. It may become obvious during this process that established objectives are not realistic and so may require change or adjustment. In many cases, the lack of necessary and adequate

Goal	Objectives	Critical Tasks
To increase diversity within the organization	To use recruitment as a method to gain an awareness and interest among potential diverse candidates	Conduct job fairs at strategic locations using the department's public education specialists
	Offer scholarships for fire academy and paramedic programs	Training division to meet with local agencies such as Job Corps and Worknet to identify scholarship monies
	To implement a structured program as a guide toward promotion	Identify and assign mentors within the organization to guide potential candidates in the officer development process
	Educate the public and students concerning the fire service within the community	Implement a committee to develop a citizen's fire academy as a marketing tool

FIGURE 4–1 An example of the strategic planning process.

resources is what most organizations will have to work through. Throughout this process, issues will be identified that do not involve an increase in financial resources to accomplish all of the goals and objectives. The focus should remain on efficiency and a better way to deliver service both internally and externally. The common theme should be an improvement to health and safety of employees and the public at large. See Figure 4–1.

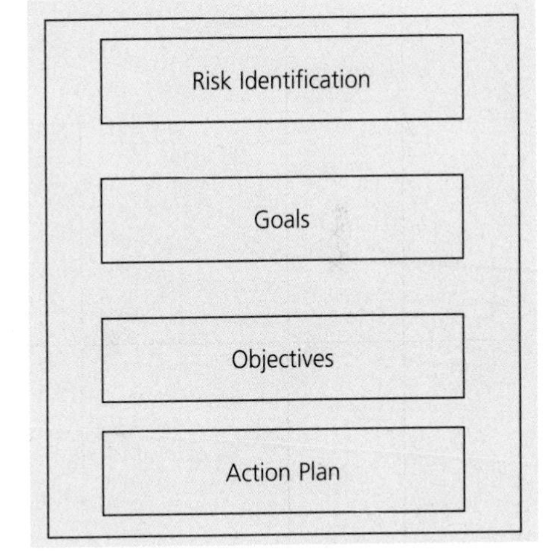

FIGURE 4–2 The sequence of events among goals, objectives, and the action plan.

Action/Strategic Planning

The action/strategic plans are written plans to tie the goals and objectives together. The next step after clarifying the goals and objectives is to develop a written list of actions or critical tasks in order to achieve the goals and objectives, generating many different options and ideas (goals and objectives). At this stage, it is important not to analyze, but to write as many ideas as possible. See Figure 4–2.

There are many components of writing an action/strategic plan. After the list of ideas is placed on paper, decisions must be made as to which ideas are effective steps in achieving the goals and objectives. Which action items or critical tasks can be dropped from the plan without significant consequences to the outcome? The remaining list should then be organized into a plan. Actions, critical tasks, and ideas should be listed in a sequence of ordered action steps. Each objective needs to have an action plan developed for it. After the sequence of action steps is determined, simplify the list as much as possible. The plan should then be executed and monitored on a continual basis.

From this point, new information should be used to adjust and optimize the plan further. The action/strategic plan should be developed by an individual or by a team that has been assigned to address the objective. The following should be included in the plan: completion times, benchmarks, resources needed, a completion date, a place to list barriers and obstructions, and the necessary steps to implement all of the objectives and critical tasks. Progress can be evaluated by the established benchmarks concerning the critical tasks. Implementation of the plan should include issues of anticipated roadblocks. Development of the final action plan should include:

- All components (parts of the plan)
- Individual assignments
- Time frame
- Budget/financial needs

Action	Action Steps	Person Responsible	Time Frame	Budget Needs	Time Allocated	Resources Needed	Support/ Roadblocks	Evaluation Method

FIGURE 4–3 Sample action plan.

- Evaluation method
- Date for final evaluation

The sample action/strategic plan is not all-inclusive; however, it provides a good road map to meet goals and objectives. See Figure 4–3.

Cost–Benefit Analysis

A cost–benefit analysis is a formal discipline used to help appraise, or assess, the case for a project or proposal. It is also an informal approach to making decisions of any kind. An action plan involves weighing the total expected costs against the total expected outcome.

In emergency services, a cost–benefit analysis will involve the study of the future reduction of risk. Local government uses cost–benefit analysis to evaluate the desirability of a given intervention; for example, a cost–benefit analysis of the purchase of filter masks for firefighters to conduct overhaul operations in an effort to reduce the amount of particulate matter firefighters might breathe, resulting in a reduction of associated lung problems and health-related issues. The aim is to gauge the efficiency of the intervention relative to the status quo.

When conducting a cost–benefit analysis, list all of the parties affected by an intervention. A monetary value is then determined concerning the effect the intervention has on their welfare as it would be valued by them. In emergency services, it is common to determine an initial outlay for a certain program to result in a future risk reduction, should money be spent on the program. The process involves monetary value of initial and ongoing expenses versus expected return.

At times, it is difficult to determine measures of the costs and benefits of specific actions. The benefits of a cost–benefit analysis may be based on estimations of improvement instead of actual results. Whereas it is very difficult, if not impossible, to determine *exact* outcomes of a project, *expected* outcomes can usually be established. Many emergency services organizations use information and results from agencies that have implemented similar projects to estimate costs and outcomes. Cost–benefit analysis is mainly used for large private and public sector projects that are less amenable to being expressed in financial or monetary terms.

In emergency services organizations, it would be common to analyze historical data, such as injury and death rates, in doing a cost–benefit analysis. Such a review aids the organization in determining whether injuries or deaths could have been prevented had a certain program, the one in question, been in place. When using cost–benefit analysis, remember that emergency services organizations are like no other organization in either the private or the public sector. Emergency services organizations cannot put a price on certain losses within the organization; for example, the loss of life is viewed as unacceptable, and certain programs must be in place regardless of a cost–benefit analysis being completed.

The accuracy of the outcome of a cost–benefit analysis depends on how accurately costs and benefits have been estimated. It is the health and safety program manager who determines and evaluates the cost-effectiveness of a cost–benefit analysis as it relates to the health and safety program. Such an evaluation should involve an analysis of the current cost of the risk, comparing that with the cost of the program and its implementation. The outcomes of cost–benefit analysis of direct and indirect costs should be treated with caution, because they may be highly inaccurate. Direct costs could be the purchase of new equipment and the cost of overtime; indirect costs could be the loss of equipment from use and the cost of hiring replacement employees.

Another challenge to cost–benefit analysis is in determining which costs to include in the analysis. This process is often controversial, as interest groups may disagree about what costs to include in or exclude from a study. Cost–benefit analysis is a powerful yet relatively easy tool for deciding whether to make a change. Management must decide whether measures will be effective from purely a cost standpoint. See Figure 4–4.

To use a cost–benefit analysis, first work out how much the change will cost the organization. Then calculate the benefit to be gained from it. Once the calculations have been completed, informal decisions can be made whether or not to move forward. It is also important to work out the time it will take for the benefits to repay the costs. Cost–benefit analysis can be carried out using only financial costs and financial benefits; however, intangible items may be included within the analysis. As a value is estimated for these intangibles, it inevitably brings an element of subjectivity into the process.

There is also a cost associated with no action concerning a cost–benefit analysis. It is always easier to implement a program and justify the costs associated with injuries that have already occurred than it is to show a cost benefit for preventing injuries that have not yet occurred. Unfortunately, most emergency services organizations do not take any action or incur further costs until the next serious injury or fatality occurs.

Cost-Benefit Analysis Goal: Replace Firefighter Gloves with a Type Offering Better Protection	Cost
Current Situation and Costs	
15 hand injuries over a year resulting in emergency room visits at $500.00 per visit and an average of 2 days (48 hours) of lost work	
Direct Costs	
Medical expenses	$7,500
Indirect Costs	
The cost of overtime to cover vacancies on shift at an hourly rate of $20.00 per hour	
48 × 15 × 2 × $20.00	$28,800
Total Cost	$36,300
Future Estimation of Hand Injuries	
5 hand injuries per year resulting in emergency room visits at $500.00 per visit and an average of 2 days (48 hours) of lost work	
Direct Costs	
Medical expenses	$2,500
Indirect Costs	
The cost of overtime to cover vacancies on shift at an hourly rate of $20.00 per hour	
48 × 5 × 2 × $20.00	$9,600
Total Cost	$12,100
First-Year Savings (Estimated Based on Research)	$24,200
Program Costs	
Direct	
Medical expenses	$2,500
New gloves at $50.00 per pair × 300 pair	$15,000
Indirect	
Training time	$2,000
Total Program Costs	$19,500
Total Cost/Benefit—First Year	
First-Year Savings Minus Program Implementation Costs	$4,700

FIGURE 4–4 Example of a cost–benefit analysis.

Training

Every emergency services organization must consider the direct correlation between training and health and safety. Often the training officer or chief officer in charge of the training division is also the department health and safety program manager. Due to a fire department's

FIGURE 4–5 A training class on safety and survival. *Courtesy of Joe Bruni.*

impact on society, its importance cannot be stressed enough. To meet the demand for services, the skills and proficiency of firefighters must continually be improved. See Figure 4–5.

Fire training allows firefighters to be placed in real-life situations and put their skills to the test. Although theoretical knowledge is important, it must be coupled with hands-on practical training and experience so that individuals can practice and apply important safety procedures and methods.

As discussed in Chapter 2, the NFPA and OSHA have established regulations and standards on the many levels of required training. Firefighters must train to work within the confines of a calculated risk due to the dangerous nature of the job. Training gives firefighters the opportunity to develop their situational awareness, a vital need on the fireground. All emergency services organizations normally mandate training so that responders can operate safely and learn to become responsive to the situations they face daily. Without adequate training, the injury and death rates in the fire service would be at a much higher level than reflected by current statistics.

In most cases, individual states require a minimum level of training for firefighters, EMTs, and paramedics. Firefighters must continually hone and practice their skills to respond in an efficient and effective manner. Many of the training requirements are directly related to the health and safety of emergency services personnel. It is vital that as the health and safety program develops, so do the training efforts. Training ensures that firefighters are proficient in the use of their equipment.

#8

The organization's policies and procedures such as an accountability system and rapid intervention team procedures may look good in theory, but are of little use if they do not work when needed. It is through training that these theories can be proven effective. Training helps to ensure emergency responders can work with the innovative tools, techniques, and equipment being developed on a continual basis.

The final reason why training is so important is that it develops teamwork. When workers prove successful during training, it boosts morale as well as knowledge, skills, and abilities. Emergency responders must develop confidence in one another.

Of the many functions within any emergency services organization, the training function and staff should be the most closely involved with the health and safety component. A fire service training officer or specialist develops programs, coordinates programs to maintain and enhance performance, develops standard operating procedures (SOPs), and

BOX 4.1 NFPA STANDARDS RELATED TO FIRE SERVICE TRAINING

NFPA 1401 — *Recommended Practice for Fire Service Training Reports and Records*
NFPA 1402 — *Guide to Building Fire Service Training Centers*
NFPA 1403 — *Standard on Live Fire Training Evolutions*
NFPA 1404 — *Standard for Fire Service Respiratory Protection Training*
NFPA 1410 — *Standard on Training for Initial Emergency Scene Operations*
NFPA 1451 — *Standard for a Fire Service Vehicle Operations Training Program*

ensures that these SOPs comply with health and safety regulations and standards. The fire service training officer is normally a mid-management-level position that answers directly to a deputy or division chief, or directly to the fire chief. Many resources, including fire and emergency medical organizations, are available for a department's training program: for example, the U.S. Fire Administration, local colleges and universities, fire and EMS professional organizations and training centers, textbooks, the World Wide Web, and the NFPA. See Box 4–1.

Recently, the Internet has become a common area in which to acquire training information, ideas, and experience from other agencies and organizations. In today's emergency services world, the Internet can provide valuable training information, experiences, and ideas from other emergency response departments and professional organizations. An organization that continually trains its personnel and focuses on the health and safety of the members will operate as a safer and more proficient organization.

The Development of Procedures and Policies

policy ■ A course of action thought to be prudent or tactically advantageous. It serves as a guide that establishes the parameters for decision making and action.

Standard operating procedures (SOPs) or guidelines (SOGs) as well as **policies** are a necessary component of meeting and addressing some of the health and safety objectives defined in the program. It is unfortunate that experts in the field cannot agree on terminology. Some feel that the term *procedures* does not afford flexibility as it relates to tasks and instructions, whereas they believe the term *guidelines* permits more flexibility and discretion involving job performance. Although liability is one of the leading causes of the debate between the two terms, terminology is less important than content and development, and the organization's willingness to follow the policies and procedures.

The ongoing process of developing and updating of SOPs/SOGs must be continually reviewed for effectiveness. Excellent resources from both the U.S. Fire Administration (USFA) and the Department of Homeland Security (SAFECOM) are available to emergency services organizations concerning the development of SOPs. The USFA publication is titled "Developing Effective Standard Operating Procedures for Fire and EMS Departments." This first-rate resource describes the five steps of the development process:

- The role and function of SOPs
- Conducting a needs assessment
- Developing SOPs
- Implementing SOPs
- Evaluating SOPs

The SAFECOM document, titled "Writing Guide for Standard Operating Procedures," focuses on interoperability and compatibility solutions. SAFECOM has worked with individuals in public safety to develop a comprehensive framework called the *interoperability continuum*. Many other resources are also available in both the public and the private sector concerning SOP development and writing.

SOPs contain important concepts, techniques, and requirements permitting personnel to be on the same page in the performance of their jobs. SOPs can be applicable to all emergency incidents or to those within a specific category. SOPs are not meant to provide step-by-step instructions concerning operations or technical guidance; they are intended to offer procedural guidance. SOPs must reflect the mission of the organization, the organizational environment, and the applicable regulations. A needs assessment should be completed that focuses on the internal and external factors that will affect an organization's SOPs.

In many cases, major changes in legal or operational requirements will dictate that an organization conduct a formal needs assessment. It should be done on a continual basis to ensure SOPs stay up to date. The process is easier if an organization forms an SOP committee to develop or change SOPs after a needs assessment has been completed. A needs assessment should answer two questions: what SOPs do we need, and do the current SOPs meet our needs? See Figure 4–6.

In order to conduct a needs assessment, the context in which the organization functions must be understood. During the process, the department's operational systems, functions, and procedures should be compared with the standard of practice (operating practices). In EMS, a similar term is the *standard of care* (see Chapter 2). This process helps to identify areas in which SOPs may need to be added, changed, or deleted. Although it is also important to evaluate laws, regulations, and standards, they will not address every organizational need. The local needs assessment or a community assessment should also be conducted. If SOPs address specific groups or functions within the organization, input from these affected groups should be solicited to offer input, assist with member buy-in, and help the SOP committee understand whether a procedure will work. People within the organization have the best understanding of the organizational strengths and weaknesses. At the same time, it benefits emergency response organizations to examine sources outside the organization. See Figure 4–7.

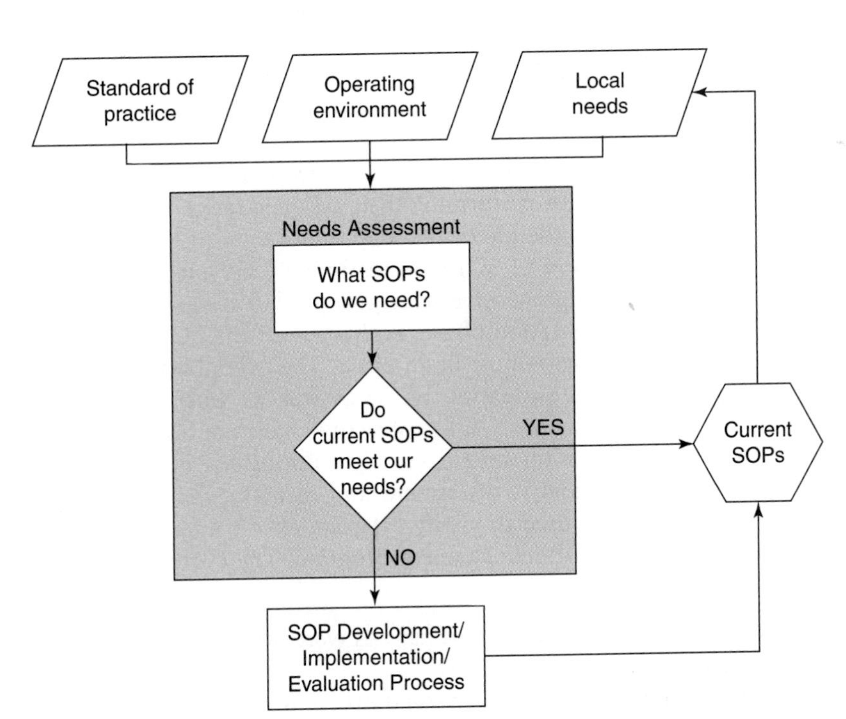

FIGURE 4–6 Sample needs assessment model. *USFA, from http://www.usfa.dhs.gov/downloads/pdf/publications/fa-197-508.pdf*

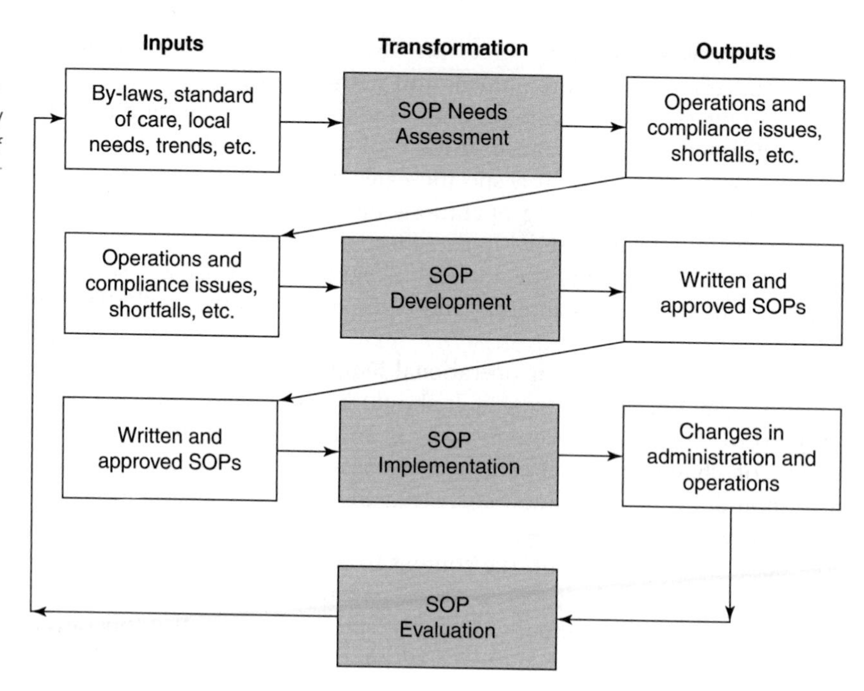

FIGURE 4–7 A systems approach to SOP management. *USFA, from http://www.usfa.dhs.gov/downloads/pdf/publications/fa-197-508.pdf*

A look outside the organization helps to identify standards and trends. A needs assessment should be completed, but all of the necessary changes will not be identified. A discussion of priorities and an action plan can also be included, and an action plan will help to ensure the necessary changes will be made.

Once SOPs have been changed or developed, all organization members should receive training. This training phase, known as the implementation phase, provides an opportunity to think through related tasks, assignments, schedules, and resource needs. The implementation phase should entail more than just a reading of the SOPs. Training needs to include a hands-on component and feedback to ensure the SOPs will be operational. SOPs that address both health and safety issues should receive a top priority during the implementation phase.

The department must ensure there is a thorough understanding of the SOPs, and that employees are capable of performing their assigned tasks. Because competence should be considered throughout the life of the SOP, more than just short-term training is needed. Effective implementation of SOPS means individuals must be accountable for their job performance, so a comprehensive strategy and plan are generally involved.

After the implementation phase, evaluation of the SOPs—the systematic analysis of operations and activities—must be in place. The overall effectiveness of SOPs should be determined during the evaluation phase as well as whether the SOPs' objectives have been accomplished. Obviously, if the objectives have not been met, changes must be made to the SOPs. This question should be asked during the evaluation phase: Are we doing the right thing? The goal is assessment and to make changes, if necessary. The evaluation phase can also be used to justify program efforts and costs, motivate personnel, and improve the overall SOP management program. The evaluation phase is the time to ask the following questions:

- What employee behaviors and actions were taking place before the SOP was implemented?
- What administrative or operational problem was the SOP designed to address?

- Was the new SOP fully implemented, or were there unexpected barriers to full implementation?
- How did employee behaviors and actions change after introduction of the SOP?
- Were the intended changes in behaviors and actions accomplished?
- Was the purpose of the SOP accomplished?
- Is the need for the SOP still current?
- Is the current SOP still the best solution?

Periodic review of SOPs should become a continual process; without a system in place to evaluate policies and procedures, organizations have a tendency toward complacency. Having a plan in place that continually reviews policies and procedures will ensure that organizations look at how they do things to determine what is best for the particular agency.

A structured system and approach will help to manage change in the fast-paced environment of emergency services today. A unified and creative effort concerning SOP design and development will ultimately lead to success.

CHAPTER REVIEW

Summary

An effective program involving health and safety must include commitment, dedication, and involvement of an organization's management. There are several essential elements necessary to formulate a successful program as well as many tools concerning program development available to emergency response organizations. The goal is to enable emergency services organizations to design effective safety programs based on community hazards and service commitment, enhance firefighter safety, and provide tools for continual evaluation of emergency response systems.

The times have changed, and emergency services organizations must adapt to these changing times. A knowledge of the laws, regulations, and standards is an important aspect concerning health and safety program development. Written policies and procedures help to ensure compliance with mandated regulations and suggested standards. Determining and designing an acceptable level of service and resource deployment should become part of the mission of the organization. The review of fatality and injury reports should play a key role concerning the health and safety of responders.

Goals and objectives must be analyzed by the health and safety program manager, as well as a safety committee. Goals and objectives must then be determined in an effort to minimize risks. Goals are broad-based statements that are measurable, whereas objectives are more specific. Each objective needs to have a step-by-step action plan developed for it.

A cost–benefit analysis should be performed for most goals and objectives and will examine direct and indirect costs. An informed decision should be made after an evaluation of the costs of the risk concerning program implementation and the expected costs of the program compared to the correct costs.

Training and safety are very closely related. Many organizations have the training division serve as the health and safety division as well. A prime focus of training should be on how to perform certain tasks with proficiency and with safety in mind. The development of SOPs concerning health and safety should also be part of the training program.

Many SOPs can be developed and evaluated by a safety committee. SOP development involves a review and analysis, as well as a review of goals and objectives. An evaluation of SOPs is the systematic analysis of operations and activities and should be ongoing for the life of the SOP. The health and safety program has several components. Using these components, the health and safety program manager must go through various processes to establish and maintain the program.

Review Questions

1. Written policies and procedures help to ensure compliance concerning mandated regulations and suggested standards.
   a. True
   b. False
2. The beginning stages of health and safety program development determine the ____________ of program development.
   a. Tools
   b. Essential elements
   c. Injury statistics
   d. Goals
3. NFPA 1500, *Standard on Fire Department Occupational Safety and Health Program*, contains the following essential elements:
   a. An infection control program
   b. A critical incident stress management program
   c. A post-incident analysis process
   d. All of the above
4. Many health and safety programs must be developed ____________ as money and resources become available.
   a. Quickly
   b. By the health and safety committee
   c. Gradually
   d. By the fire chief
5. Goals may never be reached if there is too much to do in the specified time frame.
   a. True
   b. False

6. The steps needed to get where we want to be can be viewed as:
   a. Goals
   b. Objectives
   c. An action plan
   d. Cost–benefit analysis
7. It must be kept in mind that ____________ are planning tools.
   a. Goals
   b. Objectives
   c. Action plans
   d. Cost–benefit analyses
8. A focus on generating and writing as many different options and ideas as possible is commonly known as:
   a. An action plan
   b. A cost–benefit analysis
   c. Objectives
   d. Goals
9. Every emergency services organization must consider the direct correlation between ____________ and health and safety.
   a. Goals
   b. Objectives
   c. The action plan
   d. Training
10. SOPs can be applicable to all emergency incidents and those that are within a specific category.
    a. True
    b. False

Case Study

On March 30, 2010, a 28-year-old male career firefighter/paramedic (victim) died and a 21-year-old female part-time firefighter/paramedic was injured when caught in an apparent flashover while operating a hose line within a residence. Units arrived on scene to find heavy fire conditions at the rear of a house and moderate smoke conditions within the uninvolved areas of the house. A search and rescue crew had made entry into the house to search for a civilian entrapped at the rear of the house. The victim, the injured firefighter/paramedic, and a third firefighter made entry into the home with a charged 2½-inch hose line. Thick black rolling smoke banked down to knee level after the hose line was advanced 12 feet into the kitchen area. While ventilation activities were occurring, the search and rescue crew observed fire rolling across the ceiling within the smoke. They immediately yelled to the hose line crew to "get out." The search and rescue crew were able to exit the structure safely, then returned to rescue the injured firefighter/paramedic first and then the victim. The victim was found wrapped in the 2½-inch hose line that had ruptured and without his face piece on. He was quickly brought out of the structure, received medical care on scene, and was transported to a local hospital where he was pronounced dead.

NIOSH had the following recommendations:

Recommendation #1: Fire departments should ensure that a complete 360-degree situational size-up is conducted on dwelling fires and others where it is physically possible and ensure that a risk-versus-gain analysis and a survivability profile for trapped occupants is conducted prior to committing to interior firefighting operations.

Recommendation #2: Fire departments should ensure that interior fire suppression crews attack the fire effectively to include appropriate fire flow for the given fire load and structure, use of fire streams, appropriate hose and nozzle selection, and adequate personnel to operate the hose line.

Recommendation #3: Fire departments should ensure that firefighters maintain crew integrity when operating on the fireground, especially when performing interior fire suppression activities.

Recommendation #4: Fire departments should ensure that firefighters and officers have a sound understanding of fire behavior and the ability to recognize indicators of fire development and the potential for extreme fire behavior.

Recommendation #5: Fire departments should ensure that incident commanders and firefighters understand the influence of ventilation on fire behavior and effectively coordinate ventilation with suppression techniques to release smoke and heat.

Recommendation #6: Fire departments should ensure that firefighters use their self-contained breathing apparatus (SCBA) and are trained in SCBA emergency procedures.

Recommendation #7: Fire departments should ensure that adequate staffing is available to respond to emergency incidents.

Recommendation #8: Fire departments should ensure that staff for emergency medical services is available at all times during fireground operations.

Recommendation #9: Fire departments and dispatch centers should ensure they are capable of communicating with each other without having to monitor multiple channels/frequencies on more than one radio.

Recommendation #10: Fire departments should ensure that the incident commander, or designee, maintains close accountability for all personnel operating on the fireground.

Recommendation #11: Fire departments should ensure that firefighters wear a full array of turnout clothing and personal protective equipment appropriate for the assigned task while participating in fire suppression.

Recommendation #12: Fire departments should ensure that a separate incident safety officer, independent from the incident commander, is appointed at each structure fire.

Recommendation #13: Fire departments should ensure that all firefighters are equipped with a means to communicate with fireground personnel before entering a structure fire.

Recommendation #14: The National Fire Protection Association (NFPA) should consider developing more comprehensive training requirements for fire behavior to be required in NFPA 1001, *Standard for Fire Fighter Professional Qualifications*, and NFPA 1021, *Standard for Fire Officer Professional Qualifications*.[1] See Figure 4–8.

1. Describe the possible reasons why the fire attack crew could not handle the amount of given fire at this incident with a 2½-inch hand line.
2. Describe how ventilation of this structure could have been detrimental to the interior crews.

Illinois House Fire Concerns and Issues	Yes	No
Did interior crews perform a proper risk-versus-benefit survey?		
Were hose line and fire stream selections adequate for this type of event?		
Was there an adequate incident management system in place?		
Did the interior crew provide an adequate and recent update of conditions to the incident commander?		
Did the incident commander continuously evaluate changing conditions?		
Was crew integrity maintained throughout this event?		
Did the lack of recognition of the fire conditions play a role at this event?		
Did horizontal versus vertical ventilation play a role at this event?		
Would adequate firefighter survival training have played a role at this fire?		
Was a 2½" hand line adequate for interior operations?		

FIGURE 4–8 Illinois house fire concerns and issues.

References

1. NIOSH Fire Fighter Fatality Investigation and Prevention Program (FFFIPP), *One Career Fire Fighter/Paramedic Dies and a Part-time Fire Fighter/Paramedic Is Injured When Caught in a Residential Structure Flashover—Illinois*, September 13, 2010. Retrieved December 10, 2010, from http://www.cdc.gov/niosh/fire/reports/face201010.html

CHAPTER 5

Risk Assessment and Safety Planning

KEY TERMS

critical incident stress management (CISM), *p. 112*
National Association of Emergency Vehicle Technicians (NAEVT), *p. 102*
National Incident Management System (NIMS), *p. 113*
personal protective equipment (PPE), *p. 107*
preventative maintenance (PM) program, *p. 97*
unified command, *p. 113*

OBJECTIVES

After reading this chapter, you should be able to:

- Identify the areas where risks occur prior to incidents.
- Identify the categories of the risk management plan, where risk assessment and safety planning should be a concern.
- Describe the safety considerations involved in facility design.
- List the ongoing concerns related to apparatus.
- Identify the risks in which the response takes place.
- Describe the components of the risk assessment and safety planning process.
- Explain why risk assessment, safety planning, and training are important.
- List the components of employee wellness and fitness.
- Explain why planning is important as it relates to interagency operations.

ResourceCentral

For additional review and practice tests, visit **www.bradybooks.com** and click on Resource Central to access book-specific resources for this text! To access Resource Central, follow directions on the Student Access Card provided with this text. If there is no card, go to **www.bradybooks.com** and follow the Resource Central link to Buy Access from there.

Risk assessment and safety planning deals with risk identification and planning to prevent accidents and injuries from occurring. The risks include workplace risks, vehicle and equipment risks, and the human resource risk. This chapter covers these important areas, as risk assessment and safety planning focuses on those risks that occur prior to an incident taking place or while responding to an incident. Training and knowledge that center on technology and data management make risk assessment and safety planning possible. Technology forces emergency response organizations to adapt to a continually changing world, whereas data management produces areas that need to be studied and addressed.

The health and safety program manager plays a leading role in the area of safety risk assessment and planning. Together, the on-scene incident safety officer(s) and the department health and safety officer or program manager bear the responsibility not only for monitoring and assessing safety hazards or unsafe conditions but also for developing measures to ensure personnel safety. Development of a risk management plan is critical in assessing risk and planning safety and should involve a department health and safety officer or program manager. This important individual needs to be closely involved in training in safety procedures related to all department operations and functions. The risk management plan must meet the requirements of Chapter 6 of NFPA 1500, *Standard on Fire Department Occupational Safety and Health Program.*

Risk Management Plan

Many components of the risk management plan described in NFPA 1500 can be applied to a fire department or EMS organizations. Chapter 2 of NFPA 1500 states that the risk management plan covers administration, facilities, training, vehicle operations, protective clothing and equipment, operations at emergency incidents, and other related activities. The plan should not be developed and then placed on a shelf to be used only occasionally. Instead, it should serve as a living document that continually identifies and evaluates risks for which a reasonable control plan will be put into place and followed. A risk management plan positively impacts the organization and plays an important role in risk assessment and safety planning by addressing operations, safety, and liability issues. The health and safety program manager uses the plan to minimize the potential risks that operations personnel face in the performance of their duties.

The fire chief or department director has the ultimate responsibility to make sure a risk management plan is put into place. In most cases, the chief or director delegates this task to the health and safety program manager. The health and safety program manager, the health and safety committee, or the fire chief or director can complete a review of the

FIGURE 5-1 Facility design should consider safety prior to construction as well as safety during day-to-day operations.
Courtesy of Joe Bruni.

plan, which must be done annually. At a minimum, the plan should include risk identification, risk evaluation, risk control techniques, and risk monitoring.

Facility Safety Considerations

Safety in the facility must be considered to be vital because personnel spend a great deal of time in their facility when they are not involved in incidents. Whatever the design and use of the facility, a safe working environment must be provided. Safety can be grouped into two categories: facility design and ongoing operations.

Design deals with new facility design as well as remodeling and renovations to an existing station. With design, remodeling, and renovation, safety must be considered during the planning phase. For example, an existing facility that does not have fire protection features such as a sprinkler system should have one designed into the building when it is renovated.

Ongoing operations focus upon safety and health issues concerning day-to-day operations, such as proper ventilation in engine bays and areas that pose an exposure risk to personnel from infectious disease contamination.

FACILITY DESIGN

Safety in design is important prior to construction taking place. Most facilities are considered public buildings, so the design should also take into consideration the safety of the public who will be visiting the building. Local building and fire codes are critical in the area of facility design, as the station must comply with applicable codes and standards according to NFPA 1500. Just as the design of the building is important, it is equally important to have private protection systems in place upon completion of the project. See Figure 5-1.

Specific areas should be protected with nonslip surfaces in new buildings as well as in existing buildings. Certain areas should be equipped to prevent falls, and lighting must be adequate in all parts of the building. OSHA regulations require that a section of the facility serve as a decontamination and cleanup area to prevent infectious disease. Areas for the storage of PPE must be free of ultraviolet light and well ventilated. Some

FIGURE 5-2 Example of a station design checklist.

	Yes	No
Physical Design		
Adequate storage areas provided		
Ample parking		
Adequate access to station		
Storage room for PPE with adequate ventilation, motion-sensor light switch, and no sunlight coming in		
Biohazard/disinfection room provided		
Apparatus bay adequate		
Apparatus bay doors equipped with self-reversing door switches		
Adequate nonslip surfaces provided		
Exhaust emission devices provided		
Adequate lighting provided in all areas		
Adequate drainage provided in all areas		
Poles or chutes provided with fall protection		
Low-level lighting provided in dormitory		
Routes of travel—free from obstruction, well lighted		
Safety and Security		
Automatic sprinkler system provided		
Fire alarm system provided with smoke and carbon monoxide alarms		
Smoke alarms provided		
Carbon monoxide alarms provided		
Station security provided		
Ability to shut down cooking appliances from apparatus bay		
Code Compliance		
Health		
Building/life safety		
Fire		

type of exhaust emission system must be present in the apparatus bays; several different types of emission systems are available to satisfy the health and safety of employees. A laundry area must be designed with OSHA regulations in mind to clean up clothing and materials that have been contaminated with blood-borne pathogens and body fluids. The health and safety program manager as well as a safety committee should be involved in the design of the entire facility. Figure 5-2 shows an example of a station design checklist.

Improvements during the renovation of an existing station must be prioritized with safety issues in mind first. The health and safety program manager and the safety committee must acknowledge that the personnel who will be housed at the station can provide the health and safety committee with valuable input. Their recommendations should be passed on to the organization's administration.

ONGOING OPERATIONS

The procedures and day-to-day operations that relate to risk assessment and safety are crucial to a safe facility. In many cases, it is the responsibility of the station captain to develop and enforce department policies and individual station policies to reduce hazardous situations. See Box 5-1.

BOX 5-1: EXAMPLE OF A FACILITY POLICY

STATION #3 POLICIES

Station Security

1. All station exterior doors will be kept locked at all times when not in immediate view of personnel. Exception: Shift change between 07:00 and 08:00 when the station is occupied.
2. Engine room doors are to be kept closed unless working in the area. Doors will be opened completely when used. Do not leave doors opened partially at any time. Personnel should avoid standing directly under doors when in use as a safety precaution.
3. Unsupervised children shall not be allowed in the station with the exception of project safe inquiries.
4. Civilians will not be allowed to spend excessive amounts of time in or about the station as determined by the company officer.
5. Nonessential personnel are not to be left on the premises while the station is unattended.

In Station

Many injuries in the station could have been prevented if the environment had been controlled. The station captain or individual company officers should inspect stations on a monthly basis as they continue to look for safety-related issues in-house. It is also beneficial to inspect stations on an annual basis. An annual staff inspection by the organization's administration will identify issues and health and safety concerns. Staff officers will have a narrow and focused view of safety issues within fire stations. It is not uncommon for company officers to overlook safety concerns and issues during their monthly inspection.

Fire protection systems should be up to code and inspected on an annual basis. See Figure 5-3.

FIGURE 5-3 Fire stations should meet local fire and building codes. *Courtesy of Joe Bruni.*

Smoke and carbon monoxide detectors should be checked monthly for proper operation. An inspection form should be used to complete the inspection report. It is also beneficial for a fire inspector to conduct an annual fire and safety inspection of the facility and then complete an inspection form. See Figure 5-4.

The inspection form provides ongoing documentation of hazards as well as necessary requests for maintenance related to safety. Those inspecting the facility can access Department of Labor and OSHA standards to properly identify safety problems. During their daily inspections, the on-duty crew and the company officer should address fire hazards as an immediate health and safety concern. Although not a common feature concerning ongoing operations, it should be possible to shut down the cooking area from remote locations in the event of an incident. With this feature, firefighters do not have to return to the cooking area when an alarm sounds for an emergency response. Health and safety should be endorsed in the station during ongoing operations just as it is on the incident scene. Station safety is the result of regular daily, monthly, and annual inspections by the station captain, lieutenant, or a designee. Violations found during station inspections should be reported to the fire chief or his or her designee for proper follow-up and documentation. The fire chief may use this type of documentation for budgetary purposes and to ensure the health and safety hazards have been eliminated.

FIGURE 5-4 An example of a station inspection form.

Monthly Station Inspection Report

Inspecting Officer: ______________________ Date of Inspection: ____________

Station: ________ Shift: ______ Captain: ____________

Station Interior	Office	Kitchen	Dayroom	Dormitory	Restroom	HVAC
Ceilings						
Floors						
Windows						
Lights						
Furniture						
Storage areas						
Walls						
General appearance						

Station Exterior		Safety Items		Office		Apparatus Bay	
Paint		Smoke alarms		Record storage		Floor	
Doors		Carbon monoxide detectors		Log book		Workbench	
Ramps		Caution and warning signs		Drug log		Hose rack	
Sidewalks		Tool storage		Inventories		Storage	
Parking area		Safety bulletin board		Officer's desk		Apparatus	
Yard		Combustible storage		Bulletin board		Windows	
Hose tower		Chemical storage		Public education material		Bay doors	
Flagpole		Monthly SCBA tests		Wall maps			
		Extinguishers					

Personnel								
PPE								
Accountability tags								
Driver's license								
ID card								
Proper uniform								
Grooming and cleaniness								

Risk Identification

The topic of risk management was covered in depth in Chapter 3; however, it is worth mentioning its role in risk assessment and safety planning. Concerning ongoing operations at the incident scene, the risk management plan should be used. Risk identification, risk evaluation, and risk prioritization must be completed to reduce health and safety issues at the fireground or incident scene. Risk identification involves compiling a list of all emergency and nonemergency operations and duties. The health and safety program manager should take into account all of the potential worst-case scenarios and possible events. There are many sources from which these data and information can be accessed: one of the best places is the department's loss prevention data. Organizations that are too small to have their own database or a statistically valid trend can make use of national averages and trends available from the NFPA and the National Fire Academy.

Risk Evaluation

Once risk identification has been determined, the health and safety program manager can evaluate the risks in terms of frequency and severity. OSHA refers to frequency as "incident rate," or the likelihood of a risk's occurrence. Repeated injuries will continue to occur until a job hazard or task analysis has been performed to identify causes and effective control measures.

Severity addresses the seriousness of a risk at the incident. It can be measured in a variety of ways including lost time from work, cost of the damage, cost for time and repair, cost for replacement, disruption of service, and legal costs. Calculating risk evaluation will vary from organization to organization. Risk management in emergency services is as simple as looking at potential problems and taking action during the risk assessment and safety planning phase to prevent those problems from occurring. It must be understood that the mistakes that emergency workers make will remain constant and can be predicted from the mistakes made throughout the history of emergency services.

Risk Prioritization

Risk frequency and severity are always compared with one another to determine priorities as a way to establish action. Top-priority risks have a high frequency probability and result in serious consequences; they warrant immediate action. Lower priority risks have a low frequency and severity; they should be placed near the bottom of the list of risks requiring action. Because things that go wrong at the incident scene are very predictable, they are preventable. Speaker Gordon Graham, a veteran of California law enforcement, is well known as a successful risk manager and practicing attorney. He has focused his attention on providing risk management knowledge to both the public and the private sector. Graham has developed a risk management matrix that can be applied to any emergency services organization's risk management plan.

Apparatus and Equipment Safety and Maintenance

Necessary maintenance of apparatus and equipment can be performed in-house at the facility or at a location dedicated to this purpose. Shift personnel cover many ongoing issues as part of their daily, weekly, or monthly inspections and duties. Nonuniformed personnel whose assignment and responsibility is apparatus and equipment maintenance also cover these duties. The health and safety program manager must ensure that a safe working environment is provided during day-to-day operations and that policies focus on personal protection. These policies should also incorporate the training necessary to perform maintenance and specify the types of protection required. The essential equipment should be made available in the maintenance areas to personnel who will be performing the required tasks.

The tools required for maintenance need to be in good working condition and stored properly with safety in mind. As is done with other fire department equipment, tools should be periodically inspected, and damaged tools replaced or repaired. Training in the safe use of tools should focus on proper storage and selecting the correct tool for the task at hand. It should never be assumed that employees understand safety related to tool use; knowledge of the tool manuals should be promoted.

Safely driving and responding to incident scenes is a major concern, because a large percentage of injuries and deaths occur during apparatus response and returning to quarters. The NFPA has reported that approximately 25% of firefighter fatalities have been the result of vehicle and apparatus incidents. Many of the NIOSH Fire Fighter Fatality Investigation and Prevention Program reports also reflect this critical area. Every emergency services organization's personnel should review these reports on a monthly basis. The emergency services as a whole needs to recognize that response to the incident must be accomplished as safely as possible. It is the health and safety program manager's responsibility to ensure safe operating practices are in place in the training area and during day-to-day operations. Continual training in this area is a critical component of health and safety.

Apparatus Design

#16

Numerous NFPA standards dictate the requirements for fire apparatus. Firefighter safety is the primary concern when it comes to apparatus design. The U.S. Department of Transportation (DOT) also issues requirements for fire apparatus. Ambulances will be governed by NFPA 1917, *Standard for Automotive Ambulances*, in the near future. Currently, ambulances are covered by federal regulations, but some states use their own design guidelines. See Figure 5-5 and Box 5-2.

The apparatus or safety committee should study apparatus design prior to purchase. It should focus on the minimum requirements as it complies with safety standards.

The health and safety program manager and the safety committee should lead the design process and be sure risk identification is part of it. They should keep the following features in mind: equipment storage, hose bed height, ladder placement, and apparatus visibility. For example, hose bed height should be designed with safety and effective deployment in mind. Are ladders in the correct position to be safely removed? The health and safety program manager should work with the safety committee and the specification committee to ensure the NFPA standard is met related to warning lights, color scheme, and reflective trim. In ambulance design, the level of safety should be geared toward the patient compartment. Considerations such as the following require attention: the proper placement of handrails, grab rails, and access steps; and ergonomically designed storage areas. Local needs and the conditions faced by the department, as well as safety, should also be kept in mind; for example, the type of terrain in which the apparatus will operate.

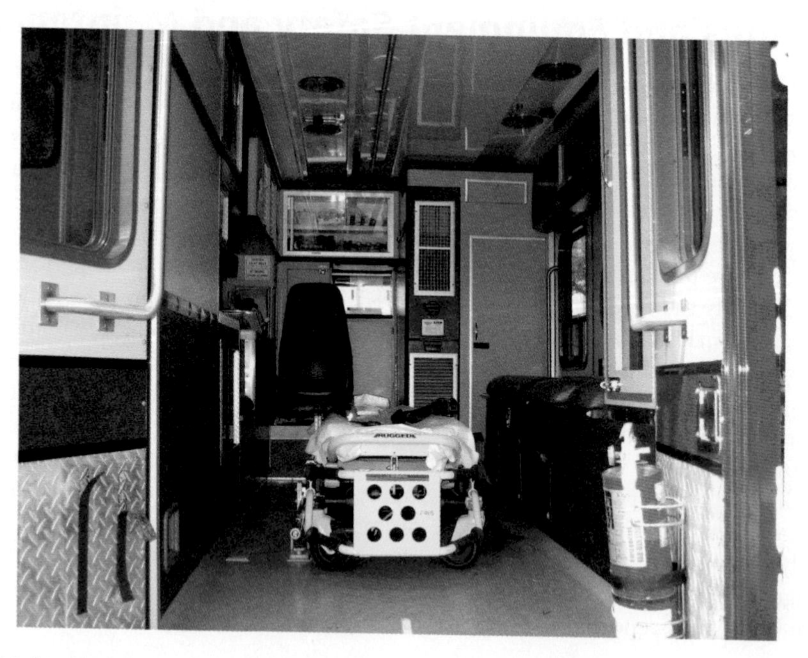

FIGURE 5-5 Ambulance design is governed by the Department of Transportation. *Courtesy of Joe Bruni.*

BOX 5-2: NFPA STANDARDS RELATED TO APPARATUS DESIGN AND ONGOING OPERATIONS

NFPA 1901 — *Standard for Automotive Fire Apparatus*
NFPA 1906 — *Standard for Wildland Fire Apparatus*
NFPA 1911 — *Standard for the Inspection, Maintenance, Testing, and Retirement of In-Service Automotive Fire Apparatus*
NFPA 1912 — *Standard for Fire Apparatus Refurbishing*
NFPA 1915 — *Standard for Fire Apparatus Preventive Maintenance Program*
NFPA 1002 — *Standard for Fire Apparatus Driver/Operator Professional Qualifications*
NFPA 11C — *Standard for Mobile Foam Apparatus*

The safety and health manager works together with the health and safety committee to complete a risk analysis and support safety considerations. Apparatus manufacturers will attempt to meet the needs of the individual department; however, changes to a stock vehicle will cost more.

Day-to-Day Apparatus Concerns

Many of the concerns regarding day-to-day use of apparatus relate to the time frame the apparatus will be in service, which is usually a significant number of years. The health and safety program director must perform a thorough risk analysis of apparatus to evaluate its safety. For example, every effort must be made to ensure hose lines and equipment are mounted and secured so that they do not fall off while the apparatus is in motion. Unsecured equipment in the cab area as well as on the outside of the apparatus will quickly turn into missiles in the event of a vehicle accident. This ongoing problem should be addressed in both new apparatus design and apparatus that has been in service for several years. Other common issues concern emergency warning lighting, traffic safety management, and roadway operations safety.

The U.S. Fire Administration (USFA) has initiated partnerships with the International Association of Fire Chiefs (IAFC), the International Association of Fire Fighters (IAFF), and the National Volunteer Fire Council (NVFC) to reduce the number of firefighters killed while responding to or returning from the emergency scene. These partnerships take the recommendations of the Emergency Vehicle Safety Initiative, a jointly sponsored project of USFA and DOT's Federal Highway Administration, and develop materials that directly target their constituency—chief officers and fire department leadership, the career fire service, and the volunteer fire service. This outreach project addresses issues such as seat-belt use, intersection safety, fire apparatus and emergency vehicle safety design, driver selection and training, policies involving alcohol and driving, and implementation of alternative response programs.

Working with the IAFC, USFA developed the guide to model policies and procedures for emergency vehicle safety, a comprehensive Web-based educational program aimed at reducing the impact of vehicle-related incidents on the fire service and the communities they protect. The program provides in-depth information for developing policies and procedures required to support the safe and effective operation of not only emergency vehicles in the fire service but also privately owned vehicles. As part of this project, the IAFF has also developed a similar innovative Web- and computer-based training and educational program.

Many problems related to ongoing operations can be reduced or eliminated with an aggressive and proactive **preventative maintenance (PM) program**. The PM program has to be established to ensure that maintenance, inspections, and repairs are performed.

preventative maintenance (PM) program ▪ This maintenance program for vehicles is ongoing and designed to catch many minor problems before they become major issues.

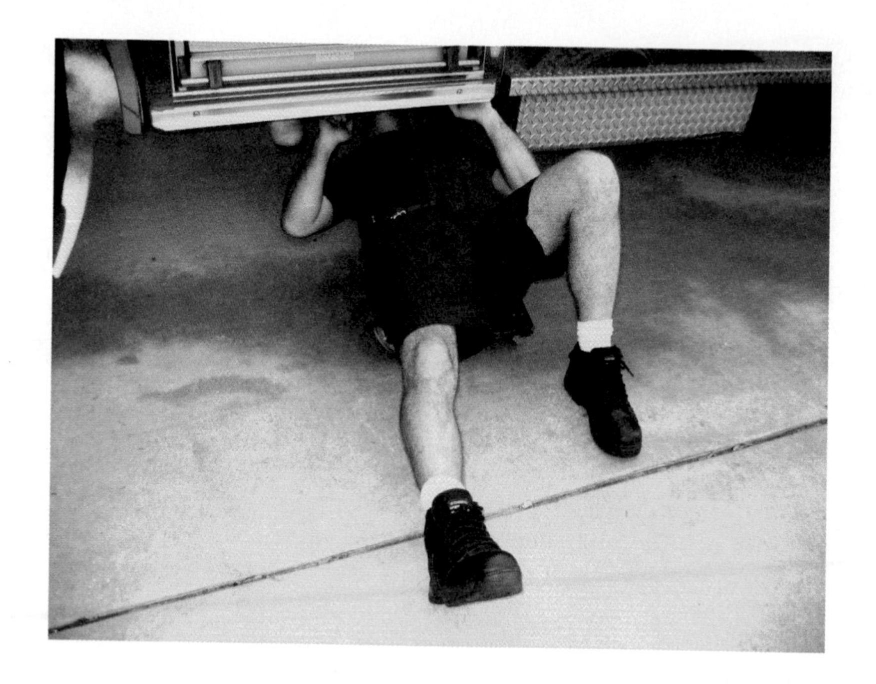

FIGURE 5-6 A preventative maintenance program is critical to having safe response vehicles. *Courtesy of Joe Bruni.*

It is a good idea to build the PM program according to the manufacturer's maintenance guidelines. NFPA 1911, *Standard for the Inspection, Maintenance, Testing, and Retirement of In-Service Automotive Fire Apparatus*, defines the minimum requirements for establishing an inspection, maintenance, and testing program for in-service fire apparatus. Preventative maintenance programs and daily apparatus checks will help in meeting ongoing safety objectives. See Figure 5-6.

In conjunction with the PM program, apparatus operators should be trained in what items to look for as they relate to maintenance problems. A key component of early problem identification and ongoing safety is the driver/operators who are very familiar with their apparatus. As a minimum requirement, driver/operators should perform a daily apparatus inspection that includes the following:

- Fluid levels
- Seat belts
- Mirrors
- All lighting components
- Windshield wipers
- Backup lights and alarm
- Tires
- Brakes

As fire apparatus becomes more complex, maintenance and testing become more critical to the safety of the firefighters and the public. Weekly inspections and inventories should also be performed so that problems and hazards can be identified early. See the checklist in Figure 5-7.

The PM program also includes an inspection at specified intervals of the brakes, chassis, fluid levels, lights, pumps, transmissions, belts, and hoses of all emergency vehicles. Lack of a PM program will result in increased costs, shorter vehicle life spans, and reduced safety of emergency responders and the public. According to NFPA 1911, pumps and ladders must also be tested annually. This NFPA standard requires the department to develop a program based on the manufacturer's recommendations, local conditions, and operating conditions. NFPA 1911 also requires the department to have a written plan in place for when apparatus is to be taken out of service. See Figure 5-8.

The NFPA 1911 standard also refers to various systems and components of fire apparatus. The criteria associated with placing an apparatus out of service should consider 49 CFR part 390, Federal Motor Carrier Safety Regulations; any applicable federal, state, and local regulations; applicable nationally recognized standards; manufacturer's recommendations; and guidelines established by the organization or its maintenance staff. The Federal Motor Carrier Safety Regulations point out that authorized personnel can place a vehicle out of service and base this decision solely on the vehicle's mechanical condition, which would likely cause the vehicle to have a breakdown or cause an accident.

DAILY/WEEKLY WALK-AROUND CHECK
FOR MOBILE FIRE APPARATUS

Fire Department Name ______________________________ Date ______________
Apparatus No. ______________________ Station No. ______________________
Start mileage ____________ End mileage ________ Start engine hours ______ End engine hours ________
Inspectors: Mon ________ Tue ________ Wed ________ Thur ________ Fri ________ Sat ________ Sun ________

Legend: X = OK
NA = Not applicable
R = Repair required (requires a comment regarding problem)
C = Corrected

OPERATIONS	Mon	Tue	Wed	Thur	Fri	Sat	Sun
ENGINE							
1. Check engine oil and transmission level.							
2. Check engine coolant level.							
3. Check for integrity of frame and suspension.							
4. Check power steering fluid.							
OUTSIDE							
1. Check for fluid leaks under vehicle.							
2. Check steering shafts and linkages.							
3. Check wheels and lug nuts.							
4. Check tire condition.							
5. Check tire air pressure.							
CAB							
1. Check seats and seat belts.							
2. Start engine; check all gauges.							
3. Check windshield wipers.							
4. Check rear view mirror adjustment and operation.							
5. Check horn.							
6. Check steering shafts.							
7. Check cab glass and mirrors.							
BODY							
1. Check steps and running boards.							
2. Check body condition.							
3. Check grab handles.							
ELECTRIC							
1. Check battery voltage and charging system voltage.							
2. Check line voltage system.							
3. Check all lights (ICC and warning).							

NFPA 1911 (p. 1 of 2)

OPERATIONS	Mon	Tue	Wed	Thur	Fri	Sat	Sun
BRAKES							
1. Check air system for proper air pressure.							
2. Check parking brake.							
3. Check hydraulic brake fluid level.							
PUMP							
1. Operate pump; check pump panel engine gauges.							
2. Check pump for pressure operation.							
3. Check discharge relief or pressure governor operation.							
4. Check all pump drain valve.							
5. Check all discharge and intake valve operation.							
6. Check pump and tank for water leaks.							
7. Check all valve bleeder/drain operation.							
8. Check primer pump operation.							
9. Check system vacuum hold.							
10. Check water tank level indicator.							
11. Check primer oil level (if applicable).							
12. Check transfer valve operation (if equipped).							
13. Check booster reel operation (if equipped).							
14. Check all pump pressure gauge operation.							
15. Check all cooler valves.							
16. Check for oil leaks in pump area.							
AERIAL							
1. Operate aerial hydraulics.							
2. Check aerial outrigger operation.							
3. Check aerial operation.							
4. Check aerial hydraulic fluid level.							
5. Visually inspect aerial structure.							

Comments ______________________________

NFPA 1911 (p. 2 of 2)

FIGURE 5-7 An apparatus daily/weekly checklist should be used as part of the apparatus safety program. *Reproduced with permission from NFPA 1911–2007,* Inspection, Maintenance, Testing, and Retirement of In-Service Automotive Fire Apparatus, *Copyright © 2007, National Fire Protection Association. This reprinted material is not the complete and official position of the NFPA on the referenced subject, which is represented only by the standard in its entirety. Classroom set for educational use only.*

Placing an Emergency Vehicle Out of Service Guide Policy

Purpose: In an effort to have department apparatus in a safe operating condition, and as a guide to assist personnel in placing a vehicle out of service based on the mechanical or physical condition of the vehicle, any of the items listed will be cause to remove a vehicle from service immediately. Notification of the on-duty company officer and shift commander is mandatory.

Braking System
- Audible or visual air leak
- Air line with leak or bulge
- Loose compressor mounting bolts
- Evidence of oil seepage
- Cracked brake drums
- Inoperative low-air warning device
- Master cylinder less than half full (if equipped)

Steering System
- Excessive free play (30 degrees before steering axle tire moves)
- Worn or faulty universal joints
- Steering wheel not properly secured
- Loose tire rod ends
- Any condition that interferes with free movement

Exhaust System
- Exhaust leak forward or below the cab

Frame
- Cracked, loose, or broken frame member

Fuel System
- Visible fuel leak
- Fuel tank not securely attached

Springs/Suspension
- Cracked, loose, or missing U-bolt or other spring to axle clamp
- Missing leaf or portion of leaf spring
- Any broken main leaf in the leaf spring (main leafs extend at both ends)
- Any displaced leafs that could result in contact with a tire, rim, or brake drum
- Broken or missing shocks
- Missing or broken axle bolts

Tires/Wheels
- Tread depth of 4/32" or less on a steering axle tire at any two adjacent major tread grooves
- Tread depth of 2/32" or less on a non-steering axle tire at any two adjacent major tread grooves
- Cut sidewall where the cord is exposed
- Flat tires
- Missing or broken lug nuts or studs

Windshield/Wipers
- Visual cracks or distortions that impair the driver's vision
- Inoperable wiper or damage that makes the driver's wiper inoperable

Lighting Devices/Warning Lights
- Any low beam head lamp missing or inoperative
- Both brake lights missing or inoperative
- Both taillights missing or inoperative
- Any turn signal missing or inoperative
- Inoperative siren
- Emergency lighting not visible from all sides

Drive Train
- Engine overheating
- Motor oil in engine (radiator) coolant
- Engine coolant in motor oil
- Broken or missing fan belts
- Coolant leak at water pump

(continued)

FIGURE 5-8 Placing an emergency vehicle out of service

- Any major coolant leak
- Automatic transmission overheating
- Transmission "Do Not Shift" indicator on
- Defective clutch components
- Defective foot throttle
- Defective charging system

Pump/Aerial Components
- Pump will not engage
- Pump panel throttle defective
- Leak in water tank
- Contaminated pump transfer case lubricant
- PTO will not engage
- Defective stabilizer system
- Cable sheaves worn excessively or defective
- Missing or damaged rungs
- Major hydraulic system fluid leak

Cab/Body Components
- Missing or broken mirrors that obstruct or limit the driver's view
- Defective door latches
- Defective defrosters

Any apparatus placed out of service should not be returned to service until these issues have been addressed and repaired.

National Association of Emergency Vehicle Technicians (NAEVT) ▪ This not-for-profit organization offers professional certification for individuals involved in service associated with emergency vehicle maintenance.

Another resource available in this area, the **National Association of Emergency Vehicle Technicians (NAEVT)**, is a not-for-profit organization that was formed in 1991. NAEVT offers information about apparatus maintenance and safe operating practices based on an overwhelming need to address issues of the everyday mechanic/technician involved with fire service apparatus. The mission of the NAEVT is to improve the quality of emergency vehicle maintenance and to promote the emergency vehicle technician profession by facilitating training, education, and legislation. NAEVT provides a national and international voice for the technician. Its certification testing for mechanics is specific to pumps, hydraulic systems, aerial apparatus, drive trains, and supervisory management. It is important for fire departments not only to make a commitment to personnel who maintain fire apparatus but also continue to provide ongoing education in this area.

Apparatus Operating Practices Safety

Approximately 20% to 25% of firefighter fatalities occur while departments are responding to or returning from alarms. Research into the National Institute for Occupational Safety and Health (NIOSH) Fire Fighter Fatality Investigation and Prevention Program's cases will produce many incidents concerned with injuries and fatalities in this area. For this reason, the area of response safety concerning apparatus as well as private vehicle response should receive a high priority with training. Classroom and hands-on practical training for members will help to reduce the potential for vehicle and apparatus crashes. This risk assessment and safety planning issue has been overlooked for far too long. A selection process and program for apparatus drivers and operators as well as training should also be in place in every emergency response organization.

NFPA 1002, *Standard for Fire Apparatus Driver/Operator Professional Qualifications*, establishes the guidelines for operators of fire department apparatus; for example, a medical evaluation to prove the fitness of a driver/operator and the proper license to operate various types of apparatus. The standard requires the fire service organization to evaluate the performance objectives in the selection of drivers. Various types of apparatus are covered by the NFPA 1002 standard: apparatus equipped with a fire pump, aerial devices, tiller operators, wildland fire apparatus, aircraft rescue and firefighting apparatus, and mobile water supply apparatus.

Many departments perform some type of testing of driver/operators to comply with NFPA 1002. Concerning performance objectives, it is a good idea to develop skill sheets that address both the NFPA 1002 and NFPA 1500 standards. Drivers should be trained in emergency vehicle operations in a vehicle similar to the one they will be operating on the job. Written procedures and policies must clearly define the guidelines and what is expected from driver/operators in their course of duty. Ongoing training programs for driver/operators should cover such operations of apparatus as various types of road surfaces, vehicle stopping distances, legal issues, vehicle maintenance, safety procedures related to the individual department, and safe driving practices. It is important to keep the geographical area as well as the environment in mind when it comes to training. A driving course is a common way to evaluate driver candidates in both the certification and ongoing formats. See Figure 5-9.

Many vocational schools and institutions, as well as insurance companies, offer driver training programs. Common driving incidents involving emergency vehicle operators include excessive speed, driving too fast for conditions, and not exercising due caution at intersections. The operator of a fire department apparatus was killed when the vehicle he was operating left the roadway and overturned. It was determined the operator was not wearing a seat belt and had a blood alcohol level over the legal limit. Driver/operators must understand the importance of vehicle control when it relates to emergency vehicle operations and adhere to driving-related issues such as apparatus speed during response, intersection safety, railroad crossing safety, school zone safety, and utilization of visual and audible warning systems. The program for driver/operators should include hands-on

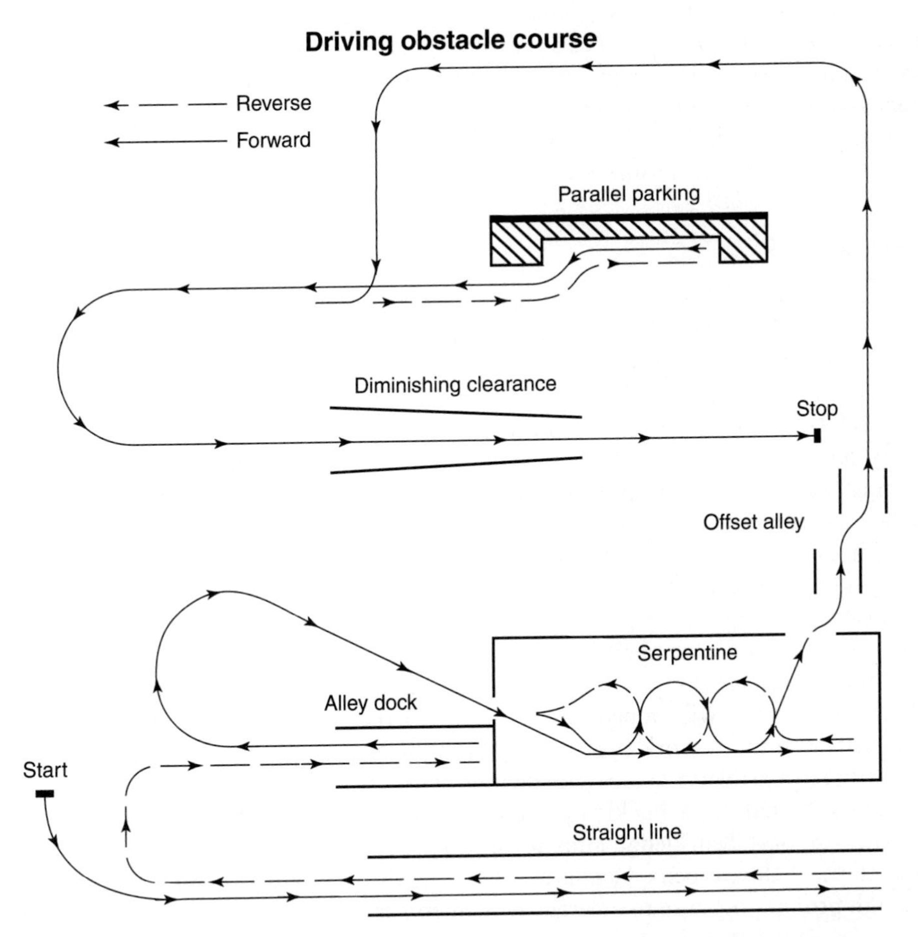

FIGURE 5-9 A typical driving course for emergency vehicle driving. *United States Fire Administration (USFA) from, http://www.usfa.dhs.gov/downloads/pdf/publications/fa-110.pdf, page C-4*

practical sessions (driving on various surfaces, such as dry roads, wet roads, and ice- and snow-covered roads) as well as classroom sessions.

Type of Response

The type of response—to emergencies versus nonemergencies—can affect the type of driving. Remember, safe response to the incident and safe return to quarters are top priorities. A response that requires emergency lights and sirens will normally heighten driver/operators' adrenaline levels and be more hazardous than normal driving during routine traffic conditions. The organization must take a hard look at the range of incidents and calls for service that require the assistance of firefighters or EMS personnel. Many of these calls do not involve life-threatening situations, and so do not require response in emergency mode. A response can be upgraded or downgraded depending upon pre-arrival information. Company officers, supervisors, and driver/operators are accountable to respond according to the information received during dispatch and while en route to the incident. Responding in nonemergency mode is appropriate for the following incidents: automatic alarms, wires down, investigation of unknown odors and smoke in the area, carbon monoxide alarms, smoke detector alarms, manual pull stations, assisting police agency, and person in a locked vehicle. Doing so will reduce the risks to responding personnel.

A review of incidents involving apparatus in accidents during emergency response will be beneficial when it relates to a reduction in accidents at intersections. A lieutenant and three firefighters were injured responding to a residential fire alarm when the operator performed a rolling stop at an intersection and was struck by a civilian in a pickup truck.[1]

During emergency response, departments should look at intersection control, because that is where most accidents occur. Traffic signal preemption control has become common, and many types of systems utilize a traffic signal controller. This controller receives a signal from the preemption device, which changes the traffic light to a green signal for the direction in which the emergency vehicle is responding. Preemption devices reduce the risk of vehicle accidents at intersections.

Seat Belts

Motor vehicle–related crashes are the second leading cause of death for firefighters. NFPA 1500, *Standard on Fire Department Occupational Safety and Health Program*, deals with the need for seat belts. NFPA 1500 section 6.2.5 addresses the issue: "drivers shall not move fire apparatus until all persons on the vehicle are seated and secured with seat belts." NFPA 1500 section 6.3.2 also states: "seat belts shall not be released or loosened for any purpose while the vehicle is in motion." It is important to develop procedures requiring seat-belt use during both emergency and nonemergency response. The lack of seat-belt use is an ongoing problem for fire service personnel and should be top priority. The company officer and the apparatus operator need to ensure seat belts are in place prior to response and movement of the apparatus. Unfortunately, standing in the jump seat during an emergency response is still a common occurrence. Department policy needs to ensure that at no time shall any personnel unbuckle or loosen their seat belt to don PPE while the vehicle is in motion. It must become part of the fire service culture that the lack of seat-belt use will impede the missions of the department. Fire service organizations must continue to work to ensure all firefighters buckle up and hold every individual personally responsible to do so—no excuses.

Across the nation, a buckle-up mind-set has come to the forefront in recent years. The National Seat Belt Pledge aims to improve fire service safety by having firefighters pledge to use their seat belts. Deaths attributed to the lack of seat-belt use are preventable. Although prompt response to emergency incidents is an organizational priority, the safety of responding personnel is always a higher priority. Unless the responding units arrive safely at the location where they are needed, they cannot deliver the required services. Unsafe operation of an emergency vehicle creates an unacceptable risk to fire department personnel, to the public, and to the individuals in need of assistance. Improving seat-belt usage is the single most effective strategy the fire service can support in reducing injury and death when a motor vehicle accident occurs.

The Environment

Risks must be identified in which the response takes place. It is the role of the company officer to assist the driver/operator to identify these hazards and act to avoid them. For example, accidents involving responding apparatus have been common. To reduce apparatus collisions, an analysis of traffic patterns and response routes should be conducted to heighten awareness of target intersections and improve radio communications. Technology such as the global positioning system (GPS) as well as other types of systems could be used for traffic preemption. Departments should develop SOPs or SOGs for emergency vehicles during emergency response. They should develop policies for firefighters to use seat belts prior to apparatus movement, and apparatus drivers should be trained in railway traffic safety. Procedures must require that a responding apparatus come to a complete stop at an intersection and not exceed the posted speed limit for response.

Personnel must never lose sight of the fact that the response environment is dynamic. Traffic patterns change during specific times of the day, and weather can be a contributing factor to accidents. Road construction should be planned for and can be monitored by the department administration. Driver/operators need to be trained in the area of response routes and continually keep abreast of any changes in these routes.

Planning Before the Incident

A safe operation on the fireground begins with an ongoing training program and planning long before an incident occurs; this area should become a top priority in the fire service. In the past, much of the training efforts have been geared toward skill development and are reactive rather than proactive. Company officers and firefighters must be encouraged to embrace the broad spectrum of safety planning prior to an incident coupled with ongoing training.

Risk assessment and safety planning is a systematic method of gathering and recording facts for the purpose of problem identification/analysis and information retrieval. Risk assessment and safety planning should consider safety during pre-incident response and during fireground operations. The health and safety program manager must work with the incident safety officers to review and identify the following: specific high-risk targets in the response district, problem areas concerning the buildings, and fire department limitations. Firefighters continue to operate in low- to zero-visibility environments with no knowledge of the structure or the environment, so target hazards need to be identified. Preplanning helps to alleviate the high-risk environment by giving firefighters more information related to the incident or structure both prior to arrival and during fireground operations. If this is not done, the lack of information and knowledge often leads to a no-win situation, and sometimes equates to nothing more than a suicide mission. If the fire service persists in sacrificing its personnel due to lack of knowledge about the building, its bloody history will continue.

Through training and an effective preplanning program, firefighters will be better prepared to conduct proper scene size-up and operate safer on the fireground. The risk assessment and safety planning process gives personnel access to the right information in a timely fashion. Preplanning allows the risks to be identified prior to an incident occurring; however, most structure fires that endanger and take the lives of firefighters occur in the residential single-family dwelling where preplanning does not take place unless it is an occupancy of vast square footage, and the owner of the property has allowed fire department personnel access to preplan the dwelling.

NFPA 1620, *Standard for Pre-Incident Planning*, addresses the components of a department risk assessment and safety planning program. The standard also deals with the various types of occupancies and information that should be collected. NFPA 1620 should serve as a guide prior to implementation of a risk assessment and safety planning program in an individual department.

Specific Areas of Concern with Preplanning

Two of the important areas concerning preplanning are life safety hazards and building construction. Risk assessment and safety planning from the life safety standpoint can be used to identify potential and specific risks. For example, a facility might utilize hazardous chemicals in its manufacturing processes. Risk assessment and safety planning would identify the hazards associated with life safety concerning employees of the facility as well as the safety hazards associated with firefighting operations. The type of construction of the building in question is closely related to both life safety of firefighters and risk management during firefighting operations.

Fire department personnel involved in the preplanning process must have a thorough understanding of how the building is designed and put together. Then personnel can predict what will happen to the building under fire conditions, which will also aid personnel in the risk-versus-benefit process and help the incident commander make better informed decisions about personnel utilization. The many areas associated with what information to include in a preplan are beyond the scope of this textbook; however, a successful preplanning program should include the following components:

- The pre-incident plan should be on a department-wide standardized form.
- Preplanning should be completed by the operations personnel who will make use of the plan. These personnel can gather specific information about and become familiar with the building.
- A system and process should be in place for updating the preplan on a specific timetable: either annually or at least once every two years in large response districts. Types of occupancies change as does ownership.
- Target and high hazards should be identified in the pre-incident planning process.

The individual preplan should include an informational text page, a site plan, and a floor plan and should provide the following information:

- Location
- Response routes
- Water locations/access
- Required rate of flow
- Nature of occupancy
- Normal hours building is occupied
- Type of building construction and features
- Number/location of people including employees with disabilities
- Life hazards/types of materials stored/special hazards to be aware of
- Utilities (location, condition)
- Extinguishing and alarm systems including fire department connections
- Methods or means of ventilation
- Ingress and egress points
- Tool and equipment needs
- All vertical and horizontal paths of fire travel
- Mutual-aid needs
- Placement of apparatus on scene

The pre-incident planning program could also be used for specific medical incidents. For example, a preplan could be designed for a mass-causality incident that might involve a structure in which large groups of people are housed or gather on a regular basis. All information should be shared on a department-wide basis, including other shifts or departments, not only to estimate and discuss the fire problem(s) but also to determine possible strategy and tactics. Once the data have been reviewed, research should be conducted in any area unfamiliar to personnel. Keep preplan information available for use at emergencies in some type of usable format such as a pre-incident plan book, onboard computers, and dispatch centers. Pre-incident planning is a very useful tool as it relates to pre-incident safety and on-scene operations; however, it focuses mainly on target hazards. Target hazards are those hazards that a department identifies as specific concerns. Examples might include hazardous materials, specific hazardous processing facilities, and large life loss facilities, such as nursing homes and places of assembly. It must be kept in mind that identification of target hazards is important, but most of the injury and deaths related to firefighting occur in single-family dwelling fires, which make up a large number of the fires in the United States.

Pre-Incident Safety and Training

Numerous fire department–related sources provide annual reports concerning firefighters injured or killed during training evolutions. Their statistics reflect that emergency response organizations must make a better proactive effort toward safety and health during training. NFPA 1403, *Standard on Live Fire Training Evolutions*, was created due to the high number of injuries and deaths related to live-fire training. Training is normally designed to simulate real emergency incidents, and command and control during training evolutions is essential to reduce injury and death, which are avoidable. The training environment must be evaluated for hazards, and any recommendations to improve safety should be made to the training division and the health and safety program manager. It is also important to keep the health and safety officer and the training division informed about weather conditions during training evolutions. Any time weather conditions would have an unsafe effect, training should be canceled. It is also vital that participants utilize full protective clothing and equipment during training evolutions and other situations that warrant their use.

During fireground operations, firefighters must remain aware of their surroundings and learn to develop situational awareness from both their experience and training. Firefighters must be taught that conditions can change rapidly; and firefighters who move too far too fast must understand the limitations of their **personal protective equipment (PPE)**, and that their escape route can be cut off. Danger signs associated with fireground and other types of emergency response and air management techniques must be explained. Reinforcing the importance of situational awareness should be done until it becomes second nature for all personnel. The NFPA standards contain many recommendations about firefighters working together as a system as they rely on each other to create a successful response.

personal protective equipment (PPE) ■ Specialized clothing or equipment is worn by employees for protection against health and safety hazards.

NFPA 1410, *Standard on Training for Initial Emergency Scene Operations*, concerns the minimum requirements for evaluating training for initial fire and rescue operations. Part of NFPA 1410's focus is on the performance of personnel during initial fire suppression and rescue operations. The standard stresses safe operation and use of adequate numbers of personnel and proper equipment. The Insurance Services Office (ISO) requires eight hours per year of facility training for companies to train together utilizing the multicompany approach. Companies that train together gain a higher level of proficiency and confidence, allowing them to operate more safely during the real incident.

The fire service should be concerned if firefighters and fire officers are not receiving the necessary training and experience to assess the risks on the emergency scene properly. Organizations that do not have a training staff or training officers should find creative ways for company officers to take a lead role with training. A reduction in structure fires means there needs to be increased training efforts related to fire behavior, building construction, reading smoke conditions, situational awareness, and risk management. Such training provides firefighters and fire officers the opportunity to learn the inexact science of firefighting. Many computer simulations and other types of media are available to help personnel continue their training and add to their knowledge base. Risk assessment and safety planning, incident safety, risk management, and incident command training can easily take place in the classroom. Post-incident analysis can be a huge benefit in teaching both personnel who were at the incident scene and those who were not. Company officers are the departmental backbone in finding ways to increase learning and understanding. Those who do not take a lead in their personnel's training are failing to perform one of their most important roles. It all boils down to what the company officer is willing to accept as the minimum standard to which his or her personnel perform. Inadequate skills and knowledge shown at the incident scene indicates that personnel need additional training in specific areas. The company officer should focus the crew's attention on performance and competency-based skills and knowledge.

Fire service leaders must provide the necessary training with promotional process opportunities to ensure that company and chief officers understand the environment their firefighters will face and the situations they will encounter. Fire service leaders must make certain that everything that can possibly be done to guarantee personnel safety is accomplished. The safety of everyone on the emergency scene must be improved. Although the fireground provides the best learning opportunities, it is a very unforgiving learning environment.

A comprehensive training program utilizing the NFPA 1500 and 1403 standards must be in place to help reduce the risk of injury and death on the fireground. These standards can be used to endorse safe operating procedures during training evolutions, to endorse the use of PPE during training evolutions to reduce the chances of injury, and to develop policies in an individual department. Rehab during training for both fire operations and EMS evolutions is important as is the appointment of a safety officer(s) during training evolutions. Personnel accountability procedures must be followed during training to ensure the same procedures will be followed during real incidents. It is time that individual fire departments realize that weak, ineffective efforts to adequately train personnel in their particular job function provide nothing more than a false sense of security and are a recipe for disaster. Unfortunately, in many fire departments the incident itself is the only training some personnel receive. Until changes are made to endorse training as an ongoing process, the fire service will not see a reduction in injuries and deaths.

The Proactive Approach to Wellness and Fitness

Firefighter wellness and fitness is a critical area of health and safety. Risk assessment and safety planning must include a component of wellness and fitness. Individual departments can develop wellness and fitness programs utilizing the expertise of personnel within their ranks, or they can be contracted using professional consultants. The IAFF and the IAFC developed the Fire Service Joint Labor Management Wellness/Fitness Initiative as a nonpunitive approach to enhance firefighter wellness and health and safety. The USFA has published a fitness coordinator's manual, which can be referenced to tailor individual programs. It is important to remember that wellness and fitness are separate areas.

#2

Wellness is directly related to medical wellness, physical fitness, and emotional wellness. Wellness also includes weight reduction, the cessation of tobacco products, proper nutrition, stress reduction, and behavior modification. Wellness encompasses a lifestyle that goes beyond issues related to on-duty practices. Wellness should be endorsed to extend beyond life in the fire station as a 24/7 commitment into which the employee must buy in. Good wellness practices allow the employee to continue learned practices and philosophies well into retirement. On the other hand, fitness relates to the ability to be physically fit to perform the job tasks required of an emergency services worker.

PHYSICAL FITNESS

A growing number of firefighters are dying in the line of duty due to cardiovascular incidents. This is due not only to stress but also to poor fitness and exposure to the products of combustion. The number-one killer of firefighters is a cardiovascular incident, which will continue until the fire service takes wellness and fitness to the mandatory level. Unfortunately, administrations and labor struggle with this hurdle, and they need help to reverse this statistic. NFPA 1500 requires fire departments to provide a physical fitness program that meets the requirements of NFPA 1583, *Standard on Health-Related Fitness Programs for Fire Department Members*. NFPA 1583 establishes five areas of focus: cardiovascular fitness, muscular strength, muscular endurance, flexibility, and body composition. The work required of a firefighter requires physical strength and cardiovascular endurance, as well as emotional stability. The medical component of an annual physical should evaluate body composition and cardiovascular fitness. Various tests measure strength and muscular fitness. One way to measure cardiovascular fitness is the use of a treadmill. Once the level of fitness is determined, a program can be developed to address overall fitness, which should include the following components:

#6

- Nutrition
- Sleep or rest
- Cardiovascular endurance
- Muscle endurance
- Muscular strength
- Flexibility

Cardiovascular fitness should be the most vital element of the fitness program. Cardiovascular exercise not only increases oxygen supply to the muscles and the brain but also reduces the chances of life-threatening diseases, such as heart disease, stroke, and high blood pressure. Organizational personnel must understand that strength, muscle fitness, and cardiovascular fitness diminish during the aging process; therefore, a regular fitness regimen becomes vital for employees as they gain seniority and grow older. It will help firefighters to handle both mental and physical stress in a more constructive manner and reduce the chances of heart attack and stroke. Firefighters of all ages need to maintain a certain level of fitness to reduce the chances of becoming a statistic.

The peer fitness program has become popular in many departments recently. The International Association of Fire Fighters (IAFF), International Association of Fire Chiefs (IAFC), American Council on Exercise (ACE), and Peer Fitness Trainer (PFT) Certification Program, which is a part of the IAFF/IAFC Fire Service Joint Labor Management Wellness/Fitness Initiative, have all become dedicated to firefighter wellness and fitness. The Peer Fitness Trainer Certification Program provides a fitness trainer standard consistent with the health and fitness needs of the fire service throughout the United States and Canada. The peer fitness program endorses a peer fitness trainer in every department who can take the lead role.

Some of the problems associated with firefighter fitness center around inadequate funding and resources, and lack of participation in all aspects of the firefighter wellness-fitness initiative. On-duty time to work out and the necessary facility to conduct a workout are needed. The department administration should conduct an analysis to show community leaders that funding a wellness and fitness program on the front side will reduce injuries and workers' compensation claims on the back side. Without this type of analysis, community leaders will not commit additional resources to this type of program.

Many department administrations and labor organizations cannot come to an agreement in the area of wellness and fitness. Because the fire service continually involves itself in standards, many believe a standard for fitness should be developed for firefighters to meet. The fire service historically has had the mind-set that punitive action should be administered when firefighters cannot meet a fitness standard. This issue will continue to be a source of debate.

Firefighter fitness does not come without consequences: some firefighters are going to sustain periodic injuries during a workout. There should be provision for rehabilitation after an injury occurs, as well as a close relationship between the department, the department physician, and a physical rehabilitation agency familiar with the work firefighters do. Many departments have a "light duty," or "restricted duty," provision for employees during the rehabilitation process. A light duty provision aids the firefighter in a safe return to work, as a way to keep a firefighter employed, and in periodic evaluations of the employee and his or her condition.

MEDICAL EVALUATION AND FITNESS

Most emergency services organizations require a preemployment physical examination; however, all emergency services organizations should have an annual physical examination in place. NFPA 1582, *Standard on Comprehensive Occupational Medical Program for Fire Departments*, addresses the requirement for a medical evaluation at the time of employment and annually. Most but not all department administrations and labor organizations agree on the importance of an annual physical for employees. For the most part, this disagreement centers around either the cost to conduct an annual physical or the privacy of the employee's medical results and condition. Every emergency services organization needs to navigate through these issues. It should be stressed that an annual physical is in the best interests of the employee, not only the organization.

An annual medical evaluation should be conducted by either a department physician or a physician well versed in the occupational health of firefighters according to NFPA 1582. Without an understanding of NFPA 1582, the physician will neither be familiar with the medical requirements of being a firefighter nor understand the work that firefighters must perform on a daily basis. The job tasks associated with firefighting are in the NFPA 1582 standard, which guides the physician in conducting the medical exam by outlining the tests and components. The purpose of the NFPA 1582 standard is to determine whether new employees and incumbent employees are at a minimum level of health and wellness. Workers' compensation has strict procedures in place outlining the medical evaluation of employees prior to their returning to full duty from injuries and illnesses resulting from an occupational illness or injury. The department should also have procedures in place concerning return to full duty for employees who have had nonoccupational injuries and illnesses See Box 5-3.

NFPA 1582 classifies problems into two categories: A and B. Category A includes conditions that would preclude a person from performing the tasks of a firefighter as they would present a significant risk to the individual or others. The department physician cannot certify or pass an individual who falls into category A. Category B includes conditions that could impede or prohibit an individual from performing the work of a firefighter and present a risk at the emergency scene and during training evolutions. Category B does not automatically preclude an individual from becoming certified as meeting the standards for duty.

BOX 5-3: MEDICAL TESTS RECOMMENDED BY THE WELLNESS-FITNESS INITIATIVE

MEDICAL HISTORY QUESTIONNAIRE

- Vital Signs
- Hands-on Physical
  - Head, eyes, ears, nose, and throat
  - Neck exam
  - Cardiovascular
  - Pulmonary—including spirometry and chest X-ray (X-rays recommended every 3 years)
  - Gastrointestinal
  - Genitourinary
  - Rectal—including digital rectal exam
  - Lymph nodes
  - Musculoskeletal
  - Body composition
- Laboratory Tests
  - White blood cell count
  - Platelet count
  - Glucose
  - Sodium
  - Total protein
  - Cholesterol
  - Differential tests
  - Liver function tests
  - Blood urea nitrogen
  - Potassium
  - Albumin
  - Heavy metal screening
  - Red blood cell count
  - Triglycerides
  - Creatine
  - Carbon dioxide
  - Calcium
  - Urinalysis
- Vision Test
- Hearing Test
- Cancer Screening
  - Breast exam
  - Prostate specific antigen
  - Testicular exam
  - Mammogram
  - Fecal occult blood testing
  - Pap smear
  - Skin exam
- Immunizations and Infectious Disease Screening
  - Vaccinations
  - Hepatitis C
  - HIV
  - Tuberculosis
  - Varicella
  - Hepatitus B virus
  - Polio

MENTAL AND EMOTIONAL FITNESS

Firefighters must learn to recognize that wellness is an ongoing process that needs continual attention. Part of the wellness program should include mental and emotional fitness. Unfortunately, physical and medical fitness has historically been prioritized over mental and emotional fitness, but priorities are changing. It is now recognized that stress is a part of life for most people, and stress can be compounded for emergency responders. Research shows that firefighters who discover a balance between physical, medical, behavioral, and emotional fitness will have the healthiest careers and retirement years. The new firefighter must adjust to becoming a firefighter; incumbent firefighters must continually seek ways to remain satisfied with their career. Employees must continually strive for well-being in their family life, and firefighters who have ended their careers must find ways to adjust to retirement. Firefighters who are under stress in any area will not perform at an effective level and then are more likely to incur an injury. Balance is the key to success.

When the balance between physical and emotional demands and periods of rest is disrupted, fatigue, poor attitude, errors in judgment, and injuries result. Attitudes, behaviors, thoughts, and emotions must remain in balance with performing the tasks of firefighting, as the mind and body are inseparable. The IAFF's Wellness-Fitness Manual gives suggestions for the assessment of medical fitness, physical fitness, and mental fitness. Sections of it provide direction for marriage and family issues, reactions to stress, training in behavioral skills, and understanding mental illness.

Many departments now have employee assistance programs (EAP) available to their employees; these are confidential, and a limited number of visits to a mental health professional are free. The employer pays for many EAP programs for fire service professionals, and the employee's family members are encouraged to participate, as many issues are related to family and home life problems.

#13

EAP programs aid organizations and their employees in establishing mental and emotional wellness. Similar to sections of the IAFF Wellness-Fitness Manual, EAP programs address the following:

- Family issues
- Drug and alcohol abuse treatment
- Tobacco use cessation
- Stress management
- Critical incident stress management (CISM)

Many company officers and higher-ranking supervisors and managers are receiving training in recognition of sudden changes in employees' work performance, because many times a sudden downward turn in work performance points to a mental or emotional issue. An employee who is disciplined may be directed to mandatory entry into the EAP program to benefit both the employee and the organization. It has been recognized that a major factor in mental and emotional wellness is **critical incident stress management (CISM).** Many CISM programs are part of an EAP program, encompassing a team approach whereby individuals with specialized training come together as a unit to deal with CISM issues as they arise. The subject of CISM will be covered in more detail in Chapter 9.

critical incident stress management (CISM) ■ This adaptive short-term helping process focuses solely on an immediate and identifiable problem to enable the individuals affected to return to their daily routines more quickly and with a lessened likelihood of experiencing post-traumatic stress disorder.

Other Agency Considerations

In most emergency response organizations and agencies, a mutual-/automatic aid agreement exists, which is vital to risk assessment and safety planning. Many agencies focus on the operational aspects of multiagency response; however, the health and safety implications must also be considered. Since the events of September 11, 2001, the concept of a comprehensive national incident management system came from the formation of Homeland Security and Presidential Directive 5: Management of Domestic Incidents. Numerous agencies have now received the training to manage domestic incidents as they relate to utilization of the **National Incident Management System (NIMS).**

National Incident Management System (NIMS) ▪ This system mandated by Homeland Security Presidential Directive (HSPD) 5 provides a consistent nationwide approach for federal, state, local, and tribal governments; the private sector; and nongovernmental organizations to work together effectively and efficiently to prepare for, respond to, and recover from domestic incidents, regardless of cause, size, or complexity.

Since September 11, agencies agree that a response to a terrorist event will require a multiagency response. In addition to the standard fire and EMS response, many agencies that normally do not respond to emergencies will respond to terrorist incidents. It will be the responsibility of the fire departments to establish an incident management structure so that all involved become part of the decision-making process. A **unified command** structure will be necessary and has proven effective in prior multiagency responses. Events involving multiagencies will require ongoing planning, review, and revision of response plans; participation by the working agencies; and joint exercises to ensure on-scene integration and coordination. The health and safety program manager in every department must become aware of the operational aspects of mutual-/automatic aid departments. Areas of concern include accountability systems, incident management systems, safety awareness, compliance with standards and regulations, compatibility with communication systems, and operational issues in compliance with NFPA 1710, *Standard for the Organization and Deployment of Fire Suppression Operations, Emergency Medical Operations, and Special Operations to the Public by Career Fire Departments.*

unified command ▪ This is a way to carry out command in which responding agencies and/or jurisdictions with responsibility for the incident share incident management.

When it comes to police departments, fire and EMS personnel must recognize the need to integrate forces in certain types of responses. An example would be working together on a civil disturbance in which the ability to communicate effectively and the integrated unified command system must be in place and used. Individual departments must establish a pre-incident understanding between agencies. Police, fire, and EMS agencies respond together to many incidents on a daily basis, such as vehicle accidents, crime scenes involving a need for EMS, and incidents requiring traffic or crowd control. Police agencies sometimes provide EMS services or medical evacuation by helicopter.

Critical to safety and health is a preplanned understanding of each agency's roles, priorities, and responsibilities prior to an incident occurring. In situations in which there is no clear understanding of responsibility, a unified command system must be used to establish an incident commander for both law enforcement and other agencies. Incident commanders must work together to bring unity and a safe operating environment. Useful information involving multiagency response must also be preplanned and continually reviewed so all agencies can work together effectively. All considerations can usually be dealt with at the local level utilizing a flexible approach, and the safety of the responders must always be a priority. The issues of safety and health, as well as effective and integrated operations, must be in the forefront when it comes to interagency response.

CHAPTER REVIEW

Summary

Risk assessment and safety planning have many components, including station safety, apparatus safety, response safety/pre-arrival, pre-incident planning, safety while training, fitness/wellness, and interagency relationships. Apparatus and station safety can be subdivided into facility design and ongoing operations. The issue of response safety includes driver issues, training programs, and response issues. Risk assessment and safety planning and programs can be used as a way for responders to develop and understand the importance of preplanning, which equates to safer operations. Preplanning affords the response agency the opportunity to identify particular hazards and characteristics about the building and the situation. Building construction and life safety hazards are the priority issues.

Training and the safety issues associated with this area create a concern with the health and safety program. Training is normally designed to simulate real emergency incidents, and command and control during training evolutions is essential to reduce injury and death, which are avoidable. Participant safety is the number-one priority when it relates to training.

The wellness of firefighters should be a top priority of the health and safety program. Wellness is an area that encompasses many components including physical, medical, mental, and emotional fitness and wellness. The issue of medical fitness can be assessed with an annual physical exam. Physical fitness assessments are necessary, and programs should be tailored for individual members of a department. The employee assistance program is essential as a way to ensure emotional and behavioral fitness. Emotional fitness must become a higher priority.

Many emergency response agencies have some type of mutual-aid agreement in place to make certain responders can work alongside outside agencies. Since the incidents of September 11, 2001, response agencies recognize the need for strong interagency coordination and effectiveness. Defined roles must be preplanned and established to reduce the chance of injury and death. The issue of interagency coordination also leads to a safer and more effective operation.

Review Questions

1. Discuss the areas of concern related to the selection of apparatus operators.
2. Discuss the reasons why personnel provide valuable information related to facility design.
3. Who should have a leading role in apparatus design?
   a. The health and safety committee
   b. Firefighters
   c. The fire department administration
   d. The Department of Transportation
4. Discuss how a traffic preemptive control device for response agencies works.
5. Discuss the importance of driver/operators and their familiarity with their apparatus.
6. List why the training for a new driver/operator must focus on response.
7. List the reasons why risk assessment and safety planning are so important.
8. An effective preplanning program will:
   a. Give responders more information
   b. Ensure firefighters are better prepared
   c. Identify target hazards
   d. All of the above
9. It is important that training take place:
   a. With a risk analysis
   b. At the facility
   c. Under controlled conditions
   d. All of the above
10. Discuss the problems associated with physical fitness.

Case Study

On July 8, 2008, a 25-year-old male volunteer firefighter (the victim) was fatally injured after being ejected in a fire truck rollover. The crash occurred as the fire truck was returning to the station after a call for a propane gas fire. The fire truck was traveling down a winding, steep grade. The paved road had a posted speed limit of 45 mph with a curve warning sign and recommended safe speed of 20 mph through the S-curve. The driver lost control of the fire truck, swerved off the left side of the road, returned to the pavement, and overturned on the right side of the road. The victim was not wearing a seat belt and was ejected out of the driver's side window. The pumper's 725-gallon water tank detached from the truck body and landed on top of the victim in the street. The victim was pinned underneath the water tank and died from injuries sustained in the crash. Key contributing factors identified in this investigation include non-use of seatbelt, inadequate driver training, driver inexperience with this specific apparatus, an older apparatus with minimal safety features, potentially incorrect installation of a replacement water tank, and difficult road conditions.

The department required that all new members be enrolled in a firefighter 1 class within one year of becoming a member. The victim (driver) in this incident had approximately 6 months' experience at this department, and had participated in three driver training sessions (number of hours unknown) with the department's most experienced driver/operator. This consisted of hands-on driver training in the local area. No other training records were available from this department.

The victim had previously been a member of another fire department for 6 years and had driven fire apparatus at that fire department, but the apparatus all reportedly had automatic transmissions. The victim had received firefighter 1 and 2 training as well as hazmat training before becoming a member of this department and was also employed by a private ambulance company as a full-time emergency medical technician.

The engine was traveling downgrade on a two-lane blacktop state highway with a posted speed limit of 45 mph. The incident occurred at an S-curve that was marked by signage indicating a recommended speed of 20 mph. The road surface was asphalt in good condition and was dry. At the time of the crash it was daylight, the skies were partly cloudy, visibility was 7 miles, and the temperature was 80°F to 85°F. Winds were 10 mph, from the west. The fatality occurred on the victim's first emergency call driving this truck. New York State law exempts emergency vehicle operators and passengers from the use of seat belts. The Federal Motor Carrier Safety Administration also exempts the occupants of fire trucks and rescue vehicles while involved in emergency and related operations from wearing seat belts.

The following recommendations are from NIOSH:

Recommendation #1: Fire departments should ensure standard operating procedures (SOPs) regarding seat-belt use are enforced.

Recommendation #2: Fire departments should provide and ensure all drivers successfully complete a comprehensive driver's training program such as NFPA 1451, *Standard for a Fire Service Vehicle Operations Training Program*, before allowing a member to drive and operate a fire department vehicle.

Recommendation #3: Fire departments should ensure that replacement fire apparatus water tanks are installed according to manufacturer's instructions.

Recommendation #4: Fire departments should ensure that programs are in place to provide for the inspection, maintenance, testing, and retirement of automotive fire apparatus.

Recommendation #5: Fire departments should consider replacing fire apparatus over 25 years old. In addition it is recommended that apparatus manufactured prior to 1991 that are less than 25 years old, that have been properly maintained, and that are still in serviceable condition should be placed in reserve status and upgraded to incorporate as many features as possible of the post-1991 fire apparatus edition.

Recommendation #6: Fire departments should be aware of programs that provide assistance in obtaining alternative funding, such as grant funding, to replace or purchase fire apparatus and equipment.

Recommendation #7: Federal and state departments of transportation should consider removing exemptions that allow firefighters to not wear seat belts.[2] See Figure 5-10.

1. Was the required driver training program in this department adequate? Explain.
2. Was the design of this apparatus safe and road worthy? Explain.

Firefighter Ejected from Pumper Concerns and Issues	Yes	No
Was vehicle speed a factor in this incident?		
Was the lack of driver/operator training a key issue?		
Was the pumper in this incident properly designed?		
Would additional safety features on the pumper have made a difference with this outcome?		
Did road conditions play a role in this incident?		
Did the driver/operator receive adequate vehicle training?		
Did the lack of recognition play a role in this incident?		
Should the law in this state regarding seat-belt use be changed?		
Should the driver/operator receive training from outside the organization?		
Would apparatus testing have produced a different outcome?		

FIGURE 5-10 Issues and concerns of a firefighter killed in a pumper rollover.

References

1. NIOSH Fire Fighter Fatality Investigation and Prevention Program (FFFIPP), *A Lieutenant Dies and Three Fire Fighters of a Career Department Were Injured When the Truck They Were Responding in Was Struck by Another Vehicle—Illinois*, August 2, 2001. Retrieved May 10, 2011, from http://www.cdc.gov/niosh/fire/reports/face200039.html

2. NIOSH Fire Fighter Fatality and Injury Prevention Program (FFFIPP), *Fire Fighter Dies After Being Ejected from a Pumper in a Single Vehicle Rollover Crash—New York*, March 20, 2009. Retrieved May 7, 2010, from http://www.cdc.gov/niosh/fire/reports/face200825.html

CHAPTER 6
Fire Emergency Safety

KEY TERMS

aircraft rescue firefighter (ARFF), *p. 139*
back draft, *p. 136*
defensive fire attack, *p. 121*
flashover, *p. 136*
freelancing, *p. 143*
heat transfer, *p. 136*
high efficiency particulate air (HEPA) cartridge filter mask, *p. 126*
incident stabilization, *p. 123*
life safety, *p. 122*
line-of-duty death (LODD), *p. 119*
offensive fire attack, *p. 121*
personnel accountability report (PAR), *p. 143*
personnel accountability system, *p. 141*
property conservation, *p. 125*
rehabilitation, *p. 146*
rollover, *p. 136*
self-contained underwater breathing apparatus (SCUBA), *p. 134*
team unity, *p. 141*
thermal layering, *p. 136*
thermal protective performance (TPP), *p. 128*
total heat loss (THL), *p. 128*
tunnel vision, *p. 123*

OBJECTIVES

After reading this chapter, you should be able to:

- Describe the purpose of the incident management system.
- Describe the key components of safer operations at fire scenes and other types of emergencies.
- Discuss the characteristics of the incident management system.
- List the three incident priorities.
- Discuss the role of the incident safety officer.
- Discuss why effective communications are essential at the fire and emergency scene.
- Define the need for risk management at fire scenes.
- List the categories of PPE for firefighting.
- Describe the hazards faced by firefighters operating at fire scenes.
- List the objectives that accountability systems should meet.

- Discuss the need for and the differences between an initial rapid intervention team and a formal rapid intervention team.
- Discuss the concept of rehabilitation and why it should be multilayered and tailored to the incident.

ResourceCentral

For additional review and practice tests, visit **www.bradybooks.com** and click on Resource Central to access book-specific resources for this text! To access Resource Central, follow directions on the Student Access Card provided with this text. If there is no card, go to **www.bradybooks.com** and follow the Resource Central link to Buy Access from there.

Firefighting operations involve a high level of physical exertion. The uncontrolled environments faced by firefighters place an extreme amount of stress on personnel both physically and emotionally. Each year, statistics continue to reflect that firefighters experience a high level of injuries, occupational-related diseases, and death. Almost half of firefighter injuries occur during fireground operations. For these reasons, firefighting is one of the most hazardous occupations. It has become common for organizations to endorse management education with their members who are striving to become administrators. Yet leadership must be the prime focus if emergency services organizations wish to be successful in promoting occupational health and safety.

Ongoing training efforts and an emphasis on becoming certified will encourage personnel to take responsibility for their own health and safety. Such efforts will direct every emergency services organization to focus on risk management.

The Common Problem Areas

#4

Areas that continue to need attention as they relate to fireground safety are personal protective clothing and equipment, fire apparatus, training, communications, and medical requirements. Risk management must be integrated with incident management at all levels. Leadership must be stepped up at all levels of management and supervision, with a high standard of accountability. All firefighters should be encouraged to take personal responsibility for their own safety on the fireground, and they must be empowered to stop any unsafe practice.

#6

A fundamental behavioral change must occur in how fire departments and firefighters view occupational health and safety on the fireground. Fire departments must continue to educate their members, as well as their administrations and their citizens, about the hazards of firefighting. The NFPA standards about medical and physical fitness are equally applicable to all firefighters, and all fire service organizations should take them seriously and put them into practice.

#5

Standards also need to be addressed and adhered to regarding training and the qualifications and certifications needed to be a firefighter; these standards should be based on the duties firefighters are expected to perform. Unfortunately, much of the certification and ongoing training for firefighters is tragically deficient due to the lack of funding and sometimes priorities. One of the highest priorities in the training of personnel for firefighting is developing firefighters' SCBA and mask confidence. Personnel must continually train with their SCBA and face mask in place to develop the necessary skills for both firefighting and safety and survival. Personnel who are not proficient in their abilities and confident in operating with an SCBA in place may easily become a hazard to themselves and others.

#8

Leadership in the fire service must continually utilize knowledgeable, experienced, and certified instructors, company officers, and senior firefighters to pass on the knowledge and endorse health and safety. Ways must be found to increase funding for training efforts, and technology must be updated to produce higher levels of health and safety. Some of the needed funding could come in the form of grant funds; obtaining this type of funding should also incorporate and mandate the implementation of safe practices for fire departments. Training and safe practices on the fireground will ultimately lead to a reduction in injuries and deaths related to fireground operations.

#10

#3

Many fire departments are in need of a written risk management plan that complies with NFPA 1500, *Standard on Fire Department Occupational Safety and Health Program*. Such a risk management plan would include strategies to meet certain objectives. It is every fire chief's responsibility to implement a risk management plan and oversee its operations. A fire department's plan should include development of an occupational health and safety program that recognizes operational risks to its members and makes a concerted effort to reduce them. The focus of a large portion of the risk management plan should be on fireground safety. Unfortunately, a major difference exists between a written occupational health and safety plan and an implemented plan. Setting up and following through with periodic evaluations and analysis is the key to eliminating this difference.

Behavioral Changes Are Needed

Study over the years has revealed that the dominant factors contributing to firefighter injury and **line-of-duty death (LODD)** are related to situational awareness, a lack of fitness and wellness, a lack of a proper risk-versus-benefit analysis, and human error. When many of these contributing factors occur together, they are viewed as cluster events—which all fire department administrators, supervisors, and firefighters should be aware of. Many times a review of any injury or LODD will show how one bad event usually leads to another at the fire scene. These events are usually under the direct control of individual company officers or chief officers. The consequences of cluster factors should be used by individual fire departments to improve their risk management plan and health and safety programs. A lack of adequate operational polices and priorities has a detrimental effect on fire scene safety. This chapter addresses many of the factors that lead to the lack of safety at fire scenes.

line-of-duty death (LODD) ■ A firefighter who dies or is killed in the performance of his or her duties as a firefighter.

Incident Management

All fire scene operations, training functions, and other emergency situations should be conducted in a similar manner because of their similar hazards. Although many textbooks have been dedicated to the incident management system, this chapter focuses on safety at fire scenes and emergency incidents as related to the incident management system (IMS) and risk management.

#3

It is of the utmost importance to have an incident management system in place that recognizes hazards and strives to reduce accidents and injuries not only to the operating forces but also to the public at large. This system must meet the requirements of NFPA 1561, *Standard on Emergency Services Incident Management System*; and the department should have written standard operating procedures (SOPs) that apply to all members operating at the fire or emergency scene. Incident management systems in conjunction with SOPs are key components to safer operations at fire scenes and other types of emergencies.

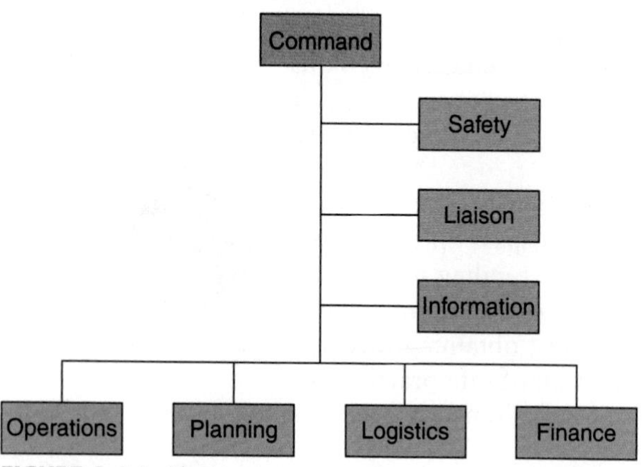

FIGURE 6–1 Incident management system, command staff, and general staff positions. *Courtesy of Joe Bruni.*

Unfortunately, some fire departments have neither a standard written set of SOPs nor experienced and qualified incident commanders within their ranks. In most cases, it takes a combination of both fireground experience and training to achieve proficiency as an effective and safe incident commander. Training that encompasses tabletop exercises, company evolutions, and drills can be a great aid in preparing personnel for the incident commander role. An incident commander mentoring program also has many benefits.

There are variations in the type of incident management systems in place; however, in 2003 a presidential directive (HSPD-5) instituted a National Incident Management System (NIMS). According to NFPA 1500, an incident management system that will be used at all emergency incidents is comprised of a command staff and a general staff. The command staff is made up of the incident commander, safety officer, liaison officer, and public information officer. The general staff working directly under the command staff consists of operations, planning, logistics, and finance. See Figure 6–1.

The staff's common goal is to utilize a system that can be adapted to different operations, such as single agency/single jurisdiction response; single jurisdiction/multiagency response; and multi-jurisdiction/multiagency response. Although variations may exist from incident to incident in the strategies and tactics used and in each jurisdiction's procedures, the incident management system should have the following characteristics:

- Shall be used at all incidents
- Shall be a modular organization
- Shall utilize management by objectives
- Shall rely on an incident action plan
- Shall have a manageable span of control
- Shall have predesignated incident mobilization center locations and facilities
- Shall have comprehensive resource management
- Shall utilize integrated communications
- Will make provisions for the establishment and transfer of command
- Shall encompass the chain of command and unity of command
- Shall have a unified command
- Will utilize an accountability of resources and personnel
- Will make provisions for deployment of personnel and resources
- Shall utilize a system of information and intelligence management
- Shall utilize common terminology

For years the fire department made use of coded radio terminology much like police agencies have done and continue to do for security reasons, and to shorten the "airtime" during radio transmissions; however, in recent years fire departments have been using "clear text" for radio communications to reduce confusion at incidents at which multiple agencies are operating. Examples of beneficial clear text and common terminology are transmissions such as "firefighter down," "emergency retreat," "firefighter lost," and "firefighter missing." Yet even with this proactive communication approach, poor communication unfortunately continues to plague the fire service, evidenced by the agencies that investigate fire scene fatalities and injuries. Effective communications are essential at the fire and emergency scene so that the incident commander's awareness of the events as they unfold helps him or her to provide effective supervision and control.

At a fire scene and other emergency incidents, the incident commander's responsibility is the overall management of the incident and the safety of operating personnel.

During the beginning stages of first arriving fire company operations, the incident commander and the first arriving company officer must make the decision whether to implement an **offensive fire attack** or a **defensive fire attack** strategy.

offensive fire attack ■ Usually this is an aggressive firefighting strategy accomplished from the interior, unburned side of the structure. Interior operations are the principal factors of an offensive strategy.

defensive fire attack ■ Such a fire attack strategy utilizes an exterior attack, with related support, to stop forward progress and then to control the fire. The first defensive tactic normally is to protect the exposures while operating around the perimeter of the involved fire building.

The role of both incident commander and safety officer can usually be handled by the company officer at an incident that normally requires a single unit to take charge of the situation; however, as an incident escalates in size and complexity, the incident commander must divide the incident into management components at the tactical level to enable a manageable span of control. Incident commanders who become overwhelmed as an incident escalates are jeopardizing the operating personnel and themselves.

The assignment of a safety officer should also be made early into an incident that is growing or one that is already large or complex. The role of the safety officer is to focus on hazards and potential hazards. As part of the command staff, he or she has the authority within the incident management system to communicate directly with the incident commander to cease, suspend, or alter any operation deemed unsafe. For this reason, the individual assuming the role of the safety officer should be well versed in fireground operations and risk management. An incident that is small in nature and requires a handful of units to mitigate the situation (sometimes referred to as a single alarm) can usually be handled by a company officer or a department's shift commander. Incidents that require greater response need the incident management system to be expanded. See Figure 6–2.

Many times at fire scenes and other emergency incidents, it is common for the incident commander and the operating forces to perceive time to be passing at a slower pace. For this reason, many fire departments have their dispatch center transmit and report elapsed time. It has been discovered that buildings built with modern lightweight construction materials are seriously compromised after 10 minutes of heavy fire involvement; this elapsed time report allows the incident commander and the operating forces to make an informed decision as to whether a transition should be made from an aggressive offensive fire attack strategy to a defensive fire attack strategy, allowing the incident commander to remove operating personnel from harm's way.

Because the incident commander has a great deal of responsibility at fire scenes and other emergencies, he or she should have a high regard for firefighters' safety and their survival at every incident. Through integrating this concern into his or her regular

FIGURE 6–2 Incident safety officer. *Courtesy of Lt. Joel Granata.*

command functions, the incident commander promotes firefighter welfare in all facets of the job. Incident commanders must keep in mind that under fire conditions they are at an extreme disadvantage to perform any additional tasks other than the role of incident commander. The incident commander's safety plan must become part of the standard command plan. As the incident commander develops an incident action plan and tries to predict the sequence of events as they play out at the fire and emergency scene, he or she should always incorporate firefighter safety into this evaluation and forecasting.

#4

It is difficult to discuss the components of the incident management process without briefly touching on risk management. It cannot be stressed enough that risk identification, evaluation, and management concepts should be incorporated into each stage of the command process. A risk-versus-benefit analysis must become part of the normal thought process of every company officer and incident commander as well as of every incident action plan.

#3

Risk management has been discussed in detail in Chapter 3. The goal at any fire scene or emergency incident is to raise the awareness of responders from the incident commander down to the firefighter level about safety-related issues. Everything from individual safety-related actions to overseeing the overall incident as an incident commander must undergo a cultural change, and the endorsement of safe operations must move to the forefront of fire and emergency scene operations.

#2

Fire Scene Safety and Incident Priorities

Throughout the course of history, the fire department has been a paramilitary organization with one key exception: in the fire department, there is no acceptable loss of life concerning fire department personnel. The fire department's number one priority is life safety, and this includes the safety of its personnel as the top priority.

INCIDENT PRIORITIES

The incident priorities at every fire scene are life safety, incident stabilization, and property conservation. These priorities form the basis of the decision-making process that firefighters, company officers, and incident commanders use to establish a strategy and implement the tactics to mitigate the incident. The responsibilities of the first-due company officer are tremendous, and often overwhelming. The incident commander must continually assess these three priorities throughout the incident.

Life Safety

life safety ■ The actions firefighters take to make certain the dangers associated with injury and death to firefighters and the general public are reduced to the absolute minimum.

Life safety is the top priority at every fire scene and incident; this never changes for any reason. The goal is to maximize life safety and minimize any threats to it. Life safety may be as simple as firefighters wearing their full protective gear at a simple trash fire to the total evacuation of a high-rise building under fire conditions. The lives of firefighters come before those that they are trying to save. This may be difficult for the general public to accept; however, placing firefighters in a situation that could get them seriously injured or killed will only add to the encountered problems at the fire scene. It is important to remember that the value of the lives of both firefighters and the general public they are sworn to protect are equal; however, the fire department must protect its own people. Firefighters must make difficult decisions when civilian lives are hanging in the balance.

The risk-versus-benefit analysis called a risk assessment that must be made during these difficult situations can add a great deal of psychological and emotional stress on firefighters and other emergency responders. As tragic as it may be and as difficult as it is to accept, one life lost is better than six lives lost. Firefighters will always make a concerted effort to afford the best chance of survival for victims while, at the same time, doing everything possible to protect their own personnel.

The leading incident priority is safety of both firefighters and the general public. Safety is considered with each decision made by company officers and the incident commander with a safety officer assigned to oversee the incident as decisions are made to mitigate the incident. As noted previously, risk management, discussed in Chapter 3, at fire scenes begins at the time of receiving the alarm. An ongoing size-up must be made to ensure personnel are operating as safely as possible in a hazardous situation. Fire scene supervisors must continually use a risk identification process prior to making their decisions so they can provide a high level of safety at the fire scene. Decisions concerning strategy and tactics should be made after incident commanders first identify the risks to responders while also considering safety. Many departments have adopted the risk management philosophy:

- We will take a significant risk to protect savable lives.
- We will take a very controlled risk to protect savable property.
- We will not risk firefighter lives to save what is already lost, be it lives or property.

In order to comply with this philosophy, company officers, firefighters, and the incident commander must undertake a risk-versus-benefit analysis. All ranks of firefighting personnel arriving on scene must also learn to gather information as quickly as possible from bystanders; this can difficult to do because of **tunnel vision,** which commonly occurs with inexperienced fire department personnel. For some fire department personnel, tunnel vision can be a difficult issue to work through. Some personnel never learn to overcome it to view the big picture. Certain bystanders may state to arriving firefighters that someone is inside the burning structure whereas other bystanders or occupants at the same scene may state that no one is left inside and all escaped the structure.

tunnel vision ■ The focus of one's attention on one specific item or event to the exclusion of everything else that is happening around him or her.

Arriving firefighters must learn how valuable information from bystanders can be early into any fire scene emergency. Many vacant or boarded-up structures may very well house a vagrant or someone who is homeless at the time of any fire; however, experienced company officers and incident commanders may need to "pull back on the reins" of firefighters who are attempting to access an extremely unsafe area and ignoring safety practices and situational awareness. Aggressive firefighters do not like to be placed in the reverse mode, and fireground supervisors must exercise a great deal of internal strength to accomplish this. Many fires in vacant or boarded-up buildings without electricity supplied to them are the direct result of arson. Incident commanders must remain cautious with their strategy and tactics during these situations because no building is worth a firefighter's life. Without the experienced senior personnel on scene, younger inexperienced personnel will not have the necessary guidance in providing safety.

Incident Stabilization

When it comes to incident priorities, **incident stabilization** occurs after all initial life safety considerations have been evaluated and considered. Although the order of incident priorities does not change, firefighters must learn that, in some instances, extinguishing the fire accomplishes life safety. Extinguishment (incident stabilization) of a fire in one room of a nursing home will accomplish life safety at the same time because evacuation of the occupants cannot be accomplished in a timely fashion. During the incident stabilization process, firefighters are concurrently attempting to mitigate the hazard or situation and to identify and solve the encountered problems at the fire or emergency scene. Incident stabilization is a process of mitigating a hazard or life-threatening situation, and it is usually coupled with an incident action plan. In many cases, incident stabilization will save the lives of people caught in harm's way. A continual size-up of the incident is required to complete and implement incident stabilization. In many cases, a quick attack on the fire is effective concerning incident stabilization, property conservation, and life

incident stabilization ■ The process of mitigating a hazard or life-threatening situation.

FIGURE 6–3 Fire knock-down. *Courtesy of Chief Mike Zamparelli.*

safety. It is through proper and safe strategy and tactics that effective incident stabilization occurs. See Figure 6–3.

Strategy and Tactics

Although this book is not on strategy and tactics, they are worth mentioning here. No matter the type of firefighting situation encountered, strategy and tactics are put into practice so that firefighters can mitigate the situation and achieve incident stabilization. The specific strategy and tactics depend upon the type of firefighting situation. Strategy and tactics on the fireground include a whole host of issues:

- Use of hose lines
- Apparatus placement
- Support work
- Ladder placement
- Rescue operations
- Ventilation
- Forcible entry
- Rapid intervention team procedures
- Salvage operations
- Overhaul

In order to operate safely, personnel must have a thorough knowledge of the strategy and tactics that will be put into practice. Such an understanding helps personnel spot unsafe practices that are not in line with the incident commander's orders and strategy; for example, a firefighter who is working on a ladder without the benefit of someone butting the ladder or tying off or taking a leg lock. At most structure fires, the incident commander is managing unseen areas of the structure and unseen personnel. Safe and solid strategy and tactics allow personnel to operate as safely as possible by continually seeking information; identifying issues that need to be addressed; and using the on-scene personnel to manage the fire incident.

Success on every fireground depends on two important issues: information and personnel. Safe and effective strategy and tactics is about information, communication, and skilled operating forces. The fire service is continually changing, being refined, and updating due to the information it receives. Personnel who do not stay abreast of updated information may get themselves or someone else injured or killed. In many cases, the only thing that stands between the saving of lives and total losses is a strong incident commander and personnel well versed in the types of strategy and tactics to be put into practice at the fire incident. Individual fire service personnel must understand they are students from the day they start their training to become a firefighter until the day they retire. They must assume a specific role on every fireground and make individual decisions to keep safe while attempting to bring the incident to a safe and successful closure. The efforts of firefighters who continually study and participate in hands-on training to learn the job required will pay off in large dividends.

PROPERTY CONSERVATION

Property conservation is the third incident priority on every fire scene. Efforts are directed toward minimizing property damage and the long-term effects an incident has on those affected. Many times, effective property conservation can be accomplished through quick, safe, and effective fire extinguishment along with effective salvage and overhaul efforts. Salvage and the efforts that make it possible need to be considered at almost every fire event from its earliest stages. Salvage operations do not begin after the fire is extinguished and the situation has been mitigated, but should begin as soon as possible to reduce damage from smoke, fire, and water both during and after the fire event.

property conservation ■ Efforts made by the fire department to reduce further loss to property including the salvage of savable property.

Effective overhaul involves checking void spaces and other areas of a structure or fire scene for hidden fire; this ensures the end to property loss. A large-scale incident may require that property conservation take place for days or weeks after mitigation of the emergency incident is completed. Property conservation is commonly referred to in the fire service as "stopping the loss." Property conservation should become a standard in every fire department to commit whatever fireground resources are needed to stop or reduce property loss.

The personnel used to conduct fire extinguishment and rescue may be fatigued when the time comes to commit resources to property conservation; this can equate to shoddy property conservation efforts and injuries. Many of the injuries during the overhaul or property conservation phase of operations fall into the category of hand, eye, and back injuries. Firefighters are normally attracted to fire control operations more than property conservation efforts. The incident commander must keep this in mind in order to reduce careless property conservation efforts and reduce injuries.

Risks After Extinguishment

The faster fire extinguishment occurs, the quicker the fire scene becomes a great deal safer; however, incident commanders and company officers must recognize that many risks can still be present. In numerous cases, firefighters and fire scene investigators must reenter the building to complete salvage, overhaul, and to conduct a fire investigation. Incident commanders and other fire scene personnel must never let their guard down in regard to safety and risk management. Something as simple as allowing personnel to breathe airborne particles without the use of respiratory protection will compromise the long-term health of personnel.

For years, the cancer rates of firefighters have been significantly higher than those of the rest of the population. Numerous toxic by-products of combustion are usually present for several hours after extinguishment; and many are carcinogens, such as acrolein, benzene, acrylonitrile, cyanide, and formaldehyde. Recently the fire service has been doing a very good job protecting the respiratory system of firefighters from the products of combustion while the fire is occurring; it must have firefighters use the same effective protection during the overhaul and investigative phases. If the fire service wishes to

FIGURE 6–4 Firefighter in a HEPA mask. *Courtesy of Joe Bruni.*

reduce its cancer rates, firefighters must start wearing their SCBA during operations inside after the fire is out. Firefighters and investigators working a "cold" fire scene should be issued and be required, at a minimum, to wear a **high efficiency particulate air (HEPA) cartridge filter mask** and a combination cartridge for the encountered gases found at fire scenes. See Figure 6–4.

high efficiency particulate air (HEPA) cartridge filter mask ▪ A filter designed to remove 99.97% of all airborne pollutants 0.3 micron or larger.

Operating fireground personnel must also be aware of potential collapse hazards after extinguishment has been accomplished. Firefighters working inside an already damaged building may very well cause an action that could result in injury or tragedy. Gravity is trying to bring down every building that is standing; fire damages every building in some fashion and may help gravity complete its job. Past experience and the knowledge of experienced key fireground personnel will continue to play an extremely important role in this area. For this reason, it is critical that experienced personnel be a part of every fire suppression unit, if possible, to ensure added safety awareness.

Personal Protective Equipment

Respiratory and personal protective equipment (PPE) is of great value to emergency and fire scene safety. There are several types of PPE available for fire and emergency scene safety: structural PPE differs from PPE used for wildland, technical rescue, hazardous materials, and aircraft rescue and firefighting. PPE for technical rescue situations and hazardous materials will be discussed in Chapter 8.

The categories of PPE for firefighting can be broken down into several areas:

- General requirement
- Purchasing and design
- Use
- Care and maintenance
- Firefighter responsibility

Firefighters in full PPE are shown in Figure 6–5.

According to NFPA 1500, the general requirement is that firefighters wear personal protective equipment and clothing appropriate to the hazards to which they may be exposed.

FIGURE 6–5 Firefighters in full PPE. *Courtesy of Lt. Joel Granata.*

NFPA 1971, *Standard on Protective Ensembles for Structural Fire Fighting and Proximity Fire Fighting*, sets forth the general requirements for the components of structural firefighting PPE. NFPA 1851, *Standard on Selection, Care, and Maintenance of Protective Ensembles for Structural Fire Fighting and Proximity Fire Fighting*, sets forth the requirements and the associated program for the care and maintenance of PPE.

Since the terrorist attacks of September 11, the role of firefighters has changed. In addition to flames, heat, and smoke, firefighters now face the possibility of dust and other airborne particulates while searching the rubble and debris from a building collapse. Additional hazards may also include human remains, hazardous materials, and secondary explosive devices. The common types of structural PPE that firefighters have worn and trained with may not meet the requirements of searching building debris. Firefighters facing long-term operations and tasks they typically do not perform must be provided with adequate PPE to handle such unexpected activities. See Figure 6–6.

FIGURE 6–6 Firefighter in PPE for building debris search. *Courtesy of Chief Mike Zamparelli.*

Managing the incident will require assurance that all personnel be properly protected. This will also necessitate proper training for and information about the circumstances and hazards firefighters face. NFPA 1500 also dictates that the department have fit testing procedures in place.

Under the guidelines of NFPA 1500, new PPE must meet the current edition of the standard. Older PPE must meet the edition of the standard when it was purchased. Protective clothing and equipment shall be used whenever the member is exposed, or potentially exposed, to the hazards for which it is provided. Therefore, whoever is responsible for purchasing PPE for an individual department should be familiar with the NFPA 1500 standard.

The safety committee or the safety program manager should have input in the department's specification writing process. Protective clothing must be used and maintained in accordance with manufacturers' instructions. The purchasing agent and the safety program manager should work together to ensure PPE for firefighting meets the standard. Both must be familiar with the standards in regard to safety. It is expected that PPE for firefighting be made up of the following components:

- Self-contained breathing apparatus (SCBA) (NFPA 1981)
- Personal alert safety system (PASS) (NFPA 1982)
- Firefighting footwear (NFPA 1971)
- Firefighting gloves (NFPA 1971)
- Firefighting protective trousers (NFPA 1971)
- Firefighting protective coat (NFPA 1971)
- Firefighting flame-resistant hood (NFPA 1971)
- Firefighting helmet (NFPA 1971)

It is of the utmost importance that both the safety program manager and the purchasing agent understand the concepts of and the relationship between **thermal protective performance (TPP)** and **total heat loss (THL)**. Such knowledge will ensure firefighters have the highest level of safety when it comes to firefighting PPE.

thermal protective performance (TPP) ▪ This test process for firefighter protective clothing determines the realistic exposure and heat transfer criteria related to human tissue tolerance to thermal exposure.

total heat loss (THL) ▪ This test method evaluates the ability of a clothing material composite to allow heat transfer to the outside environment in a manner that simulates wearing it under a specific set of environmental conditions.

USE OF PERSONAL PROTECTIVE EQUIPMENT

Every department should have a procedure or policy that defines the use of PPE. This may seem clear to most individuals; however, many firefighters are injured because the proper PPE was not worn during certain conditions and circumstances. During firefighting operations, most firefighters realize the need for full PPE; however, the gloves, protective hood, and SCBA, at times, are not in place for various reasons. As an example, a great deal of dexterity is lost while performing intricate tasks when wearing firefighting gloves. Tying knots, working with certain hand tools, and other tasks that require manipulative hands-on skills are difficult to either complete or complete in a timely fashion. Hence, the firefighter's gloves or some type of leather protective gloves are not in position when they should be. To gain proficiency with intricate tasks, firefighters should conduct training with gloves on.

In many cases, the protective hood is either not in place or, when in place, used incorrectly. Certain firefighters believe that the exposed ears and neck are to be used as a temperature gauge inside of a working structure fire; this belief and consequent behavior cause safety issues on any fire scene. Every fire department should use disciplinary measures when firefighters disregard PPE procedures and policies.

Firefighters and chief officers agree that wearing full PPE in a high heat area and in temperatures over 80°F is physically taxing for even physically conditioned firefighters. Firefighters involved in initial fire scene operations who are held at the scene to perform salvage and overhaul operations are usually reluctant to once again don their full PPE for such operations. Statistics prove that most firefighter injuries occur during overhaul operations. The only acceptable way to perform salvage and overhaul operations later into the incident is to first determine whether the atmosphere is safe. Technology now permits firefighters to conduct on-scene air monitoring; yet this may produce a false sense of security. Many of the newest gas monitors do not measure more than five separate types of gases. The monitor may also not be sampling the entire atmosphere in a given space, but just the atmosphere close to the monitor. Monitors do not have the capability to sample many of the gases present after a fire event and will not scrutinize airborne particulate matter during the overhaul and investigation stage. If firefighters will be unprotected during overhaul and investigation, it diminishes their overall protection during the fire event.

#8

The on-scene safety officer or incident commander must ensure the atmosphere is safe to enter prior to starting overhaul operations if crews are going to work without the aid of the SCBA. If air monitoring shows an interior atmosphere free of toxic levels of fire gases, firefighters should have, at a minimum, some type of cartridge filter mask in place to reduce the amount of inhaled airborne particulates and gases.

#16

The use of PPE is primarily an individual and company officer responsibility. Fire departments must continually strive to change the beliefs, mind-sets, and culture of its members who choose to function without the proper PPE. In conjunction with proper PPE, policies, guidelines, and SOPs need to be in place to ensure the highest level of safety.

CARE AND MAINTENANCE OF PERSONAL PROTECTIVE EQUIPMENT

Both the manufacturers of PPE and the NFPA standards address the care and maintenance of PPE. It is recommended that fire departments follow the manufacturer's guidelines. NFPA 1500 requires fire departments to establish a maintenance and inspection program for PPE that includes a repair procedure. The NFPA standard addressing the care and maintenance of structural firefighting PPE is NFPA 1851, *Standard on Selection, Care, and Maintenance of Protective Ensembles for Structural Fire Fighting and Proximity Fire Fighting*.

Most fire departments issue PPE to individual firefighters, and some of those leave the responsibility of care and maintenance of PPE up to the individual. PPE for structural firefighting receives the hardest wear and becomes contaminated quickly; therefore, a department needs to institute a care, cleaning, and inspection program. Fire departments must understand that PPE is technologically designed equipment for specialized job functions. Therefore, it is the individual fire department's responsibility to clean the PPE, and cleaning procedures must be suitable for the various types of contaminants as well as the type of materials from which the PPE is made.

The removal of contaminants and the protective thermal stability of PPE should be the main concerns. PPE must be cleaned annually and also when contamination occurs. It is a good practice for firefighters to inspect their issued PPE after each use. Some fire departments provide the facilities for in-house cleaning, whereas others contract with an outside agency for cleaning and maintenance purposes. PPE that has been contaminated with possible blood-borne pathogens must be cleaned in a washing machine designed for this purpose. If cleaned in-house, the cleaning should comply with NFPA 1581, *Standard on Fire Department Infection Control Program*.

Even though department members are responsible for the care of their issued PPE, the department should have a PPE inspection program included in its safety program. The NFPA recommends monthly inspection, which the station captain, station company officers, or the shift commander can conduct. See Figure 6–7.

The idea is to identify damaged and contaminated PPE. The department safety program manager should participate in this quarterly or semiannually. Documentation of these inspections should be kept on file including date of issue, inspection results, maintenance and repairs, and cleaning. The safety program manager keeps a copy of such records at the station where the individual firefighter is assigned.

Personnel should have a thorough understanding of what makes PPE unsafe. Most structural PPE is manufactured using the same raw materials, regardless of the

FIGURE 6–7 Company officer inspecting PPE.
Courtesy of Joe Bruni.

manufacturer. The major differences exist in how the material is cut and stitched together, and departments get what they pay for. PPE that has been poorly stitched together will not have an extended life span. PPE designed for structural firefighting should also have padding and protection in the wear and heat areas and the stress points, which will extend its life span.

Exposure, age, and use each play a significant role in the protective qualities of PPE. All PPE has a user life of 10 years, according to the NFPA 1971 standard. The department should institute guidelines for the proper storage of PPE both in the station and on the apparatus. PPE should be stored out of direct sunlight and in a well-ventilated area to reduce and prevent the development of mold and mildew. PPE kept in compartments may be subject to damage from contact with the apparatus or from loose equipment. Specific companies that work at a higher than average number of structure fires will need to have their PPE cleaned and replaced more frequently. Cleaning structural PPE in a washing machine designed for such purposes changes the THL and TPP qualities of the garments. By having PPE inspected by a qualified individual each time it is sent in for cleaning, small tears and problems with the liner and thermal barrier can be identified. This will indicate whether the PPE should be sent out for repair or warrants replacement. See Figure 6–8.

FIREFIGHTER RESPONSIBILITY AND PERSONAL PROTECTIVE EQUIPMENT

Firefighters must understand the importance of their involvement regarding their issued PPE. Unfortunately, many firefighters view contaminated gear as a badge of courage and a display of firefighting experience. They must be educated in the hazards of wearing contaminated and damaged gear and in what their department will not tolerate in the area of PPE use, care, and maintenance.

Complacency regarding PPE usage may become a part of department culture, especially among newer, inexperienced members of a department who are not aware of the potential hazards of fire scene operations. It is up to chief officers, company officers, and informal leaders to ensure that the newer personnel do not become complacent. Members must be taught that firefighting operations usually involve an exposure to every individual responding to the scene, even with proper PPE in place. Younger members need to understand that such exposure has its cumulative effects over the course of time and that, as individuals, they are not "bullet proof." Education, policies, procedures, and possibly a cultural change are the keys to a successful PPE safety program.

#2

Advanced P.P.E. Inspection

*Denotes to advise Training Officer (during business hours M—F) / District Chief (after hours)

*Coats/Trousers	*Helmets	Hoods	Gloves	Footwear
Soiling	*Soiling*	*Soiling*	*Soiling*	*Soiling*
Contamination (Haz Mat or Bio Agents)	*Contamination (Haz Mat or Bio Agents)*	*Contamination (Haz Mat or Bio Agents)*	*Contamination (Haz Mat or Bio Agents)*	*Contamination (Haz Mat or Bio Agents)*
Rips, Tears, Cuts, and Abrasions	*Cracks, dents, and abrasions*	*Rips, Tears, Cuts, and Abrasions*	*Rips, Tears, Cuts*	*Cuts, tears, punctures, cracking, or splitting*
Damaged or Missing Hardware	*Thermal Damage—shell bubbling, softspots, warping, or discoloration*	*Material elasticity, stretching out of shape*	*Thermal damage—charring, burn holes, or melting*	*Thermal damage—charring, burn holes, or melting*
Thermal Damage—Charring, burn holes,melting, or discoloration of any layer	*Ear flaps have thermal damage—charring, burn holes, or melting*	*Thermal Damage—Charring, burn holes, melting, or discoloration of any layer*	*Seam—broken or missing stitches*	*Exposed/deformed steel toe, steel midsole, and shank*
Moisture Barrier	*Ear flaps are ripped, torn or cut*	*Shrinkage*	*Inverted liner*	*Seam delamination*
Fit—Coat/Trouser Overlap	*Damaged or missing components of suspension and retention systems*	*Seam—broken or missing stitches*	*Shrinkage*	*Broken or missing stitches*
Seam—broken or missing stitches	*Functionality of suspension and retention system*	*Loss of face opening adjustment*	*Loss of flexibility*	*Loss of water resistance*
Integrity—UV or chemical degradation	*Damaged or missing components to face shield*		*No added accessories without mfg. approval*	*Closure system component damage and functionality*
Wristlets integrity—loss of elasticity, stretching, runs, cuts, and burnholes	*Discoloration or scratches to the faceshield limiting visibility*		*Wristlets integrity—loss of elasticity, stretching, runs, cuts,and burn holes*	*Condition of lining—tears and excessive wear*
Reflective trim	*Functionality of faceshield*			*Lining separation from outer layer*
Legibility of Name or Department	*Damage to impact cap*			*Heel counter failure*
Hook and loop functionality	*Damaged or missing reflective trim*			*Excessive tread wear*
Liner attachment systems	*No added accessories without mfg. approval*			*No added accessories without mfg. approval*
Closure system—Zipper function				
No added accessories without mfg. approval				

Name and Station:	**Payroll #**
Inspected by:	**Date:**

(Circle all items to indicate a problem exists and comment on the next page.)

FIGURE 6–8 PPE inspection record. *Courtesy of St. Petersburg Fire/Rescue.*

Advanced P.P.E. Inspection *(Continued)*

For each item circled, please comment on the problem and identify what repairs are needed, have been corrected, or have been replaced/modified.

Employee Name and Station:	***Payroll #:***
****Training Officer/DC Notified :***	***Date and Time:***
Action taken: SENT TO SHOP/SUPPLY / REPLACED / IN STATION REPAIR/CARE / NO ACTION NECESSARY	
Inspected by:	***Date:***

FIGURE 6–8 PPE inspection record. *Courtesy of St. Petersburg Fire/Rescue. (Continued)*

RESPIRATORY PROTECTION

Every fire department should have a complete respiratory protection program. Manufacturers of fire service SCBA must meet NIOSH certification, standards, and guidelines as well as requirements of NFPA 1981, *Standard on Open-Circuit Self-Contained Breathing Apparatus (SCBA) for Emergency Services.* Fire departments are required to provide SCBA that has a reasonable level of dependability if correctly used and maintained. Having a preventative maintenance program in place ensures safety of department personnel. In those cases of a reported failure of SCBA, a before-use check, a more thorough user inspection program, or a preventive maintenance program most likely would have eliminated the failure. Unfortunately, complacency and the lack of an adequate respiratory protection program frequently play a role in the SCBA failure.

Fire departments need to develop and maintain standard operating procedures compliant with the NFPA 1500 standard in the safe use of respiratory protection. Thorough training should be in place on emergency procedures while using the SCBA. Training that involves basic emergency procedures such as the use of the regulator bypass valve, dealing with damage to the face piece and breathing tube, and breathing directly from the regulator (where applicable) should be taught to, and practiced by, the individual user. Firefighters must be taught to neither compromise the integrity of the SCBA nor remove the face piece in an immediately dangerous to life or health (IDLH) atmosphere for any reason.

Every department's SCBA respiratory protection program should include emergency procedures and written policies for the removal of victims without compromising the rescuer's respiratory protection. Firefighters should be trained to remove victims, including firefighter victims, from the hazardous atmosphere as quickly as possible. Buddy breathing techniques continue to be debated in the fire service because they have the potential to compromise the wearer's safety while using the SCBA, from either removing the face piece or disconnecting the breathing tube or regulator.

Many of the newer SCBAs have the technology either to transfill a user's air tank that is low on air from another user's tank or to tie into another SCBA unit through an emergency breathing safety system. The disadvantages include two firefighters now having to leave the area together while consuming air at a faster rate, and the difficulty of two firefighters who are buddy breathing to remain close to each other while attempting to leave the IDLH atmosphere. Whereas some believe that buddy breathing should be reserved as a last resort, having several key factors in place may reduce or eliminate this need at a fire scene:

- A strong, well-administered respiratory protection SCBA program
- Emphasis on user testing and inspection of respiratory protection SCBA
- Required before-use and after-use testing and maintenance
- Functional preventative maintenance (PM) program
- Fireground management based upon safe operations with knowledge of fire development, building construction, and coordinated firefighting operations
- Quality breathing air
- Personal alert safety system (PASS) devices and portable radios for interior firefighting teams
- Thorough training in survival techniques, controlled breathing, and stress management
- Accountability for interior firefighting crews
- Physical fitness of firefighters
- Use of positive pressure SCBA that are NIOSH approved and meet the requirements of NFPA 1981, *Standard on Open-Circuit Self-Contained Breathing Apparatus (SCBA) for the Emergency Services*

An SCBA safety program that includes SCBA testing and maintenance will prepare a department's SCBA units for regular service. The program should also include air quality testing to ensure firefighters are using good clean air. Good fireground management and training for firefighters that emphasizes situational awareness on the fireground help firefighters understand fire behavior and growth, building construction, and coordinated firefighting efforts. Training should also cover self-survival techniques and methods. All members engaged in interior firefighting operations should be equipped with their own portable radio and working PASS device. An adequate accountability system to track companies and personnel is needed at every fire scene operation. Members who engage in firefighting operations must also be in good physical condition.

For years, it has been common practice for firefighters to expel their air supply in an IDLH atmosphere until the low-air warning system operates prior to exiting the atmosphere. A current philosophy focuses on training firefighters to constantly monitor their air supply and to plan to leave the IDLH atmosphere prior to their low-air warning system sounding. Trained **self-contained underwater breathing apparatus (SCUBA)** divers who practice underwater cave diving reserve more than half of their air supply to ensure they are able to exit an underwater cave system safely; this same type of philosophy should also be adopted in the fire service when it comes to IDLH atmospheres.

self-contained underwater breathing apparatus (SCUBA) ▪ This diving mode is independent of surface air supply; the diver uses open-circuit self-contained underwater breathing apparatus, which supplies air or breathing gases at ambient pressure.

Hazards Faced by Firefighters Operating at Fire Scenes

It is very important for firefighters, company officers, and chief officers to be aware of the hazards present during fire scene operations. Younger inexperienced firefighters must rely on knowledgeable company officers and line personnel to pass on information related to these hazards. Firefighting is not a hobby nor is it for the meek individual. It is a profession that requires doing a great deal of study in order to recognize and face the hazards of every fire event.

No matter the type of fire event encountered, firefighters must be well versed in fire behavior and fire dynamics. Firefighting personnel must also understand the dynamics of smoke as a potential fuel source that can rapidly ignite, causing injury and death. Personnel must be familiar with the properties of fuels and the flammable range concept involving various fuels. Such understanding not only helps predict what a fire is about to do and when it will progress into a hostile fire event but also helps reduce firefighter injury and death.

All fire departments must have a risk assessment completed, as well as incident priorities established and effective strategy and tactics put into practice. This section discusses the various types of fire situations firefighters encounter. Hazardous materials fires will be covered in Chapter 8.

STRUCTURE FIRES

Fighting fire inside of structures is a high-risk and dangerous endeavor. Many firefighters are injured and killed every year performing this type of firefighting operation. Firefighters face accidental hazards, physical hazards, chemical hazards, and biological hazards compounded by many psychosocial, economical, and organizational factors. These can be countered by having an adequate number of firefighters, experienced officers and personnel, and up-to-date training. Fireground operations should be limited to firefighters who can safely perform at the scene, which means that more experienced officers and firefighters exercise direct supervision of inexperienced firefighters. See Figure 6–9.

It is also important for firefighters to operate in teams of two or more. Firefighters working in an IDLH atmosphere must be able to communicate with each other at all times using visual, audible, or physical means. Firefighters involved in structural firefighting should use a hose line or a safety guide rope to coordinate their activities and stay in contact with each other so that they can get assistance during firefighting operations.

Even in the biggest cities with many types of structures, it takes time to develop structural firefighting experience. In most cases, firefighters responding to structural fires are unaware of all of the possible hazards, and many younger firefighters lack the training and experience in situational awareness. Structure fires can be divided into several categories:

- Single-family dwellings
- Multifamily dwellings
- Industrial occupancies
- Commercial occupancies

Although each type of structure presents hazards common to all structure fires, interior operations at fires occurring in larger commercial or industrial occupancies with open floor plans present the greatest hazard for firefighters. It is extremely important for first arriving companies to perform a risk assessment when they receive the alarm. The type of occupancy, type of construction, and conditions upon arrival all affect initial operations.

FIGURE 6–9 Incident commander supervising fireground operations. *Courtesy of Lt. Joel Granata.*

Unfortunately, many firefighters are entering the field without solid knowledge of building construction. Building construction plays an important role during the risk assessment phase of fire scene operations. Everyone involved in fire scene operations needs to understand that the attack of fire within a building will change the designed loads related to the building's design and construction. Lightweight construction practices are common. Today's design methods make use of geometric principles instead of the concept of mass. The triangle is the strongest geometric shape; architects and engineers have learned to make use of this geometric shape to lessen the mass in buildings. Unfortunately, when one leg of a triangle fails, the triangle falls apart. This spells disaster in a modern-day structure fire in a new construction building. Mass is directly related to how long a building will stand up during fire conditions, and the effect of fire on the connections of the building influences this. It is advisable for firefighters to learn the hazards associated with building construction during fire conditions early in their careers, as such knowledge will play a large role in risk identification.

Although this is not a textbook about building construction, the issue of collapse potential and collapse profiles needs to be mentioned regarding structure fires. Personnel with knowledge of building construction and design gain a better understanding of collapse potential associated with the five types of building construction:

Fire-resistive construction (type I)
Noncombustible/limited combustible (type II)
Ordinary construction (type III)
Heavy timber (type IV)
Wood frame (type V)

It also promotes safety if, at a minimum, company officers and incident commanders have in-depth knowledge of the hazards associated with truss construction, balloon-frame construction, lightweight wood construction, cantilevered walls, and metal-framed construction.

Personnel must understand the types of buildings that are most susceptible to collapse under fire conditions. Buildings built of fire-resistive construction (type I) are truly a friend of firefighting personnel as they relate to collapse potential. The fire-resistive building contains structural components that have been protected from fire. This type of construction can hold high amounts of heat under fire conditions, which can be punishing to firefighting personnel; however, its collapse potential is relatively low. The greatest threat to firefighters comes from spalling concrete, which may fall in large pieces and will eventually expose the steel reinforcing members within the concrete. These exposed steel members will eventually cause failure of all or portions of the building. However, the fire-resistive building under construction is extremely dangerous to enter under fire conditions and highly susceptible to collapse. The temporary supports used during the construction phase can be destroyed by fire, adding to the danger. Fire service personnel must become aware of such buildings under construction in their fire district and preplan them accordingly.

rollover ▪ When the fire gases, usually at the ceiling level, begin to ignite. Visually, this may be seen as flames "rolling" across the ceiling, radiating outward from the seat of the fire to the extent of gas spread. Rollover precedes flashover. Also known as flameover.

flashover ▪ In this stage of a fire, a room or other confined area becomes so heated that the flames flash over and through the vapors being produced by heated combustible contents in the space.

back draft ▪ This situation can occur when a fire is starved of oxygen; consequently, combustion ceases but the fuel gases and smoke remain at high temperature. If oxygen is reintroduced to the fire (e.g., by opening a door to a closed room), combustion can restart, often resulting in an explosive effect as the gases heat and expand.

thermal layering ▪ Heated air currents in a confined space rise and form temperature layers. These thermal layers are the separation between different temperatures and products of combustion.

heat transfer ▪ The transmission of heat through conduction, convection, and radiation.

Of the five types of building construction, the structure that has been built and classified as a noncombustible/limited combustible (type II) building is most susceptible to collapse. These types of buildings contain a vast amount of unprotected steel, which will quickly absorb high amounts of heat and fail. In many of these buildings, the roof assembly contains unprotected metal or composite wood and metal trusses along with a roof deck constructed of steel. Unprotected steel will not hold its designed shape for long periods of time when exposed to fire. Again, it is important to identify these types of structures in the community through preplanning. The noncombustible building truly warrants either sprinklers or some other type of fire protection. Firefighters should be very cautious operating in these types of unprotected structures even with the smallest amount of fire inside.

Buildings of ordinary construction (type III) are normally susceptible to early failure or collapse, as well as to fire burning up through the structure and self-ventilating itself. In most cases, these buildings are comprised of exterior masonry walls with wood framing for non-load-bearing walls and roof assemblies. The danger related to collapse of these structures comes from any building alterations and renovations done over the course of time, building neglect, and previous fires in these buildings. Additional collapse concerns are parapet walls and decorative cornices that can literally fall off the building onto firefighters working at street level.

The heavy timber (type IV) building contains structural members of large mass, so there is a great deal of structural strength and integrity. This surface-to-mass ratio means this type of building has a great deal of collapse resistance and makes it difficult to ignite; however, once ignited, heavy timber structures burn with an intense amount of heat release. Renovations or neglect play a role in early collapse of these buildings during fire conditions. In most cases, the floors will collapse in a pancake-style fashion, leaving dangerous unsupported walls.

The wood-frame (type V) constructed building consists of load-bearing and non-load-bearing walls and roofs made totally or partially of wood. Lightweight wood constructed buildings also fall into the wood-frame type V category. The following factors contribute to the collapse potential of this type of structure: its wood members do not contain a great deal of mass; it normally contains void spaces so fire can easily gain access; or it has had numerous alterations and additions. It is important to consider a collapse zone around these buildings of 1½ times the building's height because the collapse profile of these building is difficult to determine. Collapse and fire spread potential will be dictated primarily by how these building were put together.

A large number of structure fires are fought and extinguished utilizing crews that operate on the inside of the structure during hostile and dangerous conditions. For this reason, it is important for firefighters to study and understand fire behavior and collapse potential. Everyone involved in the operation must be highly aware of signs and conditions related to **rollover**, **flashover**, **back draft**, **thermal layering**, and the three methods of **heat transfer**. It takes a combination of experience, education, and continual training—as well as a great deal of resources—to remain safe and effectively combat structure fires.

FIRES OUTSIDE OF A STRUCTURE

Fire departments routinely respond to fires that do not involve structures; however, many of these fires occur close to a structure and so threaten the structure. Fires outside of a structure can be anything from the typical small grass fire, vehicle fire, dumpster fire, or trash fire, to a fire at the local landfill. Each of these fires has its own set of hazards: a fire outside of a structure along a busy highway may pose traffic hazards to firefighters; a trash or dumpster fire may contain a wide range of hazardous or biohazardous materials. See Figure 6–10.

At times, getting to a fire outside of a structure can present responders with access problems and difficulties such as rough or difficult terrain. The fire response may be an extremely labor-intensive operation because firefighters must transport the extinguishing agent to a remote fire location. Fires occurring during hostile weather conditions may place firefighters in a precarious situation involving lightning, high wind conditions, or extreme heat or cold.

Fires involving energized electrical equipment may occur in transformers of transmission equipment, power substations, and underground electrical vaults, all of which present possible electrocution hazards. Without a life hazard at these scenes, firefighters must ensure power has been disconnected to the equipment. Firefighters who attempt to enter an area to make a rescue must be aware of the high risk involved. In most cases, it is advisable to isolate the area and wait for the power company to disconnect power.

Fires outside of a structure usually do not present life hazard issues to the general public, but may present safety issues to firefighters. Chief officers, company officers, and firefighters must utilize a risk management philosophy of taking a very controlled risk to combat these types of fires to ensure a high level of firefighter safety. A controlled risk should include fighting the fire from a safe distance and respiratory protection for those who may enter the smoke around the hazard zone. Yet it is at this type of fire that firefighters may be reluctant to wear and use their SCBA. It must be stressed that firefighters should not breathe smoke, not even light smoke, for any reason.

FIRES INVOLVING VEHICLES AND TRANSPORTATION EQUIPMENT

Fires frequently occur in many types of transportation equipment such as trains, aircraft, boats, ships, and commercial trucks, and especially passenger vehicles. Although each

FIGURE 6–10 A fire outside of a structure. *Courtesy of Joe Bruni.*

type of fire presents its own specific hazards, there are general hazards associated with them all; therefore, it is advisable to do a risk assessment early into the incident.

When no life hazard exists, it is advisable to take a very controlled risk when combating a fire in a vehicle or piece of transportation equipment. Fires of this nature usually present hazards from the fuel that powers the piece of equipment, as well as the equipment itself. Fires occurring in passenger vehicles and boats normally produce high amounts of toxic smoke due to the large quantities of plastics and other hydrocarbon-based material that make up the various parts of the vehicle. Other parts of these vehicles can present explosion and hazardous material hazards. Approaching from a safe distance with full PPE and SCBA in place is advisable.

It is also recommended that firefighters apply water from a safe distance to begin the cooling and extinguishment process. They should approach passenger vehicle and truck fires from the side at a 45-degree angle, avoiding the front or rear of the vehicle while it is burning. Shock-absorbing bumpers and gas-filled lift cylinders have the potential to explode and become a flying hazard to personnel. Because many people use personal vehicles to transport hazards, gasoline cans, small propane tanks, and hazardous chemicals can often be found in the common passenger vehicle fire. See Figure 6–11.

Fires involving any type of commercial truck or passenger vehicle should be handled with a thorough risk assessment. This should take into account what type of cargo the truck may be carrying and that the truck's structural integrity may be compromised when under the attack from fire.

Many fires occurring in transportation equipment occur along a road or divided highway, making officers in charge of personnel highly concerned with traffic flowing alongside of operating crews. Smoke conditions may impede the vision of drivers attempting to navigate through the smoke. It is always advisable to place apparatus in a position to protect working firefighters. If the road must be temporarily shut down to protect personnel, then do so; in most cases, the local police agency can assist with this. Also, a vehicle fire is nothing more than a trash fire on wheels, but with the hidden potential of hazards and surprises. Do not risk safety and firefighter's lives in combating these types of fires.

Sometimes fires in transportation equipment occur in conjunction with a vehicle accident, presenting a possible medical emergency; this scenario will be discussed in Chapter 7.

FIGURE 6–11 Car fire.
Courtesy of Lt. Joel Granata.

FIGURE 6–12 Ship fire.
Courtesy of Lt. Joel Granata.

Fires occurring in ships and other large vessels while in port present specific hazards to responders equipped and trained for structural and other types of firefighting. They can find themselves in a dangerous situation very quickly. One of the common hazards for firefighters is disorientation once gaining access to the ship. See Figure 6–12.

The personnel on board many large ships are trained and equipped for firefighting. In most cases, the local fire department offers support. Structural firefighting PPE and SCBA are not designed to fit through hatches, down ladders, or through doorways. Many shipboard firefighting teams are equipped with specialized PPE and breathing apparatus for such purposes. Fires occurring below deck on a ship present a punishing firefighting operation. Attempting to gain access to the seat of the fire is essentially like going down a chimney to fight the fire. Fortunately, large-scale fires aboard ships do not frequently occur. It is better to handle this type of fire at a slower pace and be correct in your efforts, than to be overly aggressive and place people in harm's way. Local fire departments that have large ships and ports in their jurisdiction must be prepared through coordination and effective preplanning. As at other fires, the highest priority concerning shipboard firefighting should be life safety.

FIRES OCCURRING IN AIRCRAFT

The hazards of aircraft fires are similar to those of vehicle fires: both the components of the aircraft and the fuel that powers it may present flammability hazards. An **aircraft rescue firefighter (ARFF)** must meet the requirements of NFPA 1003, *Standard for Airport Fire Fighter Professional Qualifications*. According to the NFPA, the initial IDLH during an aircraft fire is the area within 75 feet (23 m) of the skin of the aircraft, and firefighters must operate in teams of two or more.[1] Whether an aircraft fire occurs at an airport or elsewhere, it has the potential to escalate rapidly into disastrous proportions, particularly with large commercial aircraft. Local fire departments responding to the scene of a crash or aircraft fire at or near a medium-size to large local airport support the operations of the on-scene local aircraft rescue firefighters. For firefighters to be successful during these responses, they must be on scene and operating within minutes, as specified in NFPA 403, *Standard for Aircraft Rescue and Fire-Fighting Services at Airports*.

aircraft rescue firefighter (ARFF) ▪ A firefighter trained in firefighting and rescue of aircraft on fire.

Although the rescue of onboard passengers is a high priority, certain common hazards associated with aircraft involved in fire require consideration. Although it is against the law to transport hazardous materials onboard aircraft, the components of certain aircraft may indeed present these hazards. Examples include aviation and jet fuels; pressurized oxygen cylinders and other compressed gases; and large quantities of hydraulic fluid. The sizable amount of fuel on larger planes may burn at temperatures too high to permit local municipal firefighters a safe and effective approach. The highly toxic smoke produced will usually be of high carbon content. Military aircraft may contain explosives under the seats for pilot ejection purposes. Local fire departments that respond to a down military aircraft off site of an airport or military base must exercise extreme caution, use a thorough risk assessment, and control firefighting personnel with a carefully planned action.

WILDLAND FIRES

A wildfire may occur in most local fire department jurisdictions, and its extent can range from as small as a single acre burning to hundreds of square miles involved. Every year these fires involve large dollar and environmental, as well as human, losses. In many rural fire departments, the wildland fire is a common occurrence, whereas some urban fire departments seldom or never have one. The most dangerous hazard associated with wildland firefighting is the exposure to personnel in unprotected positions or areas. Firefighting personnel have been seriously injured or killed during wildland firefighting operations. See Figure 6–13.

A number of NFPA standards address wildland firefighting. For example, all members operating as wildland firefighters must meet the requirements of NFPA 1051, *Standard for Wildland Fire Fighter Professional Qualifications.* All new apparatus used for wildland firefighting operations must meet NFPA 1906, *Standard for Wildland Fire Apparatus.* Fire departments that engage in this type of firefighting are required to establish standard operating procedures for the use of wildland protective clothing and equipment according to NFPA 1977, *Standard on Protective Clothing and Equipment for Wildland Fire Fighting.*

The NFPA also requires fire departments that engage in wildland firefighting to establish guidelines for members on which ensemble to wear for a given incident. Many fire departments, both rural and urban, do not issue wildland PPE and equipment to

FIGURE 6–13 Wildland fire. *Courtesy of Lt. Rob Edwards.*

their members due to the low frequency of wildland fires or the cost of the equipment. As a result, firefighting personnel for a wildland operation may use structural firefighting PPE instead. During incidents of long duration, at low-frequency/high-risk events, or in environments of high temperature and humidity, the use of structural firefighting PPE can cause unnecessary stress and safety issues for personnel. Emergency services organizations responding to wildland fires must continually evaluate and perform a risk frequency analysis to help ensure the health and safety of members. Firefighters who become physically exhausted or experience a medical emergency related to their use of the inappropriate structural PPE during wildland firefighting operations become a liability to themselves and their fire department. The issue of proper wildland PPE would reduce or eliminate this potential hazard and liability.

Because these fires have the potential of being of large magnitude and long duration, it is important to have an expanded incident management system in place, as well as a good communication system. Procedures involving mutual aid, SOPs, communications, and rehabilitation of personnel are highly recommended. Logistics can become a high priority when wildland fires are of a long duration, because food, fluid replacement, and rehabilitation are of high concern. The NFPA recommends that all personnel involved in wildland firefighting be outfitted with an approved fire shelter and trained in its use annually.

The fire department's incident commanders, training officers, and safety program manager must all become familiar with the operations and hazards of wildland firefighting, such as the effects of weather, topography, and the type of fuels. Training in wildland firefighting operations is also recommended for personnel who engage in it. Personnel should be aware of the safety recommendations concerning wildland firefighting referred to as "fire orders" and "watchout situations." See Box 6–1.

Personnel Accountability

The NFPA requires fire departments to have in place standard operating procedures for a **personnel accountability system**, according to NFPA 1561, *Standard on Emergency Services Incident Management System*, and NFPA 1500, *Standard on Fire Department Occupational Safety and Health Program*. The incident commander is ultimately responsible for ensuring accountability is in place at every fire scene. Accurate accountability becomes the responsibility of everyone working on the fireground. There are various types of accountability systems in place; however, all are made up of several layers and must utilize proper supervision to work adequately. See Figure 6–14.

personnel accountability system ■ Such a system readily identifies both the location and the function of all members operating at an incident scene.

Firefighters are responsible for their own accountability and must use the system in place to do so. Personnel need to remain with their operating crew and company officer at all times. Each fire department must establish its use of accountability by considering local conditions and characteristics. Each fireground supervisor or company officer is responsible for his or her crew and their whereabouts. The idea is to utilize a **team unity** approach that is vital to its success. On smaller incidents and firegrounds, the incident commander can handle accountability; whereas on larger incidents, an accountability officer(s) within the incident management system (IMS) system must be appointed to track operating personnel.

team unity ■ The concept that a firefighting crew will work and stay together in a hazard zone to ensure accountability so they will remain an identifiable group.

Both the IMS and accountability systems in place may vary from department to department; however, their concepts remain the same. Many fire departments use Velcro tags attached to a plastic card to track personnel, whereas others utilize a tag clipped to the helmet or a ring in the apparatus that is removed and placed at an accountability location prior to entry into the hazard zone. As technology improves and becomes more affordable, the use of computers for accountability will likely become more common. One SCBA manufacturer now utilizes a radio signal transmitted from the individual SCBA to a computer to monitor the air pressure inside the cylinder and transmit low pressure alarms

BOX 6-1: FIRE ORDERS AND WATCHOUT SITUATIONS

FIRE ORDERS

1. Fight fire aggressively but provide for safety first.
2. Initiate all actions based on current and expected fire behavior.
3. Recognize current weather conditions and obtain forecasts.
4. Ensure instructions are given and understood.
5. Obtain current information on fire status.
6. Remain in communication with crew members, your supervisor, and adjoining forces.
7. Determine safety zones and escape routes.
8. Establish lookouts in potentially hazardous situations.
9. Retain control at all times.
10. Stay alert, keep calm, think clearly, and act decisively.

COMMON DENOMINATORS OF FIRE BEHAVIOR ON TRAGEDY FIRES

1. Most incidents happen on the smaller fires or on isolated portions of larger fires.
2. Most fires are innocent in appearance before the "flare-ups" or "blowups." In some cases, tragedies occur in the mop-up stage.
3. Flare-ups generally occur in deceptively light fuels.
4. Fires run uphill surprisingly fast in chimneys, gullies, and on steep slopes.
5. Some suppression tools, such as helicopters or air tankers, can adversely affect fire behavior. The blasts of air from low-flying helicopters and air tankers have been known to cause flare-ups.

WATCHOUT SITUATIONS

1. Fire not scouted and sized up
2. In country not seen in daylight
3. Safety zones and escape routes not identified
4. Unfamiliar with weather and local factors influencing fire behavior
5. Uninformed on strategy, tactics, and hazards
6. Instructions and assignments not clear
7. No communication link with crew members or supervisors
8. Constructing line without a safe anchor point
9. Building fire line downhill with fire below
10. Attempting frontal assault on fire
11. Unburned fuel between you and the fire

Source: Fire & Aviation Management (FAM), *Standard Firefighting Orders and 18 Watchout Situations,* adapted from http://www.fs.fed.us/fire/safety/10_18/10_18.html

and PASS activation. The department health and safety officer should monitor a department's accountability system to ensure it does an adequate job of tracking personnel. Evaluation of the effectiveness of a department's personnel accountability system should be an ongoing process with fire departments making needed adjustments. Accountability systems should meet the following objectives:

- Adapt and fit the IMS system in place.
- Account for all members operating at a scene at any point in time by location and function.
- Be able to meet the needs of an incident as it expands.
- Ensure personnel are accounted for throughout an incident.
- Dictate the points of entry into a hazard zone.

It has been discovered at many LODDs that the lack of an effective accountability system contributed or played a role in the member's death. Fire departments must make certain that their guidelines set forth concerning accountability systems are followed to

FIGURE 6–14 A status board. *Courtesy of Joe Bruni.*

the letter. Everyone from the fire chief down to the newest firefighter must buy into the accountability system. However, the requirement of an accountability system and having one in place do not ensure it will adequately work. If the system is not usable or does not adequately track personnel at all incidents, it is nothing more than the fulfilling of a requirement. Training that utilizes the department's accountability system will help to ensure the system adequately tracks personnel.

A proper accountability system starts with the firefighters and company officers fulfilling their responsibilities. Firefighters must ensure their nametags are in the proper location prior to and during any incident. Company officers need to make certain their crews remain together as a team at all times and must not tolerate **freelancing**! The incident commander is responsible to know where his or her assigned companies are operating, and company officers are responsible to know where their team members are at all times. The accountability officer(s) are responsible for accountability within a geographical area at an incident. The incident commander is responsible to ensure the accountability system is in place and functioning correctly in conjunction with the IMS system. An incident commander who has to try to play catch up with an ineffective accountability system will end up managing the accountability system instead of mitigating the incident; such activity has the potential to injure or kill firefighters.

freelancing ■ Firefighters not staying with their assigned crew at an emergency scene and doing what they want to do on their own without direct supervision.

Within the accountability system is a **personnel accountability report (PAR)**, a method the incident commander uses to determine whether all members are safe and accounted for. A PAR usually takes place during an incident at regular intervals or when certain benchmarks are reached. All department members must become familiar with its use. First, when receiving assignments at a fire scene, personnel must be clear about who reports to whom. The chain of command within the IMS must be apparent to initiate a proper PAR. Company officers who are under a division or branch within the IMS may choose to do a face-to-face PAR to reduce radio traffic. A PAR should be implemented during the following conditions:

personnel accountability report (PAR) ■ This verbal or visual polling system ensures that all personnel operating at an emergency scene are safe and accounted for.

- Any report of a lost or missing firefighter.
- During the transitional mode of going from an offensive to a defensive strategy.
- During a catastrophic change in the conditions at a fire scene (e.g., explosion, structural collapse, flashover, back draft, or sudden release of hazardous materials).

- When emergency evacuation procedures have been initiated.
- During crew rotation.
- After a fire is declared under control.
- At any point the incident commander deems necessary.

Rapid Intervention Teams/Groups

Fire departments must provide a team of firefighters standing by that is dedicated to the rescue of members who are trapped, injured, or lost at an incident. Depending on the local jurisdiction, this specialized team is referred to by various names, but for the purpose of this text, it is called the rapid intervention team (RIT). RIT members must be thoroughly trained in all aspects of firefighter recovery and removal, and the specialized equipment that goes with it. The IMS and personnel accountability systems are sufficient tools to aid the incident commander with management of the incident and the tracking of personnel, but these efforts are inadequate if the RIT team is neither in place nor highly trained in the function of RIT operations. An Occupational Safety and Health Administration (OSHA) ruling requires firefighters entering an interior structure fire to have a two-person team outside the fire building as a standby team; this is commonly referred to as two-in/two-out, or the initial rapid intervention team (IRIT).

The IRIT should consist of a minimum of two members who will be in place at the time firefighters enter an IDLH atmosphere during initial operations of first arriving companies. The IRIT must be fully equipped with full PPE and SCBA if its personnel are to take quick action to assist firefighters in distress. The IRIT should be located near the point of entry with the necessary equipment and full PPE in place, ready to go to work in a matter of seconds to locate a firefighter in trouble. Fire companies arriving on the scene after initial operations have taken place can form a full RIT by joining with the IRIT members or by forming a completely different RIT. If IRIT members are permitted to assume other responsibilities on the fireground, they must be ready to abandon those duties to perform or aid in a firefighter rescue.

Initial arriving companies who assume other responsibilities that also form the IRIT must ensure those duties do not inhibit their ability to aid firefighters in need. It is critical not to jeopardize the health and safety of the team operating in the IDLH atmosphere. It is the responsibility of the incident commander to ensure both the IRIT and the RIT teams are in place. Although it has been debated that an IRIT functionally and theoretically meets the requirement of a formal RIT, it really does not. The IRIT serves only as the initial standby team. Many fire departments adopt the philosophy that the IRIT is made up of crew members of the first arriving units starting initial operations, whereas the RIT is made up of specially trained firefighters and equipment. Once a crew has entered an IDLH environment and begins operating, the incident is no longer considered in the "initial stage," and a formal RIT should be formed.

A formal RIT should be fully equipped with the appropriate protective clothing, protective equipment, SCBA, and any specialized rescue equipment specific to the operation underway. In most cases, several firefighters must rotate through the RIT company to locate, package, and remove a single firefighter in trouble. Through training and experience, many firefighters have discovered how quickly their air supply becomes exhausted in locating and packaging a firefighter for removal. The RIT should be flexible and made up of adequate numbers of firefighters based on the size and the complexity of the incident. It is vital that department members who serve on the RIT frequently train on the specialized techniques and equipment used during a firefighter rescue. An incident commander who places a RIT in place that is not adequately trained in RIT techniques is only meeting the requirement of having a RIT in place and checking a box on his or her incident command worksheet.

The Phoenix Fire Department and many others have discovered through either an LODD or training how difficult and specialized a RIT operation will become. The incident commander should evaluate the situation and the complexity of the incident to determine the number of firefighters that will be needed for a formal RIT, and if more than one RIT will be needed. The RIT must have a team leader, normally a company officer, and team members must know their job as they operate together throughout an RIT operation. The safety and program manager along with the department's training personnel must ensure members are trained, equipped, and effective.

Anything more than a firefighter rescue involving a simple drag just a few feet to an exit from a ground-level floor will become a very labor intensive and technical operation. Many fire departments consider the formal RIT to be a team of firefighters with specialized training who also have the skills of experienced firefighters with equipment and firefighting tactics. When a fire enters the stage of a "working fire," or continues to grow beyond the incipient stage, the formation of a formal RIT becomes an issue for the incident commander to consider.

Many fire departments differ in how they utilize personnel to make up a formal RIT. Some add to the IRIT to form a formal RIT, whereas others dispatch a specialized unit of firefighters and equipment on the first alarm dedicated to the RIT. Still other departments have built an additional unit into their first-alarm assignment to be the formal RIT when they arrive on scene. For example, a heavy rescue unit or a second-due truck company may be initially dispatched to form the formal RIT on all structure fires.

The size of the fire and involved building are considered in the formation of the RIT. A small single-family dwelling may require one RIT to stand by, whereas a commercial or industrial building may need to position a RIT on several sides of the building. The role of the RIT upon arrival is to think like the incident commander. Members must continually size up the situation; attempt to view the entire building on all sides; review preplans; gather necessary equipment; and monitor radio communications for the first sign of trouble. A RIT may also act to keep firefighters from getting into trouble in the first place; for example, by placing a ground ladder to an upper floor as a secondary means of egress; placing an additional backup hand line at the point of entry; or conducting forcible entry to provide additional exit and egress points. Many firefighters trained in RIT operations learn how to use the tools they normally carry at every fire to perform special techniques. See Figure 6–15 and Box 6.2.

FIGURE 6–15 RIT team and tools. *Courtesy of Lt. Joel Granata.*

BOX 6-2: THE COMMAND RESPONSIBILITIES FOR RIT OPERATIONS

- Conduct an ongoing risk assessment while determining an action plan.
- Change strategy and tactics to the rescue mode while continuing fire suppression activities.
- Monitor and encourage all personnel to monitor the stability of the structure.
- Request ALS transport units to treat and transport rescued personnel.
- Conduct a PAR. Until accurate personnel information is obtained, command cannot develop an effective rescue plan.
- Expand the IMS as necessary to keep ahead of the demand.
- Assign a secondary RIT to replace the initial RIT.
- Request specialized resources with specialized equipment to assist in the rescue process.
- Ensure a public information officer is assigned.

Rehabilitation of Personnel

Firefighters face some of the toughest demands and must be physically and mentally ready to perform at peak levels during normal and extreme weather conditions. Extreme heat or cold will adversely affects personnel. Many times, this occurs during the first 20 minutes of going to work at the incident scene as firefighters perform at the same level as an athlete while wearing approximately 70 pounds of protective gear. This generally causes a firefighter to reach his or her fatigue level very quickly into an incident. Personnel must have adequate **rehabilitation** provided both during an incident and while training to reduce the chances of further risk and injury. See Figure 6–16.

rehabilitation ■ This action consists of providing rest, medical monitoring, observation, fluid replacement, and at times nourishment at an incident scene for personnel.

NFPA 1584, *Recommended Practice on the Rehabilitation Process for Members During Emergency Operations and Training Exercises*, lists 17 important terms to use when talking about rehabilitation at emergency incidents and training exercises. See Box 6–3.

The incident management system should also provide for rehabilitation within its structure, and the incident commander must realize this function's importance. Every fire department should have a policy or SOP addressing rehabilitation at the incident scene that defines how implementation of rehabilitation occurs during an incident as well as the responsibilities of personnel. It is important for personnel to understand that they must communicate their rehabilitation needs to their supervisor, as many inexperienced firefighters will commonly work themselves to the point of complete exhaustion, which only causes them to become a liability.

Rehabilitation should be multilayered and tailored for the type, duration, circumstances, and size of the incident. At most incidents of short duration, rehabilitation can usually be informal. For example, a short-term single-alarm incident in moderate weather conditions will cause physical exertion to rise, but not to the level of complete exhaustion. At the minimum, informal rehab should consist of rest and fluid replacement. Comparatively, formal rehabilitation should require the medical evaluation of personnel and consist of a minimum of 20 minutes of rest, fluid replacement, and two sets of vital signs. Medical monitoring includes ongoing evaluation of personnel who may be at risk of the adverse effects from working the incident. It is important to document vital signs for comparative purposes, keeping in mind that documentation may be considered confidential. The vital signs to monitor include pulse rate, respiratory rate, blood pressure, temperature, and oxygen saturation. A carbon monoxide assessment of members should also be included during an evaluation of vital signs.

FIGURE 6–16 Firefighter in rehab. *Courtesy of Larry McGevna.*

BOX 6-3: SEVENTEEN TERMS CONCERNING REHAB

1. Cooling; active cooling; passive cooling
2. Core body temperature
3. Emergency incident
4. Emergency operations
5. Hydration
6. Incident commander (IC)
7. Incident management system (IMS)
8. Medical monitoring
9. Personnel accountability system
10. Procedure
11. Recovery
12. Rehabilitation
13. Rehabilitation manager
14. Sports drink
15. Standard operating guideline
16. Standard operating procedure
17. Supervisor

At a minimum, formal rehabilitation should consist of basic life support procedures, whereas having advanced life support procedures available is an important consideration for the incident commander and scene safety officer. The vital signs of personnel need to be within established criteria prior to their returning to work at the incident scene. One of the most monitored and least understood vital signs is blood pressure. According to the NFPA 1584 standard, a member should not be released from rehab if the blood pressure is at or exceeds 160 mmHg systolic over 100 mmHg diastolic. The following measures should be evidenced before a member is released from rehab: the respiratory rate within normal limits of 12 to 20 breaths per minute; the pulse rate returned to normal (60 to 80 beats per minute); and a pulse oximetry reading of 92% or more. A rapid carbon dioxide assessment tool is needed in the rehab area to measure carboxyhemoglobin (SpCO) or methemoglobin (SpMet). Pulse CO-Oximetry can serve as an adjunct to the standard pulse oximetry. Any member with a carboxyhemoglobin level greater than 15% should be treated with high-flow oxygen. Individuals in rehab with high carbon monoxide levels possibly have cyanide poisoning, which can be treated with drugs that should be available in the rehab area.

Whether rehabilitation is formal or informal, it should take place away from environmental hazards such as smoke and apparatus exhaust fumes. During periods of inclement weather, rehabilitation sites can be established in the lobbies of nearby buildings, at parking facilities, inside municipal buses, or in shaded areas.

#3

Rehabilitation during wildland firefighting conditions, where personnel may be located a great distance from other resources, requires special consideration. NFPA 1500 requires wildland firefighting personnel to be provided with a minimum of 2 quarts (2 liters) of water. A system to replace the water for the hydration of wildland firefighting personnel must also be in place. A small amount of planning and consideration in the area of rehabilitation pays off big dividends in reducing injuries and medical risks.

Summary

Each year, statistics continue to reflect that firefighters experience a high level of injuries, occupation-related diseases, and death due to the nature of their profession. It is critically important for every fire department to have a health and safety program and a program coordinator to ensure that the fire department focuses on firefighter safety during every event and function. Firefighter safety must become the priority. The incident priorities at every fire scene are life safety, incident stabilization, and property conservation; and a risk assessment must be done with these priorities in mind at all times. All fire departments should adopt the risk management philosophy that we will take a significant risk to protect savable lives; we will take a very controlled risk to protect savable property; and we will not risk firefighter lives to save what is already lost, be it lives or property.

Because a fire scene is an uncontrolled event with unforeseen hazards, fire departments must focus on incident management systems, personal protective equipment, accountability systems, rapid intervention teams, and firefighter rehabilitation. Experienced incident commanders and company officers help reduce the risk of injury and death, and an incident management system with a safety officer assigned to ensure fire scene safety is vital.

Firefighters, company officers, and chief officers must be aware of the many hazards present during fire scene operations. It takes a combination of experience, education, and continual training—as well as a great deal of resources—to remain safe and effectively combat structure fires. Personnel involved in the operation must be highly aware of signs and conditions related to rollover, flashover, back draft, thermal layering, and the three methods of heat transfer.

The requirement of having an accountability system in place does not ensure the system will work. Ultimately it is the incident commander's responsibility to ensure accountability at every fire scene.

When an unexpected event occurs, the incident commander must make certain that a rapid intervention team (RIT) can rapidly respond to firefighters in trouble. Personnel assigned to these teams must be thoroughly trained in all aspects of firefighter recovery and removal, and the specialized equipment that goes with it.

A fire scene has personnel facing some of the toughest conditions compared to other professions. Hence every fire department should have a policy or SOP addressing rehabilitation at the incident scene that defines how implementation of rehabilitation occurs during an incident as well as the responsibilities of personnel. Having rehabilitation as an integral part of the incident management plan in place will help to ensure safer operations at all times.

Review Questions

1. Explain the role of the fire department's health and safety program manager.
2. A lack of adequate operational polices and priorities plays a major role in fire scene safety.
   a. True
   b. False
3. List the three incident priorities.
4. List the four positions that make up the incident command staff.
5. What are the key components of safer operations at fire scenes and other types of emergencies?
6. Who has the ultimate role of keeping the focus on hazards and potential hazards at a fire scene?
7. What must become a priority in the incident commander's plan?
8. Firefighters involved in structural firefighting should use a[n] ______________ or ______________ to coordinate their activities and stay in contact with each other so that they can get assistance during firefighting operations.
9. Explain the differences between the IRIT and the RIT.
10. List the conditions when a personnel accountability report should be implemented during an incident.

Case Study

On March 21, 2003, in Cincinnati, Ohio, a firefighter died in the line of duty while battling a fire at a residential single-family dwelling. The victim was a member on Engine 9 as the first arriving engine company. The victim perished when he was caught in a flashover and became disoriented. The fire also injured two other firefighters just inside the front door with the victim when the flashover occurred. The fire started in the rear kitchen area of the dwelling when a pot of grease was left unattended on the stove, and the fire quickly progressed to the flashover stage within 3 minutes and 40 seconds after the arrival of Engine 9. The occupant status was unknown by firefighters at the time of arrival of Engine 9, although the 911 caller stated that all occupants were out of the dwelling. A Cincinnati district chief who was the first to arrive on scene reported heavy smoke showing. Following the flashover, the two injured firefighters were cared for in the front of the structure, and rescue efforts were initiated to recover the victim.

The initial attack line was stretched to the front door, and the engine company crew discovered a locked front door to the structure. The engine crew advanced their hand line down the delta side of the structure, where they were met by the district chief who advised the crew to return to the front on the structure and attack from the unburned side. Engine 9 then returned to the front door by stretching their hand line back down the delta side, and they were joined by a firefighter from Engine 2 who had arrived on scene. Personnel began to don their SCBA face pieces while one member of the engine company returned to the apparatus to retrieve an ax. The Engine 9 captain called for water twice as firefighters prepared to gain entry. Realizing a problem, the Engine 9 captain began addressing kinks in the hand line. The Engine 9 captain checked with the apparatus pump operator while a truck company began venting windows on the delta side. At the same time, the firefighters on the front porch area moved into the structure. Water was delayed due to the amount of kinks in the hand line. Flashover occurred shortly after the firefighting team entered the building to advance the hand line. After the flashover, two firefighters exited the building onto the front porch. Other on-scene firefighters scrambled to hose down the exiting firefighters when they noticed the silhouette of the victim flailing his arms inside the front room of the structure.

Another team of engine company firefighters made an aggressive attack through the front door in an attempt to knock down the fire and locate the victim. The district chief serving as incident commander ordered a change to a defensive strategy after occurrence of the flashover and was upset when firefighters made an aggressive interior offensive attack. The incident commander was unaware that there was a trapped firefighter. The interior crews with the aid of a thermal imaging camera located the victim toward the rear of the structure. A "Mayday" was called by an interior captain, and the victim was removed through a window on the delta side where cardiopulmonary resuscitation efforts began. Unfortunately, the victim succumbed to his injuries and perished in the line of duty. See Figure 6–17.

NIOSH had the following recommendations:

Recommendation #1: Fire departments should review and revise existing standard operating procedures (SOPs) for structural firefighting to ensure firefighters enter burning structures with charged hose lines.

Recommendation #2: Fire departments should ensure that a rapid intervention team (RIT) is established and in position prior to initiating an interior attack.

Recommendation #3: Fire departments should ensure that ventilation is closely coordinated with interior operations.

Recommendation #4: Fire departments should ensure that crew continuity is maintained on the fireground.

Recommendation #5: Fire departments should ensure that fire command always maintains close accountability for all personnel operating on the fireground.

Recommendation #6: Emergency dispatchers should obtain as much information as possible from the caller and report it to the responding firefighters.[2] See Figure 6–18.

1. How did the theory of attacking the fire from the unburned side play a role in this incident?
2. Did the incident commander have control of this incident?
3. How did communications become a factor at this incident?

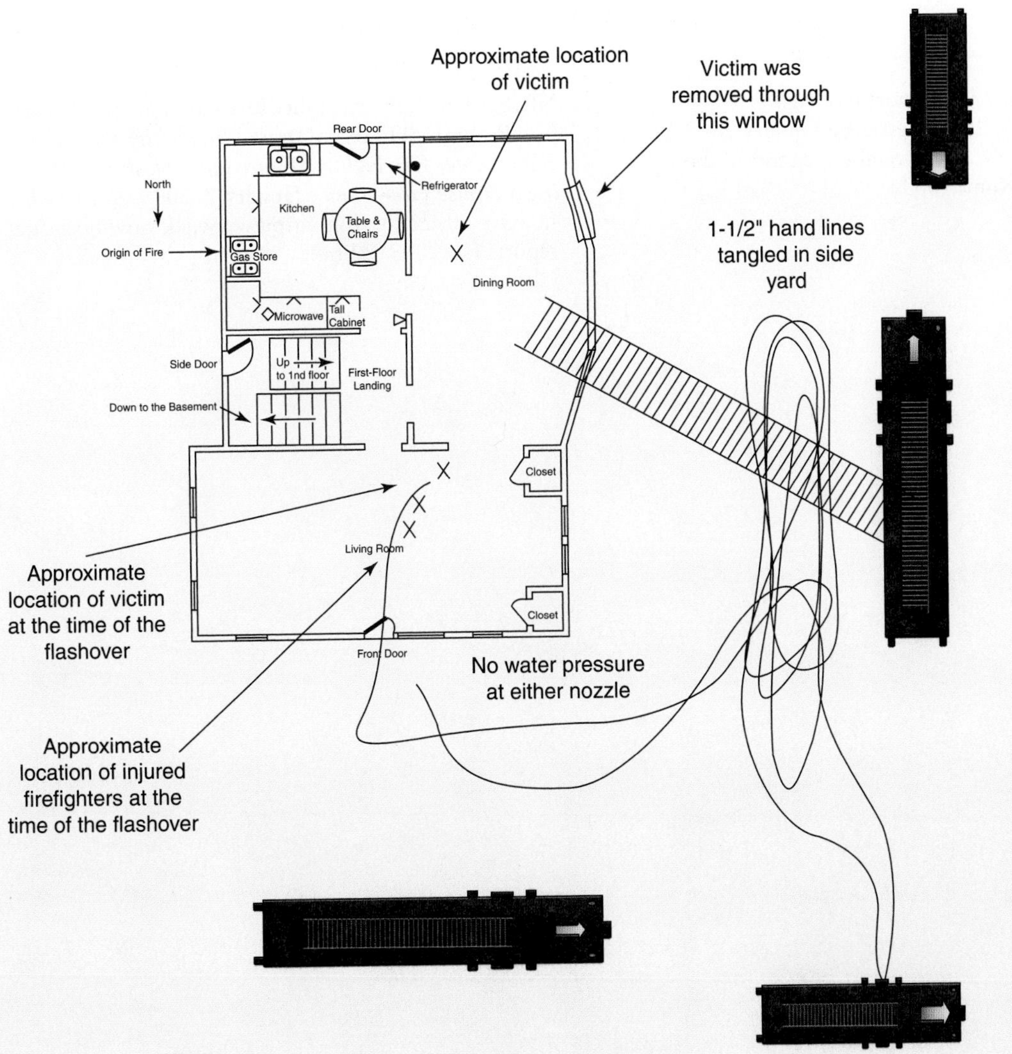

FIGURE 6–17 Aerial view of floor layout at the time of flashover.

Ohio House Fire Concerns and Issues	Yes	No
Did interior crews perform a proper risk-versus-benefit survey?		
Was hoseline and fire stream selection adequate for this type of event?		
Was there an adequate incident management system in place?		
Did the interior crew provide adequate and recent updates of conditions to the incident commander?		
Did the incident commander continuously evaluate changing conditions?		
Was crew integrity maintained throughout this event?		
Did the lack of recognition of the fire conditions play a role at this event?		
Was the established water supply adequate for this size event?		
Would adequate firefighter survival training have played a role at this fire?		
Was personal protective gear worn properly at this event?		

FIGURE 6–18 Concerns and issues of fatal house fire—Ohio. *Courtesy of Joe Bruni.*

References

1. NFPA 1003, *Standard for Airport Fire Fighter Professional Qualifications*, 2010. Retrieved January 25, 2010, from http://www.nfpa.org/aboutthecodes/AboutTheCodes.asp?DocNum=1003

2. NIOSH Fire Fighter Fatality Investigation and Prevention Program (FFFIPP), *Career Fire Fighter Dies and Two Career Fire Fighters Injured in a Flashover During a House Fire—Ohio*, January 7, 2005. Retrieved January 30, 2010, from http://www.cdc.gov/niosh/fire/reports/face200312.html

CHAPTER

7 Safety at EMS Emergencies

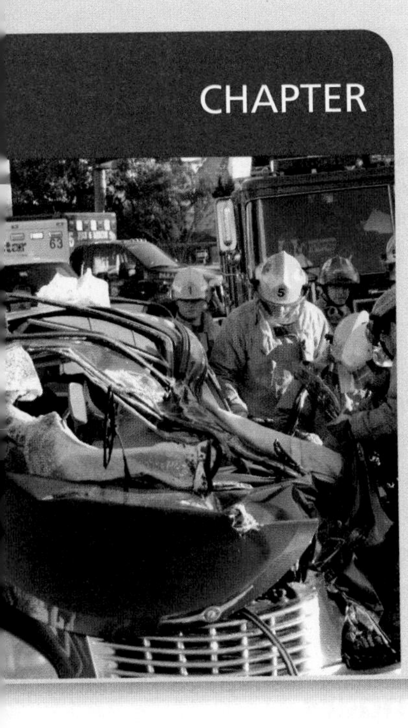

KEY TERMS

body substance isolation (BSI), *p. 167*
burnout, *p. 159*
hazardous materials (operations level), *p. 158*
Manual on Uniform Traffic Control Devices (MUTCD), *p. 163*
post-traumatic stress disorder (PTSD), *p. 159*

OBJECTIVES

After reading this chapter, you should be able to:

- Discuss the hazards faced by responders at emergency medical and rescue incidents.
- Explain the common hazards faced by responders at EMS incidents.
- Compare the similar common hazards faced by responders at both fire and EMS incidents.
- List the solutions to back injury prevention for EMS personnel.
- List the reasons associated with stress as a frequent occupational hazard for EMS workers.
- Discuss the steps responders can take to help ensure highway and roadway safety.
- Discuss why infection control is the personal responsibility of all responders.
- Discuss why an incident management system is needed at all EMS incidents.

Resource Central

For additional review and practice tests, visit **www.bradybooks.com** and click on Resource Central to access book-specific resources for this text! To access Resource Central, follow directions on the Student Access Card provided with this text. If there is no card, go to **www.bradybooks.com** and follow the Resource Central link to Buy Access from there.

Rural, suburban, and urban fire departments respond to medical and rescue types of calls. In fact, emergency medical and rescue incidents account for approximately 80% of the emergency call volume in most fire departments that provide EMS service to the community; EMS is the fastest-growing service provided by the fire service today. The different levels of EMS service are first responder, emergency medical technician, and paramedic. For the most part, the fire service utilizes cross-trained dual-role firefighters, and with these increased roles come increased risks.

The Threats and Dangers to Responders

EMS responders are routinely exposed to a wide variety of hazards. This chapter offers a fresh look at EMS incidents and the threats they pose to responders. EMS responders face many serious hazards in their jobs, placing them at high risk of occupational injury or death; these range from blood-borne pathogens, to victims who are emotionally upset or violent, to traffic hazards on the roadway, to hazards associated with disasters, to name a few. EMS responders are forced to make critical decisions in a very short time frame. Most EMS crews are made up of two individuals who respond to incidents as a team in a single unit or ambulance. In many cases, EMS responders are not afforded the same level of supervision as a fire company, which is usually led by a company officer experienced in emergency incidents. The senior member of the EMS crew normally acts as the supervisor and safety officer on EMS incidents. The senior member and his or her partner on the EMS unit must be familiar with the following:

- Medical and rescue techniques
- Exposure and infection protection
- Lifting and carrying techniques
- Department policies and procedures
- Incident scene safety
- Safe emergency vehicle operations
- Interpersonal skills

The National Registry of Emergency Medical Technicians (NREMT) currently certifies emergency medical service providers at five levels: first responder; EMT-basic; EMT-intermediate (which has two sublevels: 1985 and 1999); and paramedic. Certain states have their own certification programs and use distinct names and titles. By 2014, the various provider levels that exist across the nation will be standardized into four levels: emergency medical responder (EMR), emergency medical technician (EMT), advanced EMT (AEMT), and paramedic.

Limitations in technology associated with existing protective clothing and equipment, as well as the continually expanding roles of emergency responders drive the need to better understand the safety and health risks and protection needs of EMS responders. Exposure to infectious diseases accounts for very few actual EMS responder illnesses; however, pathogens are a growing hazard that is very difficult to protect against. See Figure 7–1.

FIGURE 7–1 All EMS incidents should be regarded as potentially infectious. *Courtesy of Lt. Joel Granata.*

The leading concerns of EMS workers are assaults, vehicle hazards, exposure to infectious diseases, and terrorism. Whereas the chances of responding to a terrorist incident are relatively low, assault incidents to EMS responders are increasing.

Many times EMS responders have little information about the potential hazards at the time of dispatch and while responding because the individuals who call in a medical emergency do not know many details. In fact, some of the leading safety hazards involve family members and bystanders because EMS incidents are emotional for them. In most cases, because people at the scene are not thinking rationally, they can cause problems for emergency responders. Uncertainty about the potential threats and hazards also frustrates coordinating protection efforts with other fire service personnel. See Figure 7–2.

As in many fire-related emergencies, EMS workers face the issue of complacency with the type of calls they respond to. Because EMS workers respond repeatedly to the same types of incidents, many allow themselves to become complacent. This can cause

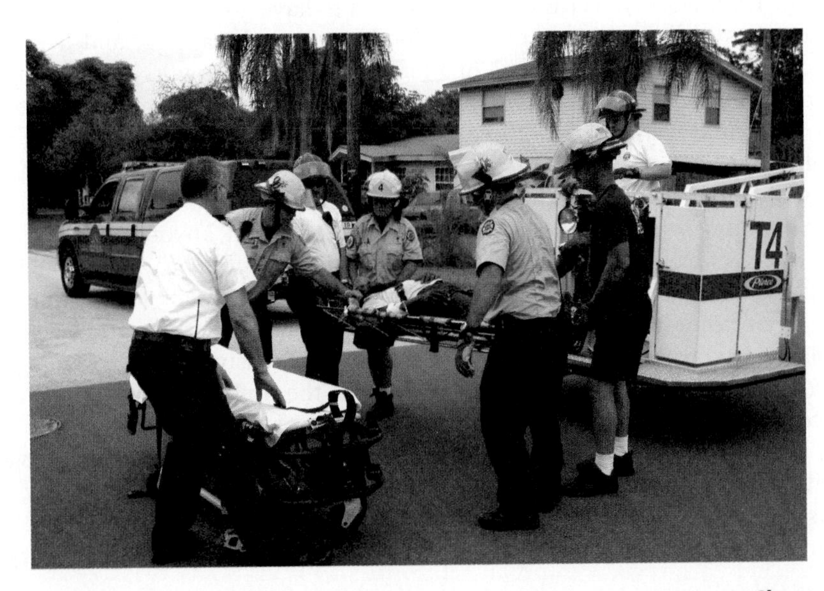

FIGURE 7–2 Every incident presents uncertain safety issues. *Courtesy of Lt. Joel Granata.*

emergency responders to let down their guard, thereby increasing their chances of injury and death.

Emergency response has traditionally been handled in a highly decentralized grass-roots manner, with solutions to problems largely being left to the local departments. Many of these issues warrant focused attention across the emergency medical responder community as the nation enters a new era in which emergency responders must be fully prepared to meet not only the challenges of routine emergencies but also those emerging from an increasingly unpredictable environment.

The Common Hazards and Perils for EMS Incidents

Responding to the scene is a leading hazard of every response. EMS personnel in the United States have an estimated fatality rate of 12.7 per 100,000 workers.[1] Other common hazards include:

- Assaults
- Infectious disease
- Hearing loss
- Lower back injury
- Hazardous materials exposure
- Stress
- Extended work hours
- Exposure to temperature extremes

The work of an EMS worker is physically and emotionally demanding. According to a study released in 2005, the number-one injury to EMS workers is assault.[2] The risk of this for EMS workers is approximately 30 times the national average. An EMS worker is more likely to be assaulted than a police officer or a prison worker. And the risk of death for EMS workers is approximately three times the national average. Unfortunately, there is insufficient training available to identify potentially dangerous situations, as well as a reluctance to provide the necessary training.

Infectious disease issues for EMS workers are developing rapidly in a complex environment of government regulations, court cases, and politics. The International Association of Firefighters (IAFF) is continually receiving new information and developing new resources. (Infectious disease will be discussed later in this chapter.)

Hearing loss affects many thousands of workers employed or formally employed as a firefighter or EMS worker. Yet EMS workers need to have good hearing to communicate with patients, as well as be able to hear blood pressure, breathing, and heart sounds. It is difficult to estimate the number of firefighters and EMS workers affected by this debilitating condition due to the fact that many in the field refuse to acknowledge they have this problem. Over a typical 20-year career, firefighters and EMS workers typically suffer significant hearing loss as a result of long-term exposure to sirens, which produce 120 decibels within a 10-foot radius.

The Environmental Protection Agency (EPA) states that anyone regularly exposed to sound levels above 70 decibels risks serious and irreversible hearing damage. OSHA regulation 1910.95 states that the employer must notify each employee exposed at or above an 8-hour time-weighted average of 85 decibels as a result of noise level monitoring.[3] Prolonged exposure to sirens is known to lead to extensive inner ear damage, which is frustrating to those who have built their career in the firefighting or emergency medical services. Yet siren noise cannot be blamed for all of the hearing loss to emergency workers, because during off-duty hours many are also using music-listening devices at a high decibel level. The key to avoiding hearing loss is education of emergency workers as well as their using hearing protection on the job. Many emergency response workers now wear hearing protection of the type used by the military and airline industry, which allows communication between members and offers a high degree of protection. OSHA regulation 1910.95 requires hearing protection for workers in high-noise environments. The bottom line: workers must be responsible for their own hearing protection if it is not provided by the employer.

Lower back injury is the most common injury suffered by EMS workers due to lifting, transferring, and otherwise moving patients. Workers' compensation claims involving a back injury can cost thousands, and the time off of work can be months to years. Compared to manually lifting or moving inanimate objects in other work-related settings, moving the human body is more strenuous, as it is heavier, more delicate, and awkward to handle. The center of gravity and distance to the patient can also change during the transfer of patients, placing the EMS worker in awkward positions. Repetition of the same movement also produces fatigue and increases the chances of injury. Physical fitness, weight, diet, exercise, personal habits, and lifestyle of the EMS worker may also affect the development of back injuries. See Figure 7–3.

#6

Workers who are not in good shape have an increased chance of injury. Excessive body weight places stress on the spine and is associated with back injuries. Here are some suggestions for back injury prevention:

- Develop a healthy lifestyle that includes proper weight control and exercise.
- Move and lift patients in teams—if possible, call for additional help.
- Utilize proper work and lifting practices.
- Use safety equipment such as power-assist stretchers, bariatric stretchers, winch cables, and ramps.

These suggestions may need to be altered to fit a particular situation, and are not the only ones possible.

HAZARDOUS MATERIALS EXPOSURE

Both EMS workers and firefighters face the possibility of hazardous materials exposure on every response. For EMS workers, hazardous materials exposure could range from spilled fluids at a traffic accident scene to a large hazardous materials incident involving victim exposure. EMS workers should learn to develop situational awareness on every response as a safeguard for one that possibly involves hazardous materials exposure. See Figure 7–4.

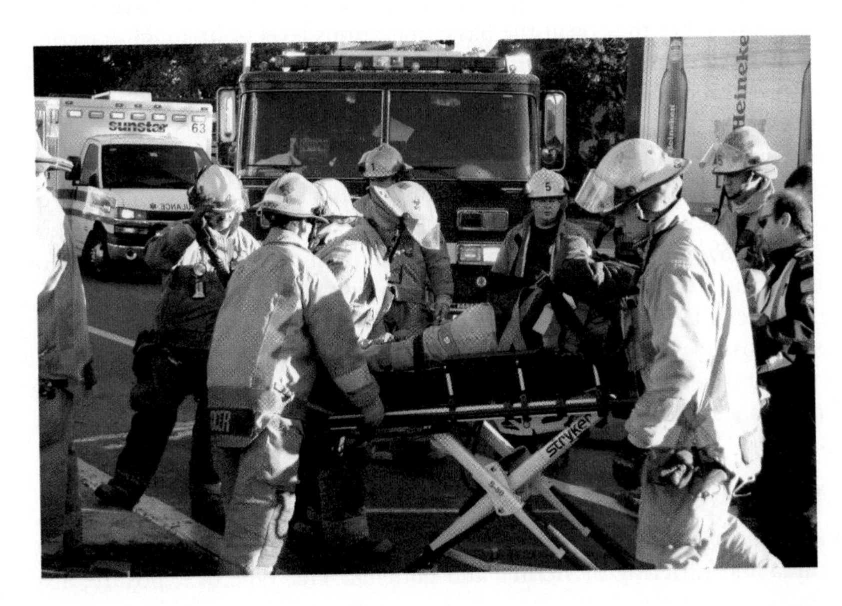

FIGURE 7–3 Responders must continually be aware of injuries related to patient movement.
Courtesy of Lt. Joel Granata.

FIGURE 7–4 Safeguarding hazardous materials at the scene reduces exposure and hazards. *Courtesy of Joe Bruni.*

hazardous materials (operations level) ▪ This training is intended for employees that will be expected to take defensive measures to stop the spread of a chemical spill, or to assist in the isolation of an area until a cleanup contractor or offsite hazardous materials team can arrive.

As a minimum, EMS workers such as firefighters should be trained in **hazardous materials (operations level)** response. The operations level helps to ensure that EMS workers understand what hazardous materials are and are able to support defensive actions at hazardous materials events. The operations level also teaches responders how to take control of the scene, call for the correct additional resources, and start the process of the incident command system. NFPA 472, *Standard for Competence of Responders to Hazardous Materials/Weapons of Mass Destruction Incidents*, now requires EMS workers to receive training at a level higher than the previous awareness level. NFPA 472 spells out the minimum competencies necessary for all responders to implement a risk-based response, with the operative word being *responder*. Previously, a hazardous material response was primarily considered a fire department event. The scope of NFPA 472 now covers all emergency responders to hazmat/WMD incidents, including EMS, law enforcement, and other agencies, as well as firefighters. Hazardous materials response now involves many more agencies and players, and new skill sets for those who respond to hazmat/WMD, whether they be at the basic operations level or at a higher technician level.

NFPA 473, *Standard for Competencies for EMS Personnel Responding to Hazardous Materials/Weapons of Mass Destruction Incidents*, addresses the response capability needs of the EMS community and organizations. The tactical and operations procedures for EMS responders must now take into account the potential threats and possible response scenarios created by hazardous materials and weapons of mass destruction incidents. Emergency responders including EMS responders must have the training and ability to effectively respond to a hazardous material incident, terrorist event, or criminal incident involving a chemical release. Proper training and the availability of the necessary equipment are crucial for personnel to protect themselves from these threats. Unfortunately, many obstacles face EMS personnel and organizations about obtaining the proper personal protective equipment, as well as the necessary training, to safely and effectively respond to these types of incidents.

THE EMS WORKER AND STRESS

Stress is a frequent occupational hazard for EMS workers, who are continually exposed to the trauma and suffering of victims and patients. This type of exposure, at times, can cause secondary traumatization to the EMS worker. EMS personnel also face personal exposure to dangerous and life-threatening situations in which they risk personal injury or even death, causing further stress to responders. Remarkably, few EMS workers suffer from **post-traumatic stress disorder (PTSD)**. A number of factors contribute to this resiliency; but the social support, bonding, and close relationships formed among firefighters and EMS workers is probably the most important reason for the low numbers of EMS workers and firefighters suffering from PTSD.

post-traumatic stress disorder (PTSD) ■ This kind of anxiety disorder can occur after workers have seen or experienced a traumatic event that involved the threat of injury or death.

Yet not all EMS job-related stress is due to traumatic incident exposure. Emergency responders who work a 24-hour shift report the stress associated with shift work, for example, lack of sleep, poor quality sleep, and disturbed sleep. Unfortunately, many EMS workers who request assignments to the busiest districts often do not recognize the stress associated with the lack of rest and rarely request reassignment when experiencing such stress. In many cases, being reassigned would equate to a separation from the people and crews to whom they have become closest.

#13

Another source of stress for EMS workers is the rigid and authoritarian organizational structure of most EMS organizations. Poor leadership in the organization, as well as management's noncommunicative style and a lack of empathy, may cause undue stress on the EMS employee and low morale.

Stress is directly related to the life-and-death decisions EMS workers make on a daily basis and the time urgency of such decisions. Repeated exposure to this form of stress is commonly referred to as **burnout**, the cumulative result of stress. Emergency workers may feel a lack of control in many areas, such as the inability not only to influence decisions that affect their job, such as which hours they will work or which assignments they will get, but also to control the amount of work that comes in. The emergency responder's work is repetitive and many times chaotic, and constant energy is needed to remain focused. When energy drain occurs, job burnout becomes a common problem. Many times the following stressors also combine to lead to burnout: needless paperwork and bureaucracy, low pay, and a lack of respect for what EMS personnel do on a daily basis.

burnout ■ A state of physical, emotional, and mental exhaustion can be caused by long-term exposure to demanding work situations.

Stress management in EMS must become an ongoing process that ranges from self-care to professional mental health services. Stress management should focus on both the worker and the organization as a whole and include proper diet and exercise, as well as training in stress reduction and relaxation techniques. Most importantly, organizations need to adopt policies and procedures that support their personnel, which will help to reduce unnecessary organizational stress. Much like the fire service, EMS organizations that are not part of a fire service organization should adopt a comprehensive wellness and fitness program. See Figure 7–5.

The promotion of the most effective employees within the organization should also be the goal as a way to ensure the continual review of stress reduction methods. Line personnel who develop a working relationship with the first-line supervisor experience greater job satisfaction and reduced stress. EMS workers must also be encouraged to take personal responsibility for their own physical and mental condition.[4]

THE EMS ENVIRONMENT

The occupational hazard of exposure to temperature extremes affects all EMS workers, as they do not have the luxury of working in a controlled environment. Instead, they work outside for prolonged periods exposed to the effects of heat and cold, so workers must take the necessary steps to protect themselves from the effects of weather with the proper clothing. Layers of clothing can be removed during periods of warmer temperatures and placed back on when necessary in the cold. Clothing includes gloves, hats, and face and ear protection, as serious cold injury can take place in a relatively short amount of time. On the other end of the spectrum, when heat gain is greater than heat loss, individual core

FIGURE 7–5 A wellness and fitness program is essential to safety and the reduction of stress.
Courtesy of Joe Bruni.

temperature rises. The body's response to the heat environment is to perspire; however, working in extreme heat can lead to mild to severe heat-related injury even while perspiring heavily. Other contributing factors may be the worker's age, weight, fitness, medical condition, and ability to acclimatize to the heat.

When the temperature and work environment cannot be controlled, every effort should be made to assist the worker with maintaining a safe and healthy body temperature. This would include periods of adequate rest, fluid replacement, and a proper rehab area to monitor vital signs. Heat-related disorders include heat rash, heat cramps, heat syncope, heat exhaustion, and heatstroke, with the latter two being the most serious. The following measures can be taken by the organization and the employee to reduce the chances of heat injury:

- Acclimatization to hot climates
- Rehydration
- Appropriate clothing
- Physical conditioning
- Engineered control—fans, air-conditioning, heat shielding
- Work scheduling—schedule more workers to rotate through the workload
- Monitoring of personnel
- Education of workers

The key to preventing heat-related injury is to understand the hazards of working in hot environments, take proper precautions to ensure good health, recognize early warning signs of heat stress, and make sure workers follow recommended safe work practices.

VEHICLE ACCIDENT AND INCIDENT SAFETY ON ROADWAYS

Traffic incidents involving fire, emergency medical services (EMS), and law enforcement personnel are routine occurrences on America's roads. Operations involving those incidents on highways and roadways pose an extreme safety risk to both firefighters and EMS personnel because of limited control of flowing traffic around and through the incident scene. Many department policies, SOPs, and guidelines are inadequate for these types of incidents. Additional concerns are weather, poor lighting, crowd control, and inadequate warning of drivers of approaching vehicles.

The issue of operating safely on the highway has received a great deal of attention in the past few years. In most cases, emergency medical incidents and vehicle accidents involve only a single ambulance or fire apparatus, and so the fire company officer or senior EMS attendant must function as both incident commander and incident safety officer. What must guide the individuals in these roles are department SOPs as well as knowledge of NFPA 1581, *Standard on Fire Department Infection Control Program*, and of bloodborne pathogens covered in OSHA Title 29 CFR 1919.1030. The incident commander/safety officer must apply his or her knowledge of medical treatment, rescue techniques, exposure protection, infection control, and lifting and carrying methods to help prevent injuries to operating personnel. The training of the incident commander/safety officer and of working personnel becomes very important as it relates to the prevention of injuries. In the often chaotic and unorganized environment of many vehicle accidents, first responders must work within a critical time frame in order to ensure their safety and that of vehicle occupants. It is often the arriving first-due emergency unit that must establish safety for the overall scene. See Figure 7–6.

The first-due unit must initially determine where to place the emergency vehicle to accomplish the following:

- Protection for responders
- Avoidance of hazards at the scene
- Access to patients
- Facilitation of transport
- Extrication and tool needs
- Easy access for later arriving emergency vehicles

Safety is the priority concern with vehicle placement. Later arriving larger fire apparatus may need to establish close access or can be used to protect the scene and responders. Down power lines, the presence of hazardous materials, and spilled fluids are all common hazards of vehicle accidents and roadway incidents. It is generally advisable to place vehicles so responders do not have to cross through oncoming

FIGURE 7–6 Responders place themselves in harm's way at every vehicle accident. *Courtesy of Lt. Joel Granata.*

FIGURE 7–7 SOPs should be in place addressing safer operations during conditions when traffic is flowing.
Courtesy of Lt. Joel Granata.

traffic to get to the scene, although this is not always possible. Of particular concern is EMS and rescue services that take place on the roadway while traffic is still flowing. See Figure 7–7.

SOPs should address conditions when traffic is flowing during operations to create a safety or buffer zone protecting responders. In most cases, emergency vehicles can be used to create this zone. Larger emergency vehicles should be considered and dispatched to the scene if the situation warrants. Treatment and transport must take place within these safety zones. Responders must not be afraid to close down the highway in order to protect responders; however, many police agencies will have an issue with closing an interstate highway and will want to get traffic flowing again as soon as possible. At a minimum, responders should wear a reflective vest that meets ANSI standards at all highway and roadway incidents, and this requirement needs to be part of an SOP on operations during traffic accidents and other roadway incidents.

All responders working in a hazard zone should have full PPE ready. It is difficult for responders to work in an area that does not afford them adequate protection and safety. On incidents in which the apparatus operators or company officer have not taken steps to ensure responder safety, the incident commander must take the necessary action. It will usually be safer and easier for a transport vehicle or ambulance to pull beyond other larger apparatus and the incident itself so it can be located in the safety zone to facilitate the loading of patients. When a lone ambulance is at the scene, it becomes vulnerable to passing motorists. Consideration should be given in those situations to having an additional unit or larger apparatus brought to the scene for the protection it can offer to the scene and responders.

Unfortunately, federal and state governments have done little to address responder safety while working in or around moving traffic. However, NFPA 1521, *Standard for Fire Department Safety Officer*, does provide for the necessity of appointing a safety officer when responders are working in or near moving traffic. The standard also states that the safety officer should evaluate motor vehicle scene traffic hazards and apparatus placement, and take appropriate action to mitigate hazards. Unfortunately, "appropriate action" is left to the interpretation of the individual reader. NFPA 1500, *Standard on Fire*

Department Occupational Safety and Health Program, notes three areas that promote traffic safety:

- Require the wearing of a garment with fluorescent and retro-reflective material when in or around traffic.
- Use apparatus to block the scene whenever possible (but NFPA 1500 does not specify when, where, or how this will occur).
- Apparatus should use warning devices, such as traffic cones and flares, to mark the emergency scene.

NFPA 1901, *Standard for Automotive Fire Apparatus*, addresses apparatus markings by requiring a minimum of a four-inch reflective stripe on the front, side, and rear of the apparatus. The 2003 **Manual on Uniform Traffic Control Devices (MUTCD)** introduced the first exposure of a temporary traffic control zone (TTC). The manual states that "anytime an incident affects the normal flow of traffic a TTC zone should be created to protect firefighters working on the scene," and "the primary function of the TTC is to provide for the reasonably safe and efficient movement of road users through or around TTC zones while reasonably protecting workers and responders to traffic incidents and equipment."[5]

Manual on Uniform Traffic Control Devices (MUTCD) ■ This manual defines the standards used by road managers nationwide to install and maintain traffic control devices on all streets and highways. The MUTCD is published by the Federal Highway Administration.

One of the best strategies as it relates to emergency operations on the roadway is to limit the exposure time on scene in conjunction with assuring the correct resources are present, therefore enabling responders to operate safely. Additional apparatus should be requested to respond to incidents on limited-access highways to ensure responders are able to work in a safety buffer zone. Preplanning of unique or specific roadways in a response district is also an added safety measure as it can be used to present responders with a graphic representation of how to correctly and safely block specific roadways or intersections.

#11

Unfortunately, some EMS responders and firefighters do not receive the necessary training involved with safety at vehicle accidents and highway safety. At a minimum, responders should be trained in basic awareness of traffic safety. One of the highest priorities is to control the traffic not involved in the incident. The responder is responsible for adequately protecting the scene, because onlookers commonly "rubberneck," which often leads to secondary incidents and emergency workers being struck by vehicles. One of the most important actions responders can take is to identify themselves at the scene and give adequate warning to approaching motorists. Responders must understand that before motorists can react to any hazard, they must first be able to see the hazard. Apparatus positioning and arrangement to guard the scene and the responders can be accomplished at every scene.

Another main hazard associated with operations on the highway is vehicle extrication. This requires gaining access to victims trapped in vehicles using hand and hydraulic tools to enable rescuers with the ability to disentangle and remove victims from the vehicles. Many times this is accomplished with tools with which the rescuers can remove the vehicle components from around the victims. In most cases, the responders involved with this type of incident focus on the operations and lose sight of the safety aspects of the incident. During incidents in which extrication is necessary, an incident management system should be in place and an incident safety officer assigned. Also, personnel involved in the extrication process that are located in the hazard zone or action circle should wear full PPE, and precautions should be in place for fire suppression. See Figure 7–8.

Vehicle accidents are not the only concern when it comes to incidents on roadways. In many cases, incidents involving assault are far too common on the highway. In these incidents, responders must be aware that law enforcement support and scene control must be in place. An incident that involves some type of violence on the roadway may also involve additional violence after responders have arrived on scene.

FIGURE 7–8 Responders working in the hazard zone should be in full PPE. *Courtesy of Lt. Joel Granata.*

#12

Environmental hazards are always a concern with highway and roadway incidents. Responders may be exposed to extreme weather conditions related to heat, cold, rain, snow, and ice. The extended time frame of some incidents during periods of extreme weather will impact the safety of responders. Rehab considerations should always be considered at every highway incident as part of the response plan.

Interagency preplanning should be considered as part of the planning process. Such incidents will require police department interaction for crowd and traffic control. Preplanning will ensure the proper equipment and resources are brought to the scene. The expectations of all agencies are also essential to ensure responder safety at every incident. Interagency preplanning will help to determine what resources and equipment that other agencies have on hand and will bring to certain incidents. Without planning, there is no certain way to ensure outside agencies and surrounding departments will meet the expectations of a department requesting assistance and resources.

BLOOD-BORNE PATHOGEN AND INFECTION CONTROL

Infection control for EMS workers at incidents is required by both federal regulations and the NFPA standards. OSHA 1910.1030 and NFPA 1581 were presented earlier in Chapter 2 in regard to infection control. OSHA 1910.1030 specifically deals with bloodborne pathogens, and NFPA 1581 addresses the requirements of a fire department infection control program. All emergency responders who treat patients and victims as part of their department's routine operations need to understand how essential the issue of infection control really is. Personnel must make an effort to reduce risks and mistakes. Infection control measures protect not only the responders but also the patient. Firefighters who perform the role of EMS must know that the duties related to EMS work require specialized clothing.

Every department is required to have an infection control officer. In some departments the health and safety program manager fulfills this role. The infection control officer's responsibility is to be a liaison between the department physician, the department health and safety officer, the infection control representative at health care facilities, and other health care regulatory agencies. The infection control officer must ensure that notification, verification, treatment, and medical follow-up occur after being notified of a possible exposure. He or she must ensure that the proper forms and paperwork have been completed correctly. The NFPA 1581 standard requires the department has established procedures for reporting and treating an exposure. The infection control officer must be notified within 2 hours of the exposure incident, and the medical facility must follow up with the infection control officer in writing within 48 hours of receiving a request as to whether there was an exposure.

In conjunction with an infection control officer, the department must also have an infection control plan to reduce the chances of the spread of disease. Such plans must detail standard operating procedures for employee and patient safety; patient confidentiality; exposure reporting and follow-up; facility considerations for storage; disposal and cleaning of infectious waste; and continuing education for personnel. An important component of any fire department infection control program is the training and education of firefighters and other emergency personnel. The infection control officer and/or a committee must review the plan at least annually and incorporate any technical or legislative changes from the previous year noted in the Federal Register, Occupational Safety and Health Administration (OSHA) enforcement guidelines, and Centers for Disease Control and Prevention recommendations. The infection control plan is based on three specific areas, which are a function of risk management:

- Engineered controls
- Personal protective equipment
- Training and education

Risk management of infection control should include the identification, evaluation, control, and monitoring of risks to the following:

- Fire department facilities
- Fire department vehicles
- Emergency medical operations
- Members when cleaning and disinfecting protective clothing and equipment
- Members from other situations that could result in occupational exposure to a communicable disease (NFPA 1581, section 4.2.2, 2010)

Responders are routinely exposed to a number of diseases. A responder's chances of being exposed to blood-borne or airborne pathogens are much greater than of being injured at a fire or other emergency. Responders who take the necessary steps to protect themselves from disease are also protecting their family and friends. See Figure 7–9.

The following list provides rules of thumb for keeping safe:

- Stay current on the TB monitoring procedures of your department.
- Avoid skin or mucous membrane contact with bodily fluids from your patient (body substance isolation).
- Consider all patients infectious and take the appropriate precautions (universal precautions).
- Wear the right size gloves, masks, and gowns; and have them available when needed quickly.
- Make sure you are fit tested annually and have the proper respiratory protection.
- Make sure you understand your department's infection control plan and know what to do if you are exposed.
- Maintain the exposed person's confidentiality.
- Clean your equipment and transport vehicle after every patient contact.
- Use disposable medical devices whenever possible.
- Use the engineered sharp's safety devices according to the manufacturer's recommendations and your department's policy.[6]

FIGURE 7–9 Responders should assume every EMS scene is an infectious zone. *Courtesy of Lt. Joel Granata.*

#4

One of the most common exposures to responders involves bodily fluids sprayed in the face. In many cases, the responder is not wearing eye and face protection, personnel do not have the face and eye protection readily available, or the protection has been damaged from being crammed into a medical bag. Responders must ensure that they are properly fit tested on an annual basis and that their protective equipment is readily available at the scene and worn for protection.

One of the best ways to reduce or prevent contamination is to prevent the means of transmission. Contamination occurs by direct or indirect contact, an intermediate host, or other vehicles. Direct contact occurs from physical contact with a patient. Indirect contact occurs when a responder touches a surface that has been contaminated by the infected person, for example, blood contamination on a surface or piece of equipment, or airborne droplets produced when a patient sneezes or coughs. Transmission via an intermediate host may occur from the bite of an insect. Transmission by other vehicles would include contact with contaminated food or water.

EMS personnel and firefighters can take other preventative measures such as immunization against hepatitis, flu shots, and testing for tuberculosis. Such preventative measures should be part of the department's infection disease plan. Among the responsibilities of each individual responder are to understand agency policies and procedures for reporting and treating exposures, to take universal precautions at every medical call, and to have the necessary protective equipment available and ensure its use when necessary. Infection control is as much a training issue as a risk management issue. Responders must have ongoing training on the modes of infectious disease transmission, as well as cleanup procedures for themselves and their equipment. The facility that houses personnel and equipment must have a cleanup/decontamination area. Units should carry some type of disinfecting gel or hand wipes for hand washing at the emergency scene and a small amount of disinfectant or bleach that can be mixed with water to sanitize equipment and footwear at or prior to leaving the incident scene.

EMS Protective Equipment

Just as firefighters and many other professions need specialized protective equipment, so do EMS personnel. NFPA 1999, *Standard on Protective Clothing for Emergency Medical Operations*, addresses PPE for EMS personnel. First responders, including medical personnel, need a greater level of protection to enable an "all-hazard" response with an emphasis on chemical, biological, radiological, nuclear, and explosive events. Firefighters who respond as medical response personnel are issued PPE suitable for many emergencies including vehicle extrication and certain rescue situations.

Whereas Chapter 6 covered PPE for firefighters, this chapter focuses on PPE specific to emergency response medical personnel involved with medical emergencies unrelated to rescue situations. EMS providers must determine what the minimum level of protection will be at particular incidents. The PPE requirements for **body substance isolation (BSI)** are defined in the respective standards and regulations. Universal precautions should be undertaken at any incident in which the responder may come into contact with body fluids. Policy and procedures should be developed for PPE in accordance with the department infection control plan. Responders should be encouraged to wear more than the minimum required at incidents involving infectious disease. EMS personal protective equipment must provide blood and body fluid pathogen barrier protection to whatever parts of the body it covers. See Table 7–1.

body substance isolation (BSI) ■ This is the practice of isolating all body substances and fluids of individuals undergoing medical treatment, particularly emergency medical treatment of those who might be infected with illnesses or diseases.

EMS responders should be issued equipment that is adequate for the possible situations they may face, including chemical protective equipment for hazardous materials incidents and incidents involving weapons of mass destruction. At a minimum, EMS personnel should be issued the following protective equipment:

- Disposable gloves
- Disposable gowns

TABLE 7–1 PPE Matrix

TASK	HAND WASH BEFORE/AFTER	GLOVES	PROTECTIVE EYEWEAR	TB MASK N95	ARM SLEEVES
Peripheral IVs	X	X	+	N/A	+
Central line/sternal IO	X	X	X	N/A	N/A
Blood glucose testing	X	X	+	N/A	N/A
Cricothyrotomy	X	X	X	X	X
Sharps disposal	X	X	N/A	N/A	N/A
Intubation	X	X	X	X	X
Insert OPA/NPA	X	X	X	+	+
Bleeding control	X	X	X	X	X
Suctioning	X	X	X	X	+
Bag valve mask	X	X	X	X	+
Extricating trauma patient	X	F	F	N/A	N/A
Possible airborne disease	X	X	X	X	N/A
Cleaning/disinfection /equip. apparatus	X	X	X	X	+
CPR	X	X	X	+	+
Handling contaminated waste	X	X	X	+	+
Childbirth	X	X	X	X	X

Key: X = Required
+ = Recommended
N/A = Not applicable
F = Full turnouts required with faceshield down or goggles in place

Source: Courtesy Tempe Fire Department http://www.tempe.gov/fire/PoliciesandProcedures/PDF%20Files/210.08B.pdf

- N100 or N95 protective masks
- Eye protection

PPE for EMS personnel must be geared to the local conditions, and the employer should also consider the following for EMS personnel:

- Hard hat or work helmet
- Hearing protection
- Hooded chemical-resistant clothing
- Full-length jacket with reflective striping
- Chemical protective gloves
- Multiuse work gloves
- Multiuse safety footwear
- Chemical-resistant foot covering
- Rain gear
- Body armor

Just as with firefighter gear, PPE for EMS personnel should be inspected, cared for, and disposed of according to the manufacturer's instructions. The main difference between a firefighter's PPE and that of EMS personnel is that much of the PPE for EMS personnel is disposable. Records and documentation should be kept for EMS personal protective equipment, just as is done with firefighter gear. In addition to record keeping, EMS personnel need training in the use of issued PPE.

EMS Incident Management

EMS incidents must be managed with an incident management system (IMS). In most cases, the senior crew member assumes the many roles of a command staff without the formal expansion of the incident management system. When EMS operations expand into a larger incident, such as those involving a large number of casualties, the incident commander should call for an incident safety officer and formally expand the management system as necessary. The incident safety officer will be required to ensure the safety of operating personnel, as well as work closely with a triage officer or medical branch to communicate incident needs to the incident commander. A large-scale incident may require more than one incident safety officer. With large incidents, EMS operations become a branch of the IMS. Just as with other emergency operations, IMS is used in some fashion for all EMS incidents. It is of vital importance that IMS be used for EMS incidents that require multiple unit responses and multi-jurisdictional responses. See Figure 7–10.

During situations and incidents that require only a response from EMS, the IMS can be tailored to fit the need for the command staff and general staff. The incident command system (ICS) has been replaced by the National Incident Management System (NIMS). Responsibilities within the command structure change only slightly depending on the type of incident. Within the incident command structure at disasters and mass-casualty incidents, an EMS or medical branch may be utilized to handle such issues as triage, treatment, and transport. Various groups may also be established to handle specific hazards at a scene or incident. The IMS may be adapted to any type or size of incident. NIMS provides a consistent, flexible, and adjustable national framework within which governmental and private entities at all levels can work together to manage domestic incidents, regardless of their cause, size, location, or complexity. The common denominator in any incident is the assignment of an incident safety officer. The safety officer is needed any time personnel will be working in a hazardous situation that is fire- or EMS-related. On a smaller EMS incident, the senior EMS crew member

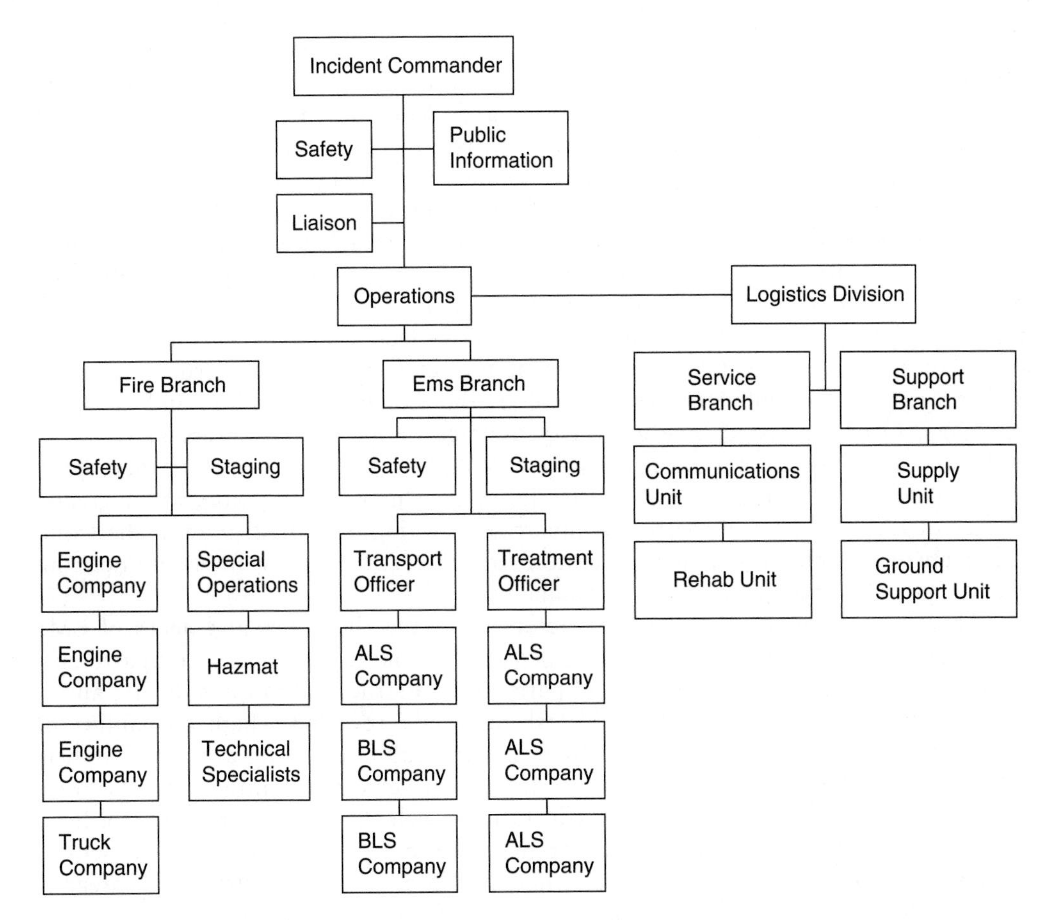

FIGURE 7–10 An expanded IMS including EMS functions. The EMS function is operational.

must assume the roles of both incident commander and safety officer and ensure that personnel utilize the proper PPE.

Accountability of personnel also becomes a key factor on large-scale EMS incidents. EMS personnel must adhere to an accountability system just as fire personnel do at fire-related incidents. Even when the EMS service is provided by a private or third-party company, it must follow both the IMS and the accountability system. The safety program manager for an independent EMS company needs to make certain the personnel within his or her company have been properly trained in both IMS and accountability so they can work within the framework of another agency. Preplanning and training combine to make certain there will be interagency coordination when the time comes to work together. Preparedness is a continuous process that involves efforts at all levels and within all organizations that will become involved in an incident. NIMS requires organizations to ensure that effective interoperable communications and information management processes, procedures, and systems exist to support a wide variety of incident management activities across agencies and jurisdictions.

Ambulance Motor Vehicle Accidents

EMS workers face a wide array of hazards, the most common being a motor vehicle accident involving an EMS response vehicle. Nearly all work-related deaths among ambulance workers are due to transportation incidents and crashes. Although there is currently no national count of ground ambulance accidents, the total number of fatal crashes involving ambulances can be ascertained from the National Highway Traffic Safety Administration (NHTSA) Fatality Analysis Reporting System (FARS). The National Institute for Occupational Safety and Health (NIOSH) is testing and identifying measures that will help to reduce injuries and deaths of EMS workers. The risk for EMS workers to be involved in a vehicle accident is two times higher than the national average for all other workers in general because of the time-urgent nature of the response.

NIOSH recently set out to reduce the number of ambulance crash-related injuries and fatalities concerning EMS workers in ambulances by partnering with the Ambulance Manufacturers Division (AMD) of the National Truck Equipment Association and the General Services Administration (GSA) to revise federal ambulance standards. The Ambulance Crash Survivability Improvement Project focused on the structural layout and integrity of the ambulance compartments, the design of hardware, and occupant restraints. Over the last decade a large percentage of the EMS workers killed in an ambulance crash were not wearing a restraint. A study of fatal ambulance crashes identified that most injuries and deaths occurred during the emergency response, with a greater chance of EMS workers in the rear compartment getting hurt than those in the front. Unfortunately, nonfatal injuries to volunteer firefighters and EMS workers are not routinely captured in occupational injury databases.

There are a number of important factors related to the injury and death of EMS workers: the lack of vehicle restraints, personnel striking equipment compartments in the treatment area, workers striking patients, and structural failures during an involved crash, to name a few. NIOSH crash tests have revealed a high probability of a responder sustaining a head injury from striking cabinets during a crash. Cabinets located in a higher position coupled with the use of vehicle restraints will help to reduce the probability of injury and death during a vehicle crash. Increasing the head clearance of compartments above workers who are in the seated position will help to eliminate a significant source of head injury. The improvement of structural integrity (e.g., padding installed in the interior of the unit) of the vehicle will also help to reduce injuries and deaths. Ambulance manufacturers need to design ambulances that provide occupant protection, increase measures to attain a higher level of survivability during ambulance crashes, and improve the overall occupational safety and health of EMS workers.

Common EMS Worker Injuries

Back and shoulder injuries are the most common ones sustained by EMS workers. Study has shown that almost 50% of EMS workers will have some type of back or shoulder injury during their EMS career. One in four EMS workers will have a career-ending back injury during the first four years of employment, most often caused by lifting. Obesity, which is up to 27% of the general population, also contributes to EMS worker injury. One research group discovered that back injuries among EMS workers often result from a reinjury rather than a sudden injury. Many reinjuries occur from cumulative wear and tear over time.

Recently, strides have been made in technology and equipment design to reduce back and shoulder injuries. Many of the newer ambulance cots are battery or air powered, thereby reducing the need to lift patients manually into the ambulance. Stair chairs and power-assisted cots have also helped reduce injury, although EMS workers must still lift the patient onto the stair chair or cot. It would be highly beneficial for fire departments and ambulance companies to endorse proper lifting techniques and exercise within their ranks; however, even this would provide only a limited solution. Without a national standard, a program that establishes, tests, and maintains fitness levels for EMS and public safety workers has uneven support. So it is up to the EMS and public safety workers to take care of their own health and fitness by making use of valuable downtime to exercise and stay fit.

One of the best solutions is to engineer the things that must be lifted and the hazards of lifting out of the job of the EMS worker. Texas was the first state to pass a law protecting health care workers (TX SB 1525, also known as the Lifting Law for Hospitals and Nursing Homes); it states, "the governing body of a hospital or the quality assurance committee of a nursing home shall adopt and ensure implementation of a policy to identify, assess, and develop strategies to control risk of injury to patient and nurses associated with the lifting, transferring, repositioning or movement of a patient." It is hoped that this type of law will also be passed in other states.

CHAPTER **REVIEW**

Summary

Emergency medical services (EMS) responders encounter many serious hazards in their jobs, placing them at high risk of occupational injury or death. Responding to the scene is a leading hazard, and other common hazards are assaults, infectious disease, hearing loss, lower back injury, hazardous materials exposure, stress, extended work hours, and exposure to temperature extremes.

The work of an EMS worker is physically and emotionally demanding. EMS responders are forced to make critical decisions in a very short time frame, leading to a great deal of individual stress. Interaction and interpersonal relations with victims, patients, bystanders, and family members also contribute to stress for the responder.

Whereas the chances of responding to a terrorist incident are relatively low, assault incidents to EMS responders are increasing. The risk of assault for EMS workers is approximately 30 times the national average, and the risk of death is approximately three times the national average. Unfortunately, there is insufficient training available to identify potentially dangerous situations, as well as a reluctance to provide the necessary training. Training in the area of infection control must become a large part of the continuing education process for response personnel. Training also plays a large role in safer operations at all incidents.

Healthy physical fitness, weight, diet, exercise, personal habits, and lifestyle choices enable the responder to work safer and help to reduce the chances of injury. Workers who are not in good shape increase their chances of injury. Physical injury as well as injuries from climatic conditions can easily occur. Physical contact with patients and victims places the responder at risk of both injury and disease. Infection control is a leading concern for personnel.

EMS incidents must be managed with an incident management system (IMS). Just as with other emergency operations, IMS is used in some fashion for all EMS incidents. During situations and incidents that require only a response from EMS, the IMS can be tailored to fit the need for the command staff and general staff. On a smaller EMS incident, the senior EMS crew member must assume the roles of both incident commander and safety officer. EMS personnel must adhere to an accountability system just as fire personnel do at fire-related incidents. Many fire-related and EMS events have the same hazards and requirements, and can be made safer by following SOPs, policies, and IMS, and wearing the proper PPE.

Recently, it has become common for EMS organizations and fire departments to provide training to their employees in the area of personal defensive tactics and measures. Technology has afforded better protective equipment and gear for the EMS worker. Workers are now provided state-of-the-art equipment, which guards against infectious disease and blood-borne pathogens. Technology also has afforded equipment as a way to enhance injury prevention. Power-assisted ambulance stretchers are available to help the EMS worker avoid back injury. Orthopedic braces are available as part of the EMS personal equipment, and area ergonomics training is offered. Needleless systems are becoming more common, as well as auto-guard intravenous catheters, which allow the needle to retract after the intravenous catheter has been placed into the patient. Emotional intelligence training is also now accessible; it teaches workers how to control emotions and recognize emotions that trigger negative responses. NIOSH is currently working to reduce injuries and deaths of EMS workers. As we advance into the 21st century, EMS organizations will continue to be proactive in the field of safety and health of their employees.

Review Questions

1. List the common hazards associated with EMS response.
2. List the reasons why family members and bystanders could become a problem at an EMS scene.
3. Explain why EMS responders should be trained at the operations level of hazardous materials training.

4. Stress management for EMS personnel should include:
   a. Proper diet and exercise
   b. Policies and procedures that support personnel
   c. A wellness and fitness program
   d. All of the above
5. The placement of an emergency vehicle at an accident scene should be out of the flow of traffic and require personnel to cross the roadway to access the scene.
   a. True
   b. False
6. Explain why complacency is a serious problem in the emergency response field.
7. Explain why law enforcement support is necessary at incidents involving violence.
8. Infection control for EMS workers at incidents is required by:
   a. Federal regulations
   b. The NFPA standards
   c. The incident management system
   d. Both a and b
9. Explain the main difference between a firefighter's PPE and that of EMS personnel.
10. Explain how a responder may become infected through indirect contact.

Case Study

A 59-year-old male volunteer firefighter (the victim) was fatally injured when a tractor trailer struck his parked privately owned vehicle (POV). The victim had responded to a weather-related single motor vehicle incident on an interstate highway. The vehicle was traveling eastbound when the driver lost control, drove through the median into the westbound lanes, and rolled over onto the north shoulder of the westbound lanes. Upon his arrival to the scene, the incident commander (IC) advised the victim to position his pickup truck upstream to warn oncoming traffic of the vehicle incident in the curve. He positioned himself upstream on the north shoulder of the westbound lanes and turned on his emergency flashers and rooftop light bar. The oncoming tractor, pulling two trailers, lost control when changing lanes, causing the rear trailer to swing counterclockwise. The operator swerved several times before the rear trailer struck the victim's pickup truck positioned on the north shoulder. The victim was not ejected from the vehicle and was found lying on the rear set of seats without his seat belt on. He was pronounced dead at the scene. Key contributing factors identified in this investigation include hazardous road conditions, the speed of the tractor trailer, and nonuse of a seat belt by the firefighter.

The victim had served 15 years as a volunteer with this fire department. The victim had completed online training for the incident command system (ICS) at levels 100, 200, 700, and 800. The victim had also received highway incident safety training from the department's insurance carrier in 2007. He was designated as the department's safety officer.

The assistant fire chief of the victim's department established himself as the IC during the incident. He has been a volunteer with this department for five years and received the same training as the victim.

The fire department has written guidelines regarding the use of crash scene clothing, personal protective equipment (PPE), and proper vehicle blocking. These guidelines focus on when to wear crash scene reflective gear, potential hazards encountered, and how to establish a safety zone and vehicle block for oncoming traffic. The victim was parked in his silver 2008 4 × 4 dual-cab, one-ton pickup truck. The victim had purchased and placed optical warning devices on this vehicle, which included strobe lights in the headlight and taillight assemblies; alternating light-emitting-diode (LED) red and white grill lights (two of each); and a rooftop light bar consisting of alternating LED red, white, and amber lights. All of these optical warning devices, including the vehicle's emergency flashers, were active during the incident. The victim's vehicle had also been used by the victim while working with a local wrecker service.

The temperature was 5°F, with steady snow and visibility in excess of half a mile. According to the highway patrol, though a layer of ice covered the highway, it was not slick due to snow and sand mixed over this layer. This four-lane interstate highway traveled west and east through varying elevations and mountain canyons. The state highway patrol crash investigator estimated the tractor trailer speed to be too fast for the highway conditions. According to law enforcement and the fire chief, the operator of the tractor trailer stated that he had been able to see the victim's vehicle.

Contributing Factors

Occupational injuries and fatalities are often the result of one or more contributing factors or key events in a larger sequence of events that ultimately result in the injury or fatality. NIOSH investigators identified the following key contributing factors in this incident that ultimately led to the fatality:

- Hazardous road conditions
- The speed of the tractor trailer being too fast for road conditions
- The seat belt not being used to restrain the victim in his seated position

According to the coroner, the cause of death for the victim was multiple blunt force traumas.

NIOSH had the following recommendations:

Recommendation #1: Companies using tractor trailers should ensure that operators drive in a manner compatible with weather conditions. According to the Federal Motor Carrier Safety Administration's (FMCSA's) Large Truck Crash Causation Study, "traveling too fast for conditions" was the single most frequently cited factor in large truck crashes. Stopping distances can be affected by the weight of the tractor and trailer(s), road surface, and terrain.

Recommendation #2: Fire department and fire service consensus committees should consider reevaluating current standards on seat-belt use to include their use while vehicles are parked and occupied at highway incidents. NFPA 1500 states,

> all persons riding in fire apparatus shall be seated and belted securely by seat belts in approved riding positions at any time the vehicle is in motion; where members are authorized to respond to incidents or to fire stations in private vehicles, the fire department shall establish specific rules, regulations, and procedures relating to the operation of private vehicles in an emergency mode; these rules and regulations shall be at least equal to the provisions regulating the operation of fire department vehicles.

The fire department had a standard operating procedure (SOP) that required the use of seat belts by fire department members. However, NFPA 1500 and the fire department SOP do not address the need for firefighters to be properly restrained in a vehicle when parked and performing such tasks as staging, blocking, or warning traffic at highway incidents. Seat belts are the single most effective means of reducing deaths and serious injuries in traffic crashes.

Recommendation #3: Fire departments should reevaluate current policies and procedures to ensure that temporary traffic control devices are available and deployed upstream of warning vehicles. The *Manual on Uniform Traffic Control Devices* (MUTCD) defines a temporary traffic control (TTC) zone as: "an area of a highway where road user conditions are changed because of a work zone or an incident through the use of TTC devices, uniformed law enforcement officers, or other authorized personnel." During this incident, cones and an apparatus block had been established for an incident on the right shoulder in a curve. Emergency signs had not been placed. The incident was minor and in the wrap-up stages. The victim would usually respond to the scene of highway incidents to assist in the setup of the temporary traffic control. When not needed at the scene, the victim would position his vehicle upstream from the incident to warn oncoming traffic of the downstream traffic control zone. This was very important during this incident because a motorist could not see the incident on the straightaway before the curve. The temporary traffic control zone was established 100 feet before the incident, and the victim positioned his vehicle 650 feet upstream from the first cone placed for the temporary traffic control zone. While operating in or around

Fatal Firefighter Vehicle Accident Concerns and Issues	Yes	No
Should the firefighter (victim) have remained near his vehicle instead of inside?		
Would seat-belt use have made a difference in this outcome?		
Was the incident safety training provided by the department adequate?		
Would a department policy regarding private vehicle use at an incident have made a difference in the outcome?		
Did the incident commander continuously evaluate the conditions?		
Would a traffic warning sign placed upstream have made a difference?		
Did the lack of recognition of the conditions on the part of the victim play a role at this event?		
Did a lack of recognition by the incident commander play a role in this incident?		
Would adequate firefighter survival training have played a role at this incident?		

FIGURE 7–11 Concerns and issues of a fatal firefighter vehicle accident—Montana.

moving traffic, one should never trust oncoming traffic or turn one's back to it. The victim did not exit the vehicle, leaving his back turned to oncoming traffic. The victim also did not have emergency road signs, cones, or flagging equipment in his vehicle. Even though the victim did not use these temporary traffic control devices, the operator of the tractor trailer stated that the victim's vehicle and optical emergency warning lights were visible while parked on the right shoulder. During the process of the tractor trailer moving over, the chain of events began, ultimately taking the life of the victim.[7] See Figure 7–11.

1. Explain why the victim had not followed safety guidelines and department policy or procedures regarding highway safety at this incident.
2. What alternative measures could the victim have taken to avoid placing himself in a vulnerable position?

References

1. Occupational Safety and Health Administration, *Best Practices for Protecting EMS Responders*, OSHA 3370-11, 2009. Retrieved January 30, 2010, from http://www.osha.gov/Publications/OSHA3370-protecting-EMS-respondersSM.pdf
2. *Defensive Tactics for Emergency Medical Services*, 2009. Retrieved April 7, 2009, from http://www.dt4ems.net/files/DT4EMS_training.pdf; NFPA 1581, *Standard on Fire Department Infection Control Program*, section 4.2.2, 2010. Retrieved April 10, 2010, from http://www.nfpa.org/aboutthecodes/AboutTheCodes.asp?DocNum=1581
3. U.S. Department of Labor, Occupational Safety and Health Administration, *Occupational Noise Exposure 1910.95*. Retrieved May 21, 2011, from http://www.osha.gov/pls/oshaweb/owadisp.show_document?p_table=standards&p_id=9735
4. Centers for Disease Control and Prevention, *NIOSH Traumatic Incident Stress: Information for Emergency Response Workers*, 2002. Retrieved April 8, 2009, from http://www.cdc.gov/niosh/docs/2002-107/pdfs/2002-107.pdf
5. U.S. Department of Transportation, Federal Highway Administration, *Manual on Uniform Traffic Control Devices* (MUTCD), 2003, p. 6A-1. Retrieved April 8, 2009, from http://mutcd.fhwa.dot.gov/HTM/2003r1/part6/part6a.htm#section6A01
6. R. O'Brien, M. Denton, and P. Kramm, "Firefighter Basics for Infection Control," *Fire Engineering Magazine*, January 1, 2003. Retrieved April 10, 2009, from http://www.fireengineering.com/index/articles/display/168764/articles/fireems/volume-1/issue-1/features/firefighter-basics-for-infection-control.html
7. NIOSH Fire Fighter Fatality Investigation and Prevention Program, *Volunteer Fire Fighter Sitting in His Parked Vehicle Warning Oncoming Traffic of a Motor Vehicle Incident Was Struck and Killed by a Tractor-Trailer—Montana*, April 9, 2009. Retrieved April 16, 2009, from http://www.cdc.gov/niosh/fire/reports/face200903.html

CHAPTER 8 Safety at Specific Types of Incidents

KEY TERMS

crew resource management (CRM), *p. 195*
Emergency Response Guidebook (ERG), *p. 179*
hot zone, *p. 178*
rapid intervention team (RIT), *p. 177*

OBJECTIVES

After reading this chapter, you should be able to:

- Discuss the safety issues related to hazardous materials response.
- Discuss the areas the written hazardous materials emergency response plan should address.
- Describe the safety issues related to technical rescue operations.
- Describe the safety issues related to water rescue/public safety dive operations.
- Describe the safety issues related to ground support during helicopter operations.
- List safety issues concerning operations during civil disturbances.
- List the basic strategic goals to accomplish at a terrorist event.
- List specific concerns when responding to natural disasters.
- During rapid intervention team (RIT) operations, list the considerations when a Mayday is transmitted.
- Describe how crew resource management is an effective use of all resources.

Resource Central

For additional review and practice tests, visit **www.bradybooks.com** and click on Resource Central to access book-specific resources for this text! To access Resource Central, follow directions on the Student Access Card provided with this text. If there is no card, go to **www.bradybooks.com** and follow the Resource Central link to Buy Access from there.

Safety during special operations includes responding to these types of events: hazardous materials, technical rescue, water rescue, helicopter landing zone operations, civil disturbances, terrorism events, natural disasters, and **rapid intervention team (RIT)** events. Some of these specialized operations are discussed in other chapters because few departments provide services in all of them; however, individual fire departments still may respond to a specialized incident infrequently.

rapid intervention team (RIT) ▪ A standby team of equipped and trained personnel is ready to respond and assist emergency workers who become disoriented or trapped.

Safety plays an important role in these specialized operations whether a fire department provides the specialized service or not. Local fire departments dictate the safety measures, policies, and SOPs that will be in place should a specialized incident occur. This chapter discusses each type of specialized incident and the safety concerns it presents.

Hazardous Materials Incidents

As has been mentioned earlier in this book, every fire incident is basically a hazardous materials event: from the smoke produced from the burning products of combustion to the large-scale fire-related or non-fire-related incident involving chemicals. All fires are hazardous to the responders; however, many fire departments view the basic fire response as simply an everyday event.

What differs significantly for the hazardous materials incident is the level of response. Some departments have full hazardous materials teams, whereas others provide a first responder level of service in which responders are trained at the hazardous materials awareness level. Still other departments take action with members trained at the awareness level who provide services at only the defensive level; they must rely on the response of a regional hazardous materials team. NFPA 472, *Standard for Competence of Responders to Hazardous Materials/Weapons of Mass Destruction Incidents*, now requires emergency responders to be trained at the operations level—a level higher than the previous awareness level. Departments that provide trained members frequently respond with other departments, so together they form a hazardous materials response regional team. The most vital aspect of hazardous materials response is that a group of trained members eventually respond in some offensive fashion to mitigate the incident.

#5

OSHA 1910.120 and Environmental Protection Agency (EPA) regulation 311 govern the response to hazardous materials incidents. Several NFPA standards and recommended practices address response to hazardous materials incidents: NFPA 472, *Standard for Competence of Responders to Hazardous Materials/Weapons of Mass Destruction Incidents*, deals with the training and competency levels of responders in all emergency organizations that respond to the hazardous materials incident, including law enforcement. NFPA 473, *Standard for Competencies for EMS Personnel Responding to Hazardous Materials/Weapons of Mass Destruction Incidents*, pertains to EMS personnel responding to a hazardous materials incident. This standard mandates that all emergency medical responders be trained to at least the core competencies of NFPA 472 (operations level).

In order to accomplish this level of training, EMS organizations must have access to and receive previously unavailable federal, state, and local funding, enabling organizations to support their equipment and training needs. Yet many agencies are able to accomplish the training and appropriate the necessary equipment without any funding from outside sources. However the particular department accomplishes the necessary funding, it must fulfill the recommendations that require a certain set of safety measures be in place before hazardous materials operations can even begin. These requirements were described earlier in Chapter 7.

The department health and safety program manager must determine the level of response for his or her individual department as it relates to the level of training of the employees or members and a written emergency response plan.

#6

In addition to a written emergency response plan, OSHA requires departments to provide a medical surveillance program to emergency response personnel. Personnel responding to emergency releases of a hazardous material must have an initial baseline physical, an annual physical, and an additional physical at employment termination. The written emergency response plan addresses the following areas:

- Pre-emergency planning and coordination with outside responding agencies
- Personnel roles
- Lines of authority and lines of communication
- Emergency recognition and prevention (what constitutes an emergency and how to prevent its occurrence)
- Safe distances and places of refuge
- Site security and control
- Evacuation routes and procedures
- Decontamination procedures
- Emergency medical treatment and first aid
- Emergency alerts and response
- Personal protective equipment and emergency equipment
- Engineering controls
- Air monitoring
- Critique of response procedures and follow-up[1]

#8

Under 29 CFR 1910.120, the Hazardous Waste Operations and Emergency Response (HAZWOPER) Standard, any employee who responds to an emergency must have pertinent training. Section Q of the standard breaks the training down into five levels tied to specific duties related to hazardous materials response. See Box 8–1.

#3

The correct response to a hazardous materials emergency is critical to both emergency responder and public safety. The department health and safety program manager must ensure not only that adequate policies and procedures are in place so personnel can operate within their level of training but also that the proper level of equipment is available. The department health and safety plan must also clearly address personnel roles, lines of authority, and communication as many risk factors are involved during response and mitigation.

hot zone ■ This area at a hazardous materials incident is where it is immediately dangerous to life or health of responders and the general public.

The OSHA regulation requires fire departments to have an incident command system in place during an incident and a backup team of hazardous materials personnel during entry into the hazardous zone, commonly referred to as the **hot zone**. Hazardous materials response has many safety issues, and the level of response and service varies depending upon the type of incident.

Many of the issues regarding hazardous materials response and offensive mitigation apply to departments that provide a specialized hazardous materials response and properly trained personnel, although the NFPA now requires that all emergency responders be trained to the operations level of response. Many of the issues and safety practices should be addressed within individual department policies and SOPs. Departments that

BOX 8-1: THE FIVE LEVELS OF TRAINING FOR HAZMAT RESPONSE

Level I First Responder Awareness. This level trains employees to be aware of any release of hazardous substances and to alert the response team. In most cases, law enforcement and EMS services are in this category. Responders are expected to recognize the presence of hazardous materials and take no further actions except to start the additional response of properly trained personnel. This includes observation, reporting, and evacuation training. Between four and eight hours of training are acceptable at this level.

Level II First Responder Operations. This defensive training applies to employees who are not authorized to stop a release but operate in a defensive manner. This level trains them to contain a release, slow the spread of hazardous material, and prevent exposure. A minimum of eight hours of training is required. Level II responders must know everything that Level I personnel know and may be required to take the complete 24-hour HAZWOPER program. Additionally, they must know how to select and use personal protective equipment, how to confine and control a simple spill, and basic decontamination procedures.

Level III Hazardous Materials Technician. This level teaches employees how to stop the release of hazardous material by patching, plugging, or repairing the vessel or container that is leaking. Training must be at least 24 hours in length. Responders must demonstrate competency and ability in additional areas. In addition to covering the same topics as Level II, the hazardous materials technicians must be trained to (1) implement the company's emergency response plan; (2) identify specific substances through the use of special instruments; (3) perform advanced containment operations; and (3) be able to identify personnel who exhibit exposure symptoms. This training level often includes at least one day of field experience.

Level IV Hazardous Materials Specialist. This specialist assists the technician in containing the spill and provides expertise in hazardous substances to be contained. The specialist also acts as the on-site liaison with government authorities. At this level, OSHA requires at least 24 hours of training. However, it is not uncommon for employees to receive 40 hours of instruction. Instruction for the hazardous materials specialist begins with Level II and III training. Specialists are trained to implement the company's emergency response plan, as well as state and local plans. Specialists must have an in-depth knowledge of the hazardous materials on site, hazard and risk assessment techniques, and hazardous materials disposal.

Level V On-Site Incident Commander. This person is in charge of the entire response, cleanup, and disposal operation, and OSHA requires a minimum of 24 hours of training. Many employers provide up to 40 hours of training. Training covers the following topics: (1) the company's incident command system (ICS); (2) the emergency response plan; (3) local, state, and federal emergency response plans; (4) personal protective equipment; and (5) decontamination of responders and equipment.

Note: At each level, training must be certified and documented. Employees must demonstrate proficiency during each annual refresher training. If an emergency response team is obligated, under a mutual-aid agreement, to respond to an off-site incident, the 24-hour emergency training and response procedures are valid during the emergency period only (i.e., rescue, containment and control, etc.). However, if an emergency response team is engaged in the cleanup of a hazardous waste site, training must comply with all regulations covering hazardous waste site remediation (29 CFR 1910.120(a)(l)(i)) and the full 40-hour training is required.

do not have a trained team of responders should ensure they have a policy or SOP in place that offers general guidelines for personnel trained at the minimal level of operations. Organizations that have not made certain their personnel are trained and equipped to the operations level of response should make this a top priority.

One area that is common to all law enforcement, fire, and EMS departments is initial response and arrival. All emergency response personnel should at least be trained to the hazardous materials operations level. Personnel should also be supplied with a copy of the Department of Transportation's **Emergency Response Guidebook (ERG)** on each emergency response vehicle to ensure responders have the resources to assess the incident and begin the proper response sequence. They must be able to recognize that they have responded to a hazardous materials incident.

Emergency Response Guidebook (ERG) ■ This guide was developed for use by firefighters, police, and other emergency services personnel who may be the first to arrive at the scene of a transportation incident involving dangerous goods. It is primarily a guide to aid first responders in quickly identifying the specific or generic hazards of the material(s) involved in the incident, and protecting themselves and the general public during the initial response phase of the incident.

Beyond the recognition phase, personnel must decide whether they have the capability, training, and equipment to handle the incident, or whether additional resources are needed. The awareness of safety at a potential hazardous materials incident is critical for all responders as well as the general public. Responders need to recognize their level of response, follow all safety guidelines and procedures, and be trained to perform an initial size-up of a possible hazardous materials incident as they would on any incident. The size-up phase should include the staging of units and personnel at a safe distance upwind and uphill of the incident. Along with the ERG, units should also be supplied with a pair of binoculars so that additional information can be gathered from a distance. Using binoculars increases the safety factor and enables initial responders to offer pertinent information to any additional specialized personnel who are responding. See Figures 8–1 and 8–2 and Box 8–2.

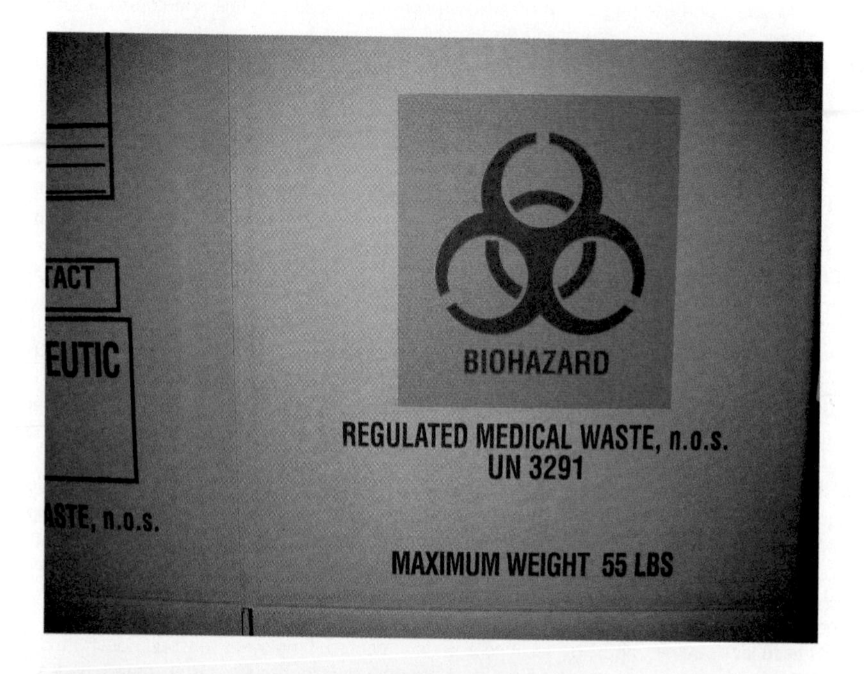

FIGURE 8–1 Labels and placards help responders identify the type of hazardous material. *Courtesy of Joe Bruni.*

FIGURE 8–2 The ERG is a valuable resource for initial responders to hazardous materials incidents. *Courtesy of Joe Bruni.*

BOX 8-2: POTENTIAL INITIAL AND PERTINENT HAZARDOUS MATERIALS INFORMATION GATHERED BY INITIAL RESPONDERS

Occupancy type and location. Occupancy refers to the purpose of a structure or location. Knowledge of the occupancy/location can provide vital information about the *potential* for the presence of hazardous materials. Location is the specific geographic area, address, and installation of the incident.

Relay wind direction and slope of the terrain. This information will enable additional responders to ensure they approach and stage their units in a location out of harm's way to begin mitigation and possible evacuation efforts.

Identify the type of products involved. The type of products involved can be identified through placards and labels, shipping papers, product names, and material safety data sheets (MSDS) for fixed facilities. Fixed facilities that store hazardous materials are required to maintain MSDS, which will provide product identification information, hazards, and procedures during emergencies. The form of the chemical or product is also important information. Whether the chemical is a liquid, powder, or gas will aid in the research process. Fixed facilities may also have the NFPA 704 placard in place on the building itself. The placard is a diamond divided into four color-coded areas for the hazard present and assigned a numerical rating from 0 to 4, with 0 representing no hazard and 4 indicating an extreme hazard. Red color-coding represents flammability, blue for health, yellow for stability, and white for special information. Employees or occupants at the site may also provide valuable information and clues concerning product information. Employees and occupants may not reveal critical information for fear of law violations. Employee or occupant information should be verified and additional research conducted.

Type and condition of container involved. Various types of containers are used to transport, handle, and store specific materials, both hazardous and nonhazardous. A container's appearance tells a lot about the product it holds and the related hazards. Information may include the container's shape, size, and composition; transport vehicle type; texture of the exterior surface; and visibility of related valving or piping. These clues can give some indication of the hazard class, type of product, level of pressurization, and amount of product. It is important to note the condition of the container. Damage to the container can sometimes be viewed from a distance, and there may be clues that identify whether the container is leaking or not.

Identify anticipated rescue or evacuation issues. Rescue and evacuation activities may already be in progress when the initial responders first arrive on scene. Initial responders should not become involved in any situation that is beyond their capability and training. Wait for specialized equipment and responders to arrive to begin rescue and evacuation in situations where the risk versus benefit outweighs the safety to initial responders. Do not add to or become part of the problem. Establish the proper control zones and summon the correct additional assistance.

Use the Department of Transportation* Emergency Response Guidebook *(ERG). The ERG enables the user to possibly identify the hazardous material, establishes initial isolation distances, and guides the initial actions of personnel. The ERG also contains a placard and label section, as well as a container type section. Important phone numbers are also within the ERG concerning chemical information and specific agencies such as CHEMTREC.

Specialized Technical Rescue

Technical rescue is the application of specialized knowledge, skills, and equipment to safely resolve unique or complex rescue situations. Technical rescue operations normally involve an extremely high risk for the rescuer with the need to rapidly stabilize the situation to provide for rescue and, at times, recovery efforts. Personnel in these operations must understand that in certain situations there may be little probability of victim survivability, and incident commanders must not unnecessarily jeopardize the health and safety of personnel. NFPA 1670, *Standard on Operations and Training for Technical Search and Rescue Incidents*, identifies levels of functional capability for conducting operations at technical search and rescue incidents while minimizing threats to rescuers. NFPA 1006, *Standard for Technical Rescuer Professional Qualifications*, addresses the qualifications needed to be a technical rescuer. The disciplines involved in technical rescue follow:

- Rope rescue
- Structural collapse
- Trench collapse and extrication
- Wilderness search and rescue
- Water rescue
- Confined space rescue
- Vehicle and machinery rescue

Much the same as with hazardous materials, technical rescue involves a group of highly trained responders and individuals with specialized equipment and skills. (Although water rescue is part of the technical rescue discipline, it will be discussed later in the chapter.)

The NFPA has established the three training levels for responders involved in technical rescue. Because such operations present a high risk to responders and victims, safety issues become a top priority. See Box 8–3.

The fire department administration must determine the level of response and training for its individual department as required by NFPA 1670 and NFPA 1006. Much the same as with hazardous materials response, the fire chief, department director, and health and safety program manager establish the level of response as it relates to the level of training of the employees or members. A department that decides to involve its members in technical rescue training and response must also satisfy a number of general requirements (see Figure 8–3):

- An advanced life support component
- Training in basic life support including cervical/spinal immobilization and vertical and horizontal packaging
- Crush injury syndrome including recognition, evaluation, and treatment

BOX 8-3: THE LEVELS OF FUNCTIONAL CAPABILITY AND TRAINING OF TECHNICAL RESCUE PERSONNEL

Awareness level. This represents the minimum level of capability of responders. As part of their normal job functions and duties, initial responders may find themselves first on scene of a technical rescue incident. This level may involve rescue, search, and recovery operations. The awareness level is the level of training that all fire departments should receive as it enables initial responders to recognize the type of incident and establishes the activation of appropriate resources. The awareness level provides initial first responders with general knowledge to work on the scene of technical rescue incidents during the initial stages and the risks associated with each type of incident.

Operations level. This represents the capability of organizations to respond to technical search and rescue incidents and to identify hazards, use equipment, and apply limited techniques specified in this standard to support and participate in technical search and rescue incidents. Personnel are involved in search, rescue, and recovery operations, but usually operations are carried out under the supervision of technician-level personnel.

Technician level. This represents the capability of organizations to respond to technical search and rescue incidents and to identify hazards, use equipment, and apply advanced techniques specified in this standard necessary to coordinate, perform, and supervise technical search and rescue incidents.

Note: Levels sometimes overlap to build a specific discipline.

FIGURE 8–3 Technical rescue involves specialized equipment and highly trained responders. *Courtesy of Lt. Joel Granata.*

Much the same as with response to fires, technical rescue response as well as its safety issues varies with each incident. There are the threats of drowning at swift-water rescue incidents, secondary collapse, and hazards encountered during high-angle rope rescues. Every technical rescue incident presents high-risk hazards that must be evaluated for risks versus benefits and safety to responders. Technical rescue has recently advanced from what was a narrow field of vehicle extrication, high-angle rescue, and water rescue operations to a become broader discipline, which is bound to continue.

Through the innovative efforts of dedicated and proactive members, the field of technical rescue has become a highly skilled and equipped group of individuals who respond to and train for one of the most dangerous fields in the fire and EMS services.

WATER RESCUE/PUBLIC SAFETY DIVE OPERATIONS

Although water rescue and public safety dive operations fall under the category of technical rescue, certain departments choose to make this a separate operation involving a specialized team of highly trained and equipped responders. The four separate water-related disciplines of water rescue are (1) dive, (2) ice, (3) surf, and (4) swift water. For the purposes of this textbook, water rescue and public safety dive operations will not include water rescue situations from swimming pools.

Much the same as with hazardous materials and technical response, the department administration must determine whether its individual department will become involved in water rescue/dive operations. In most cases, the number of incidents occurring in a specific community determines this decision. If the department administration determines it will conduct water rescue/dive operations, the health and safety program manager must ensure that adequate policies and procedures are in place so that personnel operate within their level of training and that the proper level of equipment is available. The area of water rescue/dive operations falls under the NFPA 1006 standard, and three levels of training are required of responders to these types of incidents. See Box 8–4.

Many communities have a need for a water rescue component in addition to basic fire response and rescue operations. The purpose of such operations is to rescue victims who are unable to self-rescue from water emergencies. The training of responders in water

BOX 8-4: THE TRAINING LEVELS OF WATER RESCUE RESPONSE

Awareness level. This level of training enables responders to recognize existing and predictable hazards. As part of their everyday duties, responders may find themselves as initial responders to an incident involving water rescue or public safety dive operations. The awareness level involves basic water safety and rescue training. Departments that are mandated to respond to water rescue emergencies should be trained and equipped to this minimum level. The awareness level provides initial first responders with general knowledge to work on the scene of water rescue incidents during the initial stages and the risks associated with each type of incident. The focus of the awareness level trained responder should be site control and scene management. Responders at the awareness level should be trained at identifying the resources necessary to conduct safe and effective water operations. The awareness level responder must also be able to determine rescue versus body recovery.

Operations level. Departments conducting water rescue operations in which rescuers enter the water for hands-on or swimming rescues, or use boats, as part of the rescue operation should have individuals qualified to the operations level. There are four separate water-related disciplines for the operations level: (1) dive, (2) ice, (3) surf, and (4) swift water. Departments functioning at the operations level of one or more specific disciplines must meet the requirements of the NFPA 1670 standard. Personnel operating in the hazard zone of a water rescue incident must also meet the minimum PPE requirements. Organizations operating at the operations level at water search and rescue incidents must develop and implement procedures for operations at a water rescue incident.

Technician level. Departments conducting water rescue operations at the technician level must meet the requirements of the specific disciplines at the operations level. Recognizing the unique hazards associated with the specific discipline is required at the technician level. Technicians must be trained to meet the requirements of the NFPA 1670 standard within the specific discipline the technician will operate in.

rescue is often at the operations level; however, responders involved with a specialized team are normally certified at the higher technician level.

Water rescue technicians frequently become certified as boat operators in order to maintain their operational readiness for any rescue operation. One of the most versatile tools in the water rescue technician's cache is the boat. Boat operators are trained to use the boat's capabilities to assist persons in need of rescue in remote areas or open water. Boats extend the rescuers' operational periods by limiting their exposure to cold or rough water. Water rescue involving boat operations should be conducted with the least amount of risk to the responder necessary to rescue the victim. Responding personnel and the incident commander must establish the risks versus benefits with the rescue operation.

#4

Similar to other rescue operations, water rescue often takes place during extreme weather and water conditions. Responders must have the necessary PPE in place, including personal flotation devices, during such operations.

Water rescue responders must be supported at all times during the operation, including the assignment of a safety officer trained to the technician level. If at all possible, the hazards in the water rescue area should be secured, but if they cannot be, command must monitor the hazards at all times. An alternative action plan should also be developed should planned operations not go as planned, and communicated to all personnel operating in the rescue area.

Public safety dive teams endure some of the most demanding work conditions of any special operations group. In many cases, divers are forced to operate in low to zero visibility conditions in contaminated water or under ice, searching for not only victims but also evidence related to crime. Departments that involve responders in this type of special operation must recognize it is well beyond the capability of those with recreational SCUBA diver training.

The underwater demands of public safety diving place responders in situations that at times are hazardous and dangerous. In most cases, incidents require search and recovery and special operations underwater. Personnel trained as divers should be certified by a nationally recognized agency that specializes in public safety dive operations. All members involved at the technician level of dive operations of a technician level organization should have an annual fitness test, watermanship skills test, and basic SCUBA skills evaluation supplied by an organization such as the International Association of Dive Rescue Specialists (IADRS). It is vital to maintain public safety diver capability. The department must also develop and implement procedures related to the following:

#11

- Equipment
- Various types of underwater environments
- Supervision of dive operations
- Hand and rope signals
- Search techniques
- Low-air and out-of-air emergencies
- Operations utilizing electronic communications
- Utilizing full-body encapsulation
- Equipment used in rescuing an entangled diver
- Medical monitoring of divers recovering evidence
- Protocols
- Documenting the scene

One individual on scene must be a rescue diver trained in first aid, cardiopulmonary resuscitation, and hyperbaric recognition. All divers, dive operations, and responders must comply with the current NFPA 1670 and 1006 standards. Similar to technical rescue operations, dive operations require a pre-dive safety briefing.

The department health and safety program manager should work closely with members of a public safety dive team to develop policies and SOPs that minimize the risk involved and increase the safety of all members. Because public safety diving and surface water rescue activities are by nature very dangerous, the importance of appropriate training, equipment, and experience cannot be stressed strongly enough.

Swift-Water Rescue

Sometimes referred to as whitewater rescue, swift-water rescue is a subset of technical rescue. Specially trained personnel utilize ropes and mechanical advantage systems as part of the operation to remove individuals who have been stranded or trapped in floods or swift water. In most cases, the ropes and equipment used during swift-water operations are heavy duty in design due to the added weight of moving water. There is no easy way to overcome the power of moving water, but the goal is to utilize or deflect the moving water. Much the same as with a hazardous materials incident, hot, warm, and cold zones are instituted at the incident scene, and the incident management system as well as a safety officer trained in this area should be established.

Swift-water and surface water operations are outlined in the 2008 edition of the NFPA 1006 standard. Two separate chapters of this standard define the required knowledge, skills, and abilities for surface water rescue (Chapter 11) and swift-water rescue (Chapter 12). The three terms *awareness*, *operational*, and *technician* have been replaced by two levels of qualification: Level I and Level II. A Level I technical rescuer applies to individuals who identify hazards, use equipment, and apply limited techniques specified in this standard to technical rescue operations. A Level II technical rescuer applies to individuals who identify hazards, use equipment, and apply advanced techniques specified in this standard to technical rescue operations.

The safety of the rescuer and the victim is of prime concern during swift-water and surface water operations. Special operations personnel trained in swift-water operations use a low- to high-risk algorithm for the implementation of various rescue methods. Its purpose is to keep safety in mind at all times to prevent a rescuer from endangering himself and the victim. The algorithm provides a sound, step-by-step approach when effecting a rescue. As the algorithm progresses to the higher end, the safety of the rescuer becomes more compromised. The algorithm is "talk, reach, throw, row, go and tow, and hello." Obviously, "talking" a victim into performing a self-rescue keeps special operations personnel at the safest possible level.

The threat to personnel rises rapidly once they enter the water. The greatest amount of rescues are performed utilizing a "throw bag" containing rope to throw to the victim. Emergency services personnel inadequately trained in this type of special operation should not attempt to enter the water to effect a rescue because they could drown. All personnel working near the water during this type of operation should be equipped with a personal flotation device.

All rescue operations involving swift water and surface water demand vigilance in regard to safety. IMS indicates a trained safety officer should be present to monitor and address all safety issues. If personnel are forced to enter the water, decontamination may become a health and safety issue, especially in contaminated floodwater. Yet decontamination should be of concern long before personnel enter contaminated water. Technical rescue teams taking on swift-water rescue must give it the same care as other technical rescue disciplines.

Surface Ice Rescue

Much the same as technical rescue involving dive operations, surface ice rescue is designed to develop the necessary knowledge and skills for operations within the public safety diving arena. Individuals who have fallen through ice into frigid water must be rescued within minutes if they are to survive. A speedy rescue and recovery effort focuses on knowledge, skills, equipment, and health and safety for emergency response personnel. Failure to train and follow proper techniques during a surface ice rescue emergency may result in the death of both the victim and the rescuer.

Safety during a surface ice rescue incident is a top priority for the incident commander and responders. The IMS system and a safety officer trained in this special operation must be in place early into the incident to ensure the health and safety of responders. In situations that involve the rescue of an animal, the incident commander who decides to go forward with the operation must be able to ensure the safety of the rescuers. Only trained personnel meeting the requirements of NFPA 1006, *Standard for Technical Rescuer Professional Qualifications*, should be involved in any cold water and surface ice rescue attempt.

Similar to a hazardous materials incident, a surface ice rescue incident requires the use of a backup team as a safeguard measure for the initial team attempting the rescue. Personnel who will be working as support members during this special operation should be equipped with personal floatation devices to ensure their safety. At no time should a rescue attempt be made without a cold water suit, backup personnel in place, and the rescuers tethered to shore with a rope system at a minimum. The number of victims, ice conditions, and access points should all be evaluated prior to attempting a contact type of rescue.

Preparations should be made to conduct a contact rescue shortly after arrival at the incident scene. While rescuers are attempting to either "talk" a victim through a self-rescue or perform a rescue using a rope throw bag, a minimum of two individuals should be preparing to perform a contact rescue. A delay in this preparation could mean certain death for the victim. It is also advisable to have a cold-water dive team en route to the location should the victim slip below the water or ice surface prior to the actual contact taking place. In addition to rescue procedures and skills, self-rescue and survival procedures in cold water must be emphasized to personnel.

The same as with any technical rescue operation, a great deal of specialized equipment is needed for a surface ice rescue operation, and involved personnel should be thoroughly trained on and familiar with its operation. Provisions also need to be made at the incident scene to ensure the medical monitoring of personnel and adequate warm sheltering. The latter can be as simple as a warm vehicle or a nearby structure that has been heated.

In today's modern emergency services organizations, it has become common to identify and address target hazards. Both swift-water and surface ice situations can be classified as target hazards. Although occurrences in this area may be infrequent, because they are considered to be low-frequency/high-risk incidents, they must be given top priority as they relate to skills, training, equipment, and preparedness.

This concept of identifying target hazards and developing preplans to manage the incident should be expanded to include not only occupancies and physical structures but also areas within a community such as lakes, rivers, streams, and other bodies of water, that pose a threat or danger to the public engaged in activities around or on the water or ice. This process of identifying physical hazards as well as activities that place persons at risk is called threat analysis. Organizations that are prepared, trained, and adequately equipped for water-based types of incidents consider their priorities and address the health and safety of their members and the general public as a top priority. Emergency response organizations should perform this threat analysis within their communities to determine potential incident sites and to develop preplans for them. Training for personnel in this specialized area of water-based incidents should include not only the principles of rescue operations and use of equipment but also the safety and survival principles for emergency responders.

Helicopter Operations

There are times when emergency responders are called upon to perform or assist with helicopter operations, such as:

- Working with a helicopter during rescue operations
- Setting up a landing zone
- Rescue hoist operations
- Short haul operations
- Swift-water rescue
- Mountain rescue operations
- Sling-load operations
- Longline operations
- Providing ground support for wildland firefighting
- Bambi-bucket operations
- Deploying responders into water

Many of these specialized operations require a great deal of training and coordination. No matter the operation being performed by the helicopter, general safety guidelines exist for working around these machines. For the purposes of this textbook, we focus on safety involving ground support operations and approaching the aircraft. Ground support operations that fire departments most commonly help with involve establishing a landing zone.

LANDING ZONE

Concerning helicopter operations, it is extremely important to select an appropriate landing zone for the pilot. Nighttime operations require a larger landing zone than do daylight operations. The size of the landing zone will largely depend on the size of the aircraft;

however, the general guidelines are 60′ × 60′ for a daytime landing zone, 100′ × 100′ for a nighttime one, and 120′ × 120′ for a military helicopter.

Once a landing zone has been selected, an individual appointed as the landing zone officer will have complete control of the landing zone, including all movements of the aircraft and ground support personnel. The landing zone officer should establish communication with the pilot of the aircraft by either radio or hand signals, and be the only person in contact with the aircraft during landing, while on the ground, and during takeoff. If radio communication is established with the aircraft, the landing zone officer utilizes a radio channel for scene operations that differs from the one used by the incident commander.

The landing zone officer has full authority to abort the landing or takeoff should it be deemed necessary. He or she controls all personnel entering the landing zone and should assume a position in the middle of the outer perimeter of the landing zone with his or her back to the wind. Because the pilot will prefer to approach and land with the aircraft coming into the wind, the aircraft will land facing the landing zone officer, ensuring eye contact between the landing zone officer and the pilot. One of the landing zone officer's most important tasks is keeping an eye on the helicopter tail rotor during landing because the pilot cannot see behind the aircraft.

Ground support personnel may resist assuming the landing zone officer position because of the numerous hand signals involved in directing a helicopter; however, the pilot in command is ultimately responsible for safety and will put the helicopter down in the area he or she deems safest. The landing zone officer communicates with the pilot of the aircraft using hand signals, known as marshaling. There are a total of 12 signals, but the landing zone officer will need to use only two important ones:

- Both arms outstretched forward and pointing to indicate the landing zone
- Crossing and uncrossing (waving) the arms above the head to wave off landing or takeoff

During nighttime conditions, landing zone markers must be lit, and generally the lights will be blue in color. In some cases, ground support personnel mark the landing zone at night utilizing vehicle headlights. Although the use of vehicle headlights is an acceptable practice, they may cause glare for the pilot at touchdown. It is a better to use blue marker lights, flares, or traffic cones. Local policies and SOPs determine how many markers to use when lighting the landing zone for the pilot. Many jurisdictions require four to five markers for the landing zone: four markers for the corners of the landing zone and the fifth to mark the center on the windward side to show the pilot wind direction at ground level. The landing zone should also be established so the pilot does not have to fly over the treatment area. Ground support personnel need to avoid dusty locations if possible. If that is not possible, ground support personnel should consider wetting down the area with a hose line before the helicopter lands.

It is the responsibility of the responders performing ground support operations to select an appropriate site as a landing zone. The slope of the land is also an important consideration for responders operating in a ground support role. A slope greater than 8 degrees is risky; a slope greater than 10 degrees should not be used. Because hotter weather will have a negative effect on lift for the aircraft, personnel on the ground during warmer weather should consider a landing zone that allows the aircraft to gain forward ground speed, thus helping lift the aircraft on takeoff.

The pilot is the best judge of the helicopter's ability to land in a given location, but personnel on the ground must identify all obstructions and hazards. The landing zone area should be free of small objects, which can be blown around by rotor wash, as well as metal objects, loose clothing, or blankets. Ground support personnel should check for overhead wires, poles, towers, and similar obstructions and report anything of concern to the pilot.

It is a good practice to have fire suppression forces and apparatus available at the landing zone in case of a landing or takeoff incident. Fire suppression forces assigned to the landing zone should park their apparatus so that it does not become part of the situation should an emergency occur. It is best to utilize the unit as a barrier between the suppression crew and the aircraft. The suppression crew should be prepared with full PPE, eye protection, and hearing protection in place, and be ready to deploy a foam-capable hose line in the event of a crash or sudden fire event.

Crowd Control

In many cases, the aircraft will land and keep the main prop and rotor running to prevent a delay in takeoff. This is one reason to maintain a safe distance between nonessential personnel and bystanders and the landing zone area: at least 100 feet from the aircraft at a minimum; however, 200 feet is a safer option. In certain situations, on-scene responders serve in watch and guard positions. Many times, the landing of a helicopter draws bystanders and observers, so it is important to establish vehicle and crowd control. Even though crowd control is generally a police department function, responders will normally become involved in it at the landing zone.

To best secure the safety of responders on scene and observers, the landing zone officer should have two assistants assigned as left and right perimeter guards. They are positioned midpoint on the respective perimeter lines after the helicopter has landed. Assistants to the landing zone officer establish a clear view of the aircraft, including the tail rotor. In most cases, the pilot assumes a position at the rear of the aircraft to ensure safe operations near the tail rotor. When the pilot does not assume this position, the landing zone officer becomes responsible for safety at the rear of the aircraft, as mentioned earlier. Contact with the pilot is extremely important, and permission should be received from him or her to approach the aircraft at all times.

Approaching the Aircraft

In most cases, responders on the ground have to approach the aircraft to load the patient or for other reasons. Responders should wait for a signal from the pilot to approach the aircraft. Ground support personnel and other responders should approach the aircraft in a safe zone in full view of the pilot—between 10 o'clock and 2 o'clock of the nose of the helicopter—and always be sure the pilot sees them and waves them forward. Responders approaching the aircraft should keep a crouching, low body profile to indicate to others on scene that they are aware of the rotor hazard. Responders should never carry anything that extends over their heads, including IV fluids. All long objects must be carried parallel to the ground. If the aircraft is on a slope, approach and depart from the downhill side in view of the pilot. Some helicopters are designed to load patients from the rear; however, the aircraft should never be approached or departed from the rear.

Responders and ground support personnel should not attempt to secure doors and hatches on the aircraft; this is the responsibility of the flight crew. A door or hatch that cannot be secured or becomes damaged will prevent the aircraft from flying.

The landing zone officer notifies the pilot when the landing zone is clear of all ground support personnel and maintains the same safety procedures for takeoff and departure as for landing. The landing zone officer should keep the landing zone clear for a short period of time after departure of the aircraft in the event of a mechanical emergency requiring the pilot to quickly return the aircraft to the landing zone. The landing zone officer should also observe the aircraft during liftoff for unsecured hatches, smoke or fire from the aircraft, or loose engine coverings.

Civil Disturbances

A civil disturbance is defined as a situation involving several random or specific acts of violence directed at people, emergency response personnel, law enforcement, or property.

Civil disturbances are primarily a law enforcement matter; however, in most cases, fire and EMS departments and responders become involved at some point.

During a civil disturbance, injured people will require EMS services, and arson-related fires normally occur at some point. All emergency response personnel are responsible to be alert for potential or actual hazards due to a civil disturbance. Responders must understand that even a minor incident can spontaneously escalate into a significant disturbance. The topic of interagency coordination was discussed in Chapter 6 but is also worth mentioning here.

A civil disturbance requires close coordination between law enforcement and the other agencies involved. The use of IMS and unified command to maintain effective control and safety is necessary. Keeping safety as a top priority, law enforcement, fire department, and EMS personnel must be familiar with their roles.

Responders need to recognize that even during normal operations personnel may be the target of violence. Department SOPs should make it clear that responders need to report acts of violence to their supervisors and department communications center immediately. A policy on response to incidents involving violence should be established, so that responders know how to operate when violence has been committed against them. Once a random act of violence against responders has occurred, it is a good practice to request police assistance and retreat as safely as possible from the area. As a safe practice, it is advisable to set up a perimeter, and responses to future incidents in that area should have a police department escort.

All agencies involved need to acknowledge that fire and EMS responders must not become involved in disturbance control or intervene against perpetrators of a civil disturbance or other criminal activities. Decisions made during such an event must balance reasonable degrees of safety for responding personnel and equipment against the fire and EMS responsibility to provide a service to the public. During incidents of this type, removing all exterior-mounted tools and equipment should be considered as a health and safety measure.

#12

A unified command post should be established outside of the established perimeter. Off-duty fire department staff and administration should report to the command post to serve as command support and set up the necessary liaison functions. Implement the command and general staff positions as outlined in the National Incident Management System (NIMS) to manage the incident. The possibility of calling back off-duty personnel to staff apparatus that may be kept in reserve should also be looked into. When this is not possible, a ranking police officer should be requested to report to the command post to serve as the liaison for the police department. Coordination between the agencies involved will determine the appropriate level of response for the area in question, which is best accomplished by a meeting between commanders of all involved personnel. Once that decision has been made, communication with responders and an expansion of the IMS is vital. At a minimum, a command post must be established at these events, which may include a mobile command center or a building set up as a central point where operations will be coordinated by the agencies involved.

A staging area for responders and apparatus must also be instituted for coordination and safety purposes. The command post staff needs to keep responder alertness and tension in mind as it relates to personnel located in a staging area for extended periods of time. It may be possible to rotate personnel through the staging area every few hours. Responders who enter into a hazard zone must follow general safety guidelines. See Box 8–5.

Responders entering the hazard zone for the medical treatment of people must acknowledge that the patient should be removed from the area as quickly as possible to provide treatment outside of the hazard zone at a designated treatment area. During an active fire situation, responders must focus on savable property only. Any property that does not involve a life hazard or an exposure problem, such as burning vehicles, should be left to burn. Emphasis should be on fast attack, utilizing water onboard the apparatus

BOX 8-5: THE GENERAL SAFETY GUIDELINES FOR ENTRY INTO A HAZARD ZONE DURING CIVIL DISTURBANCES

- No single company or unit responses will be permitted in the hazard zone. The task force team concept should be used for all responses.
- Police department escorts will be required for all entries into the hazard zone.
- All fire department personnel will respond to and from all emergencies in full protective clothing including helmet in place, and personnel will remain in full gear until returned to staging or their assigned fire station.
- The use of sirens and emergency lights should be avoided while operating in the hazard zone. It may be necessary to use emergency lights for response and retreat out of the hazard zone; however, personnel should avoid making themselves highly visible while working an incident in the hazard zone.
- Apparatus must be placed in a manner that will allow for rapid, unobstructed retreat from the area. Apparatus must also be parked in a manner that best protects the crew.
- All tools and equipment located on the exterior of apparatus must be removed and placed in interior compartments.
- Responders must remain aware of outside speakers on apparatus. The public may hear radio transmissions of a critical nature. Consideration should be given to the use of cellular phones for sensitive communications to the command post.
- Exercise discretion in all situations in the hazard zone. Be aware that all actions by responders are critical. Responders should back away from potentially violent situations to avoid a major disturbance from erupting.
- Body armor may be necessary and should be addressed by command.
- Unnecessary personnel should not be permitted to ride along into the hazard zone.

for a rapid knockdown and blitz attack. The use of hand lines should be limited, and personnel should not try to establish a sustained water supply. Yet if a sustained water supply is necessary due to a large volume of encountered fire, personnel must have a plan in place to disconnect and abandon the supply line should it be necessary to leave the area quickly. Routine ventilation, salvage, and overhaul practices should be suspended during times of civil disturbance.

All responding apparatus should be brought up to full strength for civil disturbance response. This means all fire units enter the hazard zone in groups of personnel and apparatus, travel in these same groups, operate in groups, and return in groups. Personnel working in the hazard zone may also become exposed to chemical agents used by law enforcement; therefore, it would be necessary to provide a decontamination area for responders.

Response to civil disturbance emergencies must be viewed as a change in normal operations to one that involves the safety of all responders. An IMS and joint command post must be established as soon as a civil disturbance incident begins to unfold.

Terrorist Events

Terrorism can be described as nonmilitary violence by individuals or organizations to attain political ends. A terrorist incident is a violent act, or an act dangerous to human life, in violation of criminal laws, with the aim of intimidating a government and the civilian population in furtherance of political or social objectives.

Most terrorist events occur with little to no warning, so organizations must prepare their responders for such events. Recently, the Department of Homeland Security and the federal government have addressed terrorism by offering grant monies and training to local governments to combat it. A terrorism event must be viewed as a nonroutine emergency and, many times, a disastrous incident. Responders must be made aware of their responsibilities when responding to a nonroutine event of a potentially terrorist nature. It is crucial to follow certain general guidelines and safety procedures regarding terrorism.

#10

Organizations need to establish uniform procedures consistent with those of other agencies that may have a role in events of this nature. Policies and SOPs including a disaster operations plan must be developed to guide fire department responses to incidents of known or suspected terrorist actions. In this way, responders can provide the greatest level of protection to themselves and the general public affected by the incident. The department health and safety program manager must become directly involved in this important area. The types of terrorist incidents can vary greatly, and the response objectives will align with the diversity of possible scenarios. Although events can range from biological to nuclear, incendiary, chemical, and explosive incidents, one critical aspect remains the same.

#3

In most cases of terrorism, the mass casualties of victims will be the common problem for responders. Initial arriving responders must begin by evaluating the terrorist event to determine its nature and intended goal before implementing a course of action. A typical goal of terrorists is hysteria, mass panic, or mass injury; and responders must take precautions to ensure they neither become victims nor get caught up in the confusion during or after the event.

The rescue of victims will be a priority if a risk-versus-benefit analysis permits safe entry and rescue to take place. Early action is vital. Responders must secure the area, coordinate with law enforcement early into the incident because the event will also be a crime scene, and call for additional resources to ensure responder safety and success. The IMS system must be quickly put in place to manage the incident effectively.

There are basic strategic goals to accomplish at a terrorist event. Casualty management largely depends on the type of incident and weapon used; the bombing or shooting massacre has historically characterized the terrorist event. The importance of triage to the proper management of mass casualties from terrorist bombings and shootings should be focused on. In many cases, there will possibly be hundreds of victims with scarce medical resources available. Traditional triage of patients is based on giving care to the most critically ill persons first. In a disaster situation or mass-casualty event, triage would change to caring for those most likely to survive first because of the scarcity of health care personnel and resources in a situation in which thousands of persons may be ill or injured. Mass-casualty collection points would be the sites of triage to prevent hospitals from being overrun by patients. So, medical responders must focus on training in the area of proper, effective, and rapid triage to ensure success. See Box 8–6.

BOX 8-6: THE INITIAL STRATEGIC GOALS AT A TERRORIST EVENT

- Be aware of personnel safety at all times.
- Establish perimeters for safety and crime scene preservation.
- Evacuate any non-contaminated members of the public.
- Initiate triage and treatment of injured individuals.
- Consider the need for specialty resources, such as hazardous materials and technical rescue personnel.
- Consider the need for mass decontamination of large groups of people.
- Recognize that any incident may have been designed to include responders in a secondary attack.
- Consider placement of the command post.

Mass-casualty disasters require a major paradigm change from our everyday approach to emergency medical care with adequate and limitless resources. A large part of the planning process should ensure all responders at all levels are equipped with the proper PPE to handle the wide range of possible terrorist events.

Prior to any incident, fire and EMS agencies must meet with local agencies including law enforcement to reach an agreement about sharing vital information and intelligence during critical events. Law enforcement, fire, and EMS share the same priorities during these critical incidents, so planning and interagency cooperation for any planned or unplanned event should be a top priority.

Natural Disasters

When a natural disaster occurs anywhere in the nation, first responders are the earliest to arrive on scene. Natural disasters include hurricanes, tornadoes, earthquakes, floods, wildland fires, and severe thunderstorms. One characteristic of these rare and many times widespread dynamic events is the rapidly evolving complexity individuals face in trying to effectively manage all of the organizations and people, operations and tasks, equipment and supplies, communications, and the safety and health of all involved.

The department health and safety plan should address this important issue because responders will be involved in the events. Risk assessment is critical at natural disaster events. Similar to terrorist events, firefighters must train and prepare for the complex nature of specific disasters in their region of the country. The department health and safety program director should focus efforts in these specific areas. For example, training and preparation of a firefighter in southern California should be for an earthquake, whereas in Florida it should be for a hurricane.

Many of the hazards will be similar as they relate to disaster response. As with terrorist events, fire departments work together with law enforcement, emergency management, public works, utility companies, and sometimes the military. The important issue is getting the big picture and summoning the necessary resources as early into the incident as possible.

Planning with other agencies will help prepare for the type of common event that might take place. Responders must understand their roles, as well as those of other agencies and organizations. Natural disaster events will be large and long term, and generally occur in hot environmental conditions. The concepts presented elsewhere in this textbook, such as IMS, rehabilitation, and accountability, must also take a leading role in disaster management operations. The focus should be on preparedness (especially planning and training) and management in controlling and reducing the hazards emergency responders face during and after the event, and during the cleanup phase.

Rapid Intervention Teams

Specialized events, other than fire events, may involve the use of rapid intervention teams, which were covered in Chapter 6, but are worth noting here. During normal specialized operations, a standby team should be available for unforeseen emergencies and events. A rapid intervention team may be utilized for events involving fires, hazardous materials, technical rescue, and water rescue. The assignment of a rapid intervention team complies with NFPA 1500 and the OSHA respiratory protection regulation 29 CFR 1910.134.

A rapid intervention team, required by NFPA 1563, should be trained to the maximum level of the situation they may have to work in, such as working structure fires, technical rescue incidents, and hazardous materials incidents. When a rapid intervention team requires deployment, the IMS must be expanded and additional resources brought to the scene.

When a Mayday event occurs, the incident commander must immediately seek additional resources. It is advisable to assign a branch, division, or group RIT leader upon deployment and then a backup RIT once the original RIT is deployed into action. The actual deployment of a RIT creates chaos. Fireground commanders must immediately gain control of the event and establish order to ensure all aspects of the special operation continue to be handled.

RIT members who have been deployed face possible injury, which makes them another part of the problem. A Mayday call during normal operations motivates responders on scene to go to the area where fellow responders are in trouble. Command officers must maintain control of their personnel during this specialized event and ensure their safety.

Accountability at the incident scene is of the utmost importance once a RIT deployment is initiated. Because an incident commander cannot manage a rescue effort of responders in trouble and normal operations at the same time, a separate rescue branch and an expanded IMS will be necessary. Having accurate information is paramount for the incident commander and the rescue branch to be able to ensure safety and effectiveness. The RIT can become so focused on the rescue situation that team members, including the RIT leader, may not realize their limitations. When a RIT loses its focus, safety may become quickly compromised, making team members a part of the problem. For this reason, the RIT branch, division, or group requires the leadership of a disciplined and knowledgeable individual trained to the highest level of the special operation. Those RIT team members and branch or division leaders who have been inadequately trained can easily compromise the health and safety of the RIT team and the members in trouble.

As an example, RIT members at a trench rescue operation must have the awareness of unique hazards such as cave-ins, the weight of soil on victims, and the possibility of secondary collapse. Consequently, RIT members at a confined space rescue must be trained in and aware of poor air quality, fire hazard potential, visibility, biological hazards, electrical hazards, and process-related hazards such as residual chemicals, to name a few. In most situations, it is advisable for the RIT to perform a 360-degree analysis of the incident scene to enhance the health and safety of all involved. In 2008, two Ohio firefighters were killed in a residential occupancy. One of the key contributing factors identified during the investigation was an incomplete initial 360-degree size-up.[2] When a Mayday is called, the following should be considered:

#1

- IC should immediately perform a personnel accountability report (PAR).
- Consider changing RIT operations to an alternative secondary tactical channel. This will help the RIT director communicate with the RIT and not get walked over by other traffic.
- Immediately deploy the RIT utilizing search and rescue techniques that can be rapidly carried out.
- The RIT should identify the needs of the victim and communicate this to the RIT branch, division, or group officer.
- Progress reports should be provided to the RIT branch, division, or group officer.
- The RIT branch, division, or group officer should request EMS units to respond and be available to receive victim(s) removed by the RIT.
- After removal of the victims, the RIT branch, division, or group officer should request another PAR.

The command responsibilities of RIT operations are very involved, as the incident commander can easily become overwhelmed. See Box 8–7.

One of the most important functions a RIT officer/leader can perform is size-up of the building, including a 360-degree view. This size-up should also involve viewing preplans or quick access plans. Once the deployment of the initial RIT occurs, the incident commander must immediately assign a RIT branch, division, or group and call for additional

BOX 8-7: THE COMMAND RESPONSIBILITIES FOR RIT OPERATIONS

- Conduct an ongoing risk assessment while determining an action plan.
- Change strategy and tactics to the rescue mode while continuing fire suppression activities.
- Monitor and encourage all personnel to monitor the stability of the structure.
- Request ALS transport units to treat and transport rescued personnel.
- Conduct a PAR. Until accurate personnel information is obtained, command cannot develop an effective rescue plan.
- Expand the IMS as necessary to keep ahead of the demand.
- Assign a secondary RIT to replace the initial RIT.
- Request specialized resources with specialized equipment to assist in the rescue process.
- Ensure a public information officer is assigned.

BOX 8-8: RIT OFFICER/LEADER CONCERNS AND RESPONSIBILITIES

- Recon the building as a team—360-degree view.
- Where is the fire? Where is it going? How is it going to get there?
- Remove as needed any type of security bars, gates, and plywood over openings.
- Evaluate the type of roof in case of a partial or total collapse onto interior crews.
- Evaluate the number and type of doors and windows.
- Ensure there is a secondary means of egress and open as necessary.
- Place ground ladders to the upper floors.
- Turn off utilities as necessary.
- Recon nearby apparatus for tools, ladders, and backup hand lines.
- Assign tasks to RIT entry crew (thermal imaging camera, additional air, tools, search ropes, backup hand line).
- Consider carry versus drag techniques (regarding obstructions, heat, and visibility).
- Execute the search plan and ensure team stays together.
- Provide timely reports to command or RIT/rescue branch.
- Remain alert and monitor the RIT air supply.
- Prepare to change out RIT members with fresh members.
- Stay focused on rescue, not firefighting.

resources. The RIT branch, division, or group leader will have to ensure he or she keeps adequate personnel and resources in the staging area as backup RIT members at all times. See Box 8–8.

An involved RIT operation will use up a great deal of resources in a short amount of time as members place themselves in harm's way and often work until the point of total exhaustion. Understandably, "one of our own" is in trouble; however, the RIT officer/leader and the RIT branch, division, or group leader must rotate RIT members before personnel exhaust themselves and their air supplies. When operating at large buildings or high-rise buildings, consider the staging of additional RIT at all points of entry or on all floors where responders are operating.

Crew Resource Management

crew resource management (CRM) ■ This management system makes optimum use of all available resources—equipment, procedures, and people.

This textbook would be incomplete without the mention of **crew resource management (CRM)**. CRM began as a workshop studied by NASA in the late 1970s. This concept was adopted into the military, and adopted by United Airlines; it has made its way into the U.S. fire service. CRM is simply the effective use of all resources, including software, hardware, and human capital. CRM is not intended to disregard the chain of command or undermine the authority of fire officers; it also does not mean management by committee, but rather enhancement of authority.

The goal of CRM is to provide better teamwork, communication, and problem-solving skills; more effective team member input; and proactive accident prevention. Human error was discovered to play a large role in accidents in the aviation industry due to lack of communication, decision making, and leadership. The same problems have plagued emergency services for years. In fact, the various models that have arisen from CRM have been adapted to different types of industries and organizations. Situational awareness is the main benefit of adopting CRM in the fire service. CRM training encompasses a wide range of knowledge, skills, and abilities.

CRM's philosophy encourages a culture in which team members are respectfully permitted to question authority. In order for CRM to be accepted and endorsed, appropriate communication techniques must be taught to supervisors and their subordinates. Supervisors must understand that the questioning of authority need not be threatening, and subordinates must learn the correct way to question authority. The International Association of Fire Chiefs (IAFC) offers a crew resource management handbook on its Web site (http://www.iafc.org/displaycommon.cfm?an=1&subarticlenbr=20).

The primary goal of CRM is not enhanced communication, but enhanced situational awareness. The fire service must understand that human nature contributes to errors and becomes the foundational cause of threats to occupational health and safety. Many emergency services leaders now realize that a lack of situational awareness during incident scene operations leads to injuries and death of team members. Contributing factors to the lack of situational awareness are the macho nature of firefighters and aggressiveness caused by adrenaline as well as inexperience and a lack of training.

Emergency services organizations and their administration must be trained to understand that cognitive skills and situational awareness are closely related. Cognitive skill is the awareness that each worker uses for solving problems and making decisions. Interpersonal skills are also closely linked with situational awareness as they are used to develop communication and a range of behavioral activities associated with teamwork. All of these skills must overlap not only with each other but also with the required technical skills.

This overlapping of skills helps to implement effective CRM. Much the same as in the airline industry, emergency services organizations should embrace CRM training as a health and safety effort to help reduce human error. CRM requires a change in interpersonal dynamics and organizational culture as team members are permitted to voice their concerns. CRM marks a paradigm shift approach to error, injury, and fatality prevention. Unfortunately, many emergency services organizations do not accept paradigm shifts easily.

An in-depth review of the NIOSH firefighter fatality reports reveals that communication failure, poor decision making, a lack of situational awareness, poor task allocation, and leadership failures contribute to LODD injuries and death. Such issues are related to all areas of incident response, not just fire situations. Both external and internal barriers will affect these contributing factors. The external barriers are physical in nature, whereas the internal barriers include opinions, prejudice, attitudes, and stress. CRM must take the necessary steps to eliminate these barriers and their negative effect on the fire service. CRM will not be effective in any emergency services organization unless the entire organization is brought on board. The greatest advantage CRM offers to emergency services organizations is empowerment.

Summary

Specialized incidents encompass hazardous materials, technical rescue, water rescue, helicopter operations, civil disturbances, terrorism, natural disasters, and rapid intervention team response. The response to any specialized incident requires a particular group of properly trained responders. Any specialized event requires a concern for the safety of all who may become involved in the response as well as an initial response by the local agencies that may not have the level of expertise and equipment required. In many cases, the initial response requires the correct and safe actions of the first responders to properly size up the situation and begin the sequence of additional specialized responders.

The response to a hazardous materials incident is regulated by federal laws that require responders to be trained at the minimum level. At a minimum, responders to hazardous materials incidents must use the proper PPE, operate in teams of two, have site-specific plans in place, and use an IMS. Training is based on five levels of response. It is the health and safety program manager's job to ensure responders have an adequate level of training for the type of incident.

Technical rescue normally involves an extremely high risk for the responders. Technical rescue entails rescue from structural collapse, below-grade incidents, high-angle situations, wilderness rescue, water rescue, confined spaces, vehicles, and machinery. Technical rescue requires responders to have a high level of training, specialized equipment, and skills. Each individual department must determine the level of response and training as it relates to the needs of the community.

The area of water rescue and public safety dive operations falls under the umbrella of technical rescue; however, certain departments choose to make this a separate operation involving a specialized team of highly trained and equipped responders. The same as with other technical rescue operations, the number of incidents occurring in a specific community will determine the need for a specialized and separate team to perform water rescue and dive operations.

Many departments are using helicopters to provide ground support for wildland firefighting; however, helicopter operations also involve medical evacuation of victims setting up a landing zone, rescue hoist operations, and short-haul operations. There are general safety guidelines for working around helicopters in all disciplines involving helicopter operations. Safety during ground support operations are the most common issues that will need to be addressed, and the selection of an appropriate landing zone is of extreme importance.

The response to a civil disturbance requires close coordination and control of personnel. Responders must be aware that they can quickly become victims of violent acts. During periods of civil disturbance, IMS and interagency coordination play a leading role. Individual department policies and SOPs should be in place to guide responders. A civil disturbance requires close coordination between law enforcement and the other agencies involved.

Terrorism also requires close coordination between the fire service responders and other agencies. Most terrorist events occur with little to no warning, so organizations must prepare their responders for such events. Responders must be trained to not become part of the problem. There are some general guidelines and safety procedures that must be followed during a terrorist event. Similar to a civil disturbance, policies and SOPs including a disaster operations plan must be developed. Incidents of this nature are designed to bring media coverage and may involve mass casualties. Responders must continually be aware of a secondary device during an incident involving terrorism.

Natural disasters include hurricanes, tornadoes, earthquakes, floods, wildland fires, and severe thunderstorms. Similar to terrorist events, firefighters must train and prepare for the complex nature associated with specific disasters in their region of the country. Management of all the organizations and individuals involved in a natural disaster event is one of the most difficult challenges. IMS, rehabilitation, and accountability must take a leading role with these events.

RIT is commonly assigned to trained individuals as a standby safety and backup team to responders. A RIT should be trained to the maximum level of the situation it may have to work in, such as working structure fires, technical rescue incidents, and hazardous materials incidents. When a RIT is deployed, the common event becomes a specialized incident, the IMS must be expanded, and additional resources must be requested and brought to the scene. There is a high likelihood RIT members could become injured or part of the problem. Command officers must maintain control of the RIT, and accountability is of the utmost importance.

The fire service of today has become a group of highly trained responders who no longer respond only to common fire and EMS incidents. The key issue of safety must continually be a top priority during any incident, especially the specialized incident.

Crew resource management can play a large role concerning the health and safety of emergency responders. The primary reason to adopt CRM is to help improve situational awareness on the fireground. The foundation of CRM is the effective use of all resources.

Review Questions

1. Hazardous materials response is governed by:
   a. OSHA
   b. NFPA
   c. EPA
   d. Both a and c
2. All fire service responders should be trained to the ______________ level of hazardous materials response.
   a. Awareness
   b. Operations
   c. Technician
   d. Specialist
3. A fire department that responds to an active chemical leaking from a 55-gallon drum and performs actions to stop the leak is operating at the ______________ level.
   a. Awareness
   b. Operations
   c. Technician
   d. Specialist
4. Describe the safety issues related to technical rescue operations.
5. As related to technical rescue response, who must determine the level of response and training for the individual department as required by NFPA 1670?
   a. The department health and safety program manager
   b. The incident safety officer
   c. The fire department administration
   d. None of the above
6. Responders trained in water rescue are many times trained at the ______________ level.
   a. Awareness
   b. Operations
   c. Technician
   d. Both b and c
7. List the safety issues concerning operations during civil disturbances.
8. During nighttime landing zone operations, the landing zone lights should be ______________ in color.
   a. Orange
   b. Red
   c. Blue
   d. White
9. List the responsibilities of the landing zone officer during helicopter operations.
10. List the seven areas considered to be technical rescue.

Case Study

On August 28, 2005, a 50-year-old male volunteer firefighter/rescue diver (the victim) died after nearly drowning during a fire department sponsored night-dive training exercise at a quarry the night before. The victim had performed a total of three training dives the day of the incident (August 27, 2005) as part of the requirements for Professional Association of Diving Instructors (PADI) certifications for advanced open water diver and night diver. After the students completed the exercises for the "night dive," they were instructed to complete the training dive with a "partner dive." During the partner dive, the victim's partner reportedly signaled to him that he wanted to surface, and the victim signaled back, "OK, let's surface." After the partner surfaced, he looked around and did not see the victim. The partner reportedly looked down and saw the victim still below him waving his light from side to side in a distress motion. The partner dove back down and found that the victim did not have his regulator in his mouth. The partner tried to donate his alternate air source, but at that point the victim's underwater flashlight dropped and he went limp. The partner brought him to the surface and yelled for help. At this time the master SCUBA diver instructing the course (Instructor #1) and his partner were below the surface on the partner dive in another part of the quarry. Another dive instructor (Instructor #2) and a diver on shore unrelated to this training heard the calls for help and immediately went to provide assistance. The victim was towed to shore and provided cardiopulmonary resuscitation (CPR). Emergency 911 was called and arrived within 15 minutes. The victim was transported to a local hospital where he died the following day.

Department

The volunteer fire/rescue department involved in this incident has 4 stations and 80 uniformed firefighters, and serves a population of approximately 12,000 in an area of about 9 square miles. One of the stations houses a paid ambulance service.

Dive Team

The incident department's dive team was established in 1994 and has an 18-member roster. The dive team provides mutual-aid dive services for the northwest region of Pennsylvania. The team consists of three levels: (1) entry level; (2) operations level; and (3) technical level. Of the 18-member team, six members are considered to be trained to the technical level by the department, and are issued equipment for dive team response to emergency situations. The victim was considered to have reached the operations level, and was permitted by the department to participate in selected recovery or search incidents based on level of training and actual diving experience required by the department. The training courses being completed at the time of the incident were part of the training necessary to achieve the technical level designation by the department.

Victim's Training and Experience

The victim had been diving for approximately 5 years and completed requirements for certifications in Basic Public Safety Diver, Equipment Specialist, and Dry Suit Diver. The victim participated with the department in three search dives for state, county, and local police departments from November 2000 to July 2003.

In addition to dive training and experience, the victim had completed a documented 64 hours of training related specifically to firefighting since becoming a probationary firefighter with this department 15 years earlier.

NIOSH had the following recommendations:

Recommendation #1: Fire departments should develop, implement, and enforce standard operating procedures (SOPs) or protocols regarding diver training

Operational protocols, minimum equipment, personnel requirements, qualifications for team membership, and issues of training, drills, health, and safety should all be addressed in fire department SCUBA team SOPs. Operational protocols should address specific needs such as a designated safety boat, backup divers, ninety-percent-ready diver, and emergency medical personnel to immediately respond to an emergency incident. SOPs should be reviewed in-house on an annual basis, at a minimum, to see whether any changes are necessary. Every team member should have a copy of the SOPs, and each member should sign a statement indicating that he or she has read, understands, and agrees to abide by them. Although this department had operational SCUBA team operational protocols, there were no specific protocols (designated safety boat, backup divers, ninety-percent-ready diver) to be followed for or during dive training.

Recommendation #2: Fire departments should ensure that each diver maintains continuous visual, verbal, or physical contact with his or her dive partner.

Effective underwater communication refers to the capability to communicate between divers and from a diver to the surface. The divers present at this incident were able to communicate by using recognized dive signals such as a "thumbs-up" to indicate they were okay. Fire departments should follow OSHA safety standard 29 CFR 1910.424(c)(2) by ensuring that a diver be line-tended from the surface or accompanied by another diver in the water who is in continuous visual contact during the diving operations. The victim was a volunteer firefighter and was not covered by OSHA regulations. However, following OSHA standards would provide additional protection for firefighters who face unique environments and hazards associated with training or technical rescue operations.

Effective communication and continuous visual contact are two ways in which divers can convey any equipment or medical problems they may be experiencing.

Recommendation #3: Fire departments should ensure that a backup diver and ninety-percent-ready diver are in position to render assistance.

Public Safety Diving states that

> in addition to having the normal duties of divers, a backup diver must be ready to act as a replacement if the primary is unable to perform for any reason, and he must be ready to render assistance if the primary diver

runs into trouble. Because of the complex nature of diving, it's always possible that the backup diver will experience a problem when called. Following a policy of having contingency plans in place, it's best to have a second backup diver available, wearing an exposure suit and with his gear fully checked and functioning. If the backup diver is called on to make the descent, the ninety-percent-ready diver completes the dressing process so that he is fully ready to enter the water. With a ninety-percent-ready diver in place, the redundancy and safety of an operation increase dramatically.

In this incident, there were only the four divers present and participating in the training. No assigned backup diver or ninety-percent-ready diver was present.

Recommendation #4: Fire departments should ensure that positive communication is established among all divers and those personnel who remain on the surface.

Effective underwater communication is imperative. Specifically, diver-to-diver and diver-to-surface communications should be established and maintained during the entire dive(s). Underwater electronic devices are available to establish diver-to-diver and diver-to-surface communications. In this incident, there were no personnel at the surface to perform monitoring, and underwater diver-to-diver electronic devices were not used. All communication was either through surface face-to-face verbal communications or underwater hand signals.[3] See Figure 8–4.

Fatal Dive Incident Concerns and Issues	Yes	No
Would a backup dive team have made a difference at this incident?		
Would better department policies and procedures have resulted in a different outcome?		
Was the incident safety training provided by the department adequate?		
Would following the OSHA regulations have produced a different outcome?		
Would underwater electronic communications equipment have made a difference?		
Would a line tender have made a difference at this incident?		

FIGURE 8–4 Fatal dive incident concerns and issues.

1. Should the victim have been permitted to train after nearly drowning during training on the previous day? Explain.
2. What additional safety measures or actions should have been in place at this incident?

References

1. Texas Department of Insurance, Division of Workers' Compensation, *Emergency Response Planning for Hazardous Materials*, February 2008. Retrieved April 18, 2009, from http://www.tdi.state.tx.us/pubs/videoresource/stperplan.pdf
2. NIOSH Fire Fighter Fatality Investigation and Prevention Program, *A Career Captain and a Part-Time Fire Fighter Die in a Residential Floor Collapse—Ohio*, July 29, 2009. Retrieved February 8, 2010, from http://www.cdc.gov/niosh/fire/reports/face200809.html
3. NIOSH Fire Fighter Fatality Investigation and Prevention Program, *Volunteer Fire Fighter/Rescue Diver Dies in Training Incident at a Quarry—Pennsylvania*, July 28, 2006. Retrieved February 15, 2010, from http://www.cdc.gov/niosh/fire/reports/face200529.html

CHAPTER 9

Safety After the Incident

KEY TERMS

critical incident stress debriefing (CISD), *p. 209*

employee assistance programs (EAPs), *p. 207*

post-incident analysis (PIA), *p. 204*

OBJECTIVES

After reading this chapter, you should be able to:

- Determine the safety and health considerations during the termination of an incident.
- Describe the process of incident termination.
- Identify the areas of incident debriefing.
- Describe the demobilization process.
- Describe the need for and the process of a post-incident analysis.
- Explain why critical incident stress management should be part of the health and safety program and its benefits.
- List the events that may lead to critical incident stress.

Resource Central

For additional review and practice tests, visit **www.bradybooks.com** and click on Resource Central to access book-specific resources for this text! To access Resource Central, follow directions on the Student Access Card provided with this text. If there is no card, go to **www.bradybooks.com** and follow the Resource Central link to Buy Access from there.

The department safety and health program should include a plan for the activities and analysis that follow an incident. This chapter covers terminating the incident, post-incident analysis, training, and critical incident stress management. It is the health and safety program manager's responsibility to ensure that the health and safety of responders is addressed after incidents have occurred. Shift commanders, company officers, and firefighters oftentimes do not deal with the issues at a scene that did not go as planned. In many cases, responders have a "business as usual" mind-set when they return to quarters after an incident that did not go well. Such behavior does not benefit training efforts or the health and safety of the members of the organization. Department safety officers and supervisors must be trained to assess incidents thoroughly to improve safety. Department administration also needs to be committed to this part of the health and safety program to ensure responder compliance.

Conclusion of an Incident

The conclusion of an incident is the end of emergency response and the beginning of the cleanup phase. The reasons for incident termination include (1) responder safety, (2) legal requirements, (3) documentation (exposures and actions), (4) cost recovery, and (5) improvement of future responses.

The process of incident termination consists of actions responders should take or have taken, and documentation they need to compile. The primary purpose of an incident coming to a close is to ensure responders can safely leave the scene.

The OSHA regulations (HAZWOPER) require incident termination for hazardous materials incidents through the use of documentation. Documenting the response at a hazardous materials incident could help in defending against any legal actions that might arise. In any event, the lessons learned from the process of post-incident management help responders do their jobs better.

Incident Debriefing

Incident debriefing is an informal process done immediately after the incident to make sure responders know what happened, when, and to whom. Normally, the responders who were involved with the incident are those who will be included in the debriefing. After the completion of a debriefing, the information can be shared or presented with the entire department. The emphasis is on responder safety and on sharing pertinent information related to the operations at the incident. Incident debriefing permits responders to identify all hazardous substances or conditions and informs all department responders about the incident hazards. Incident debriefing also identifies areas that will need to be addressed, such as cleanup procedures, critical incident stress management, and any other problems. In certain instances, there may even be a legal mandate to conduct incident debriefing. Debriefing measures should take place immediately after an event has concluded in a suitable location that is relatively comfortable. Incident debriefing may take place at the incident scene, although it is more common to conduct it elsewhere. Its location should be away from the media and the general public to prevent sensitive information from being overheard. All responders who were at the scene need to participate in the debriefing process. A facilitator should be appointed to conduct the debriefing, a time limit should be established, and a means for follow-up should be provided.

Demobilization

On larger incidents, the incident commander may decide that a demobilization plan is necessary. This will take coordination between the incident commander and the planning section, as demobilization is the point when apparatus and responders begin to leave the scene. The planning section gathers basic information on incident resources. Planning can match such information with resources and responders that remain on scene. Depending on the type of incident, demobilization may focus on getting on-scene resources back into service as soon as possible.

The many command staff officers each play a role in demobilization. The liaison officer is responsible for handling transportation availability, communications, maintenance, and continuing support. The liaison officer should be aware of the terms of agreements involving the use and release of another agency's resources. The safety officer considers the physical condition of personnel, personal needs, and adequacy of transportation. The operations officer knows the continuing needs of various kinds of tactical issues. The finance administration officer processes any claims, time records, and costs of individual resources, all of which are factors in determining the release of resources.

The communications or dispatch center also plays a role in demobilization, as its personnel give high priority to resources that need to return to service. The National Incident Management System provides a documentation form (ICS –211) to assist in the demobilization. The department should prepare for the demobilization process well in advance, often at the same time the mobilization process is taking place. Early planning in this area will help to ensure accountability and aid in the efficiency of the transportation of resources. Much the same as with mobilization, the demobilization process must ensure the adequate tracking of resources.

The incident commander needs to consider many health and safety issues during the demobilization process that are normally part of the department's SOPs and policies. One of the major issues is the order in which resources are released. The major elements to be addressed during the demobilization phase are the policies related to demobilization, responsibilities, release priorities, and release procedures. Numerous fire departments take pride in the fact that it is "their" fire scene and fire district. In fact, many first-due companies insist on remaining at the incident scene so they can be the "last out"; however, this unhealthy philosophy may compromise the safety of responders on larger incidents. The incident commander needs to take this historical philosophy into consideration as he or she decides when to demobilize and release units. In some cases, it may be necessary to release the units that were first to arrive and worked the hardest and longest. By summoning outlying companies to the incident, to complete physically demanding tasks such as overhaul, the first-due units can then be easily relieved and returned to quarters for rest and rehabilitation. In most departments such philosophical change requires a cultural change within the department, and as we know, cultural change does not come easily.

The incident commander should include the safety officer and the planning officer in the initial stages of demobilization. The safety officer provides key information concerning overworked and stressed personnel, and should also communicate this to the person in charge of rehab, who is directly involved with the crews and their physical and emotional status. The planning officer or section chief collects, evaluates, and disseminates information related to the operations of the incident. This officer maintains information and intelligence related to the current and the forecasted situation. The planning officer also maintains the status of resources that have been assigned to the incident. The preparation of incident action plans and incident maps are the responsibility of the planning officer. The incident commander should also be concerned with the status of the rapid intervention team (RIT). When a RIT is no longer needed, the team should be released from service as soon as possible.

It is not uncommon for injuries and even death to occur after the initial mitigation of the incident has been completed. Many injuries occur during the overhaul phase of an incident. During the overhaul phase, firefighters are exposed to hazards that commonly cause eye, hand, and back injuries as they work to make the fire building safe and ensure the fire is completely extinguished. The incident commander must not be surprised that safety concerns continue even when responders consider the incident to be over. It is also important for the company officer(s) and supervisors to evaluate the crew's readiness to respond to future calls. Supervisors should evaluate crew members after the incident for cases that may need critical incident stress debriefing as well as for exposure to infectious disease. If an exposure has taken place, crews should follow decontamination procedures prior to returning to their units, and then the proper documentation of any medical concerns should be done.

The company officer or supervisor must make the ultimate decision as to whether and when the crew and units are ready to return to service. The safety issues cannot be overlooked, as it is common for crews to check their equipment, restock, and think they are ready for the next incident. The safety of personnel is the primary consideration during the demobilization process.

Returning to Quarters

Many times, the crew does not get a chance to discuss the incident prior to returning to the station. The company officer or supervisor should be keenly aware of the crew's needs prior to returning to the station, and alert to any signs of critical incident stress among the crew upon return to the station.

As has been mentioned earlier in this textbook, apparatus accidents while returning to the station play a leading role in firefighter injury and death. If the apparatus operator is experiencing critical incident stress, the company officer or supervisor may not be aware of it until the crew returns to the station; it is en route when an incident can occur on the roadway. There are physical limits involved with the safe operation of emergency response apparatus. A physically and mentally exhausted driver should not be permitted to operate the apparatus back to quarters. Therefore, it is critical that the company officer determine the condition an apparatus operator is in to drive before leaving the scene. In those instances when the on-duty crews have worked to the point of sheer exhaustion, personnel should be replaced at the incident scene with individuals who are fresh to avoid apparatus accidents when returning to quarters.

Evaluation of the crew for injury and illness that may be related to the incident is vital. In most cases, responders will not complain of minor injuries and may even try to hide them. The company officer or supervisor needs to keep a close eye on the crew and ask pertinent questions that may uncover an injury or psychological issue. Once the apparatus and crew is readied for the next call or assignment, the company officer or supervisor should sit down with the crew and discuss the previous incident in an informal manner. This informal discussion leads to the next step: post-incident analysis.

Post-Incident Analysis

post-incident analysis (PIA) ■ The two-step method involves informal and formal discussion concerning incident operations.

The emergency response organization should use the **post-incident analysis (PIA)** as a tool to identify strengths, deficiencies, and needed areas of improvement in how the department handled the incident. The process involves research and the gathering of information to conduct the PIA, and the development of lessons learned and incorporating them into training lesson plans and departmental procedures. The process involves formal and informal discussion. The organization should consider conducting a post-incident analysis on every significant incident that has had a multiunit response, when the incident was not the "typical" operation. It is especially critical to complete a PIA on any incident that involves injury or death. The term *critique* has lost favor in the fire service as it has been interpreted as criticizing the event. However, the purpose of the PIA is not to find fault or place blame, but to determine what went right and what went wrong.

The department may utilize the PIA to identify other areas of need, such as for equipment, for alleviating staffing deficiencies, and for further training. The information collected may also be helpful to justify funding in future budgeting requests. The PIA is a valuable document that can be used to improve the overall operations of the fire service. See Box 9-1.

The responsible incident commander should complete the PIA and use an open-forum format to share it with organization leadership and personnel. Forms should be submitted to the shift commanders, chief of operations, and deputy chief for review, and then to the fire chief for final review and filing.

It is beneficial to complete the PIA in a short time frame. Once it is reviewed, departmental procedures should be evaluated and changed to better serve the department. This process requires the cooperation of all department personnel.

The incidents undergoing analysis may involve fire incidents, technical rescue, hazardous materials, mass-casualty incidents, and vehicle accidents involving entrapment. The incident commander or incident command team must informally analyze every incident to improve personnel, unit, and system performance.

The post-incident analysis should be based on facts, not unsubstantiated opinions. The PIA is meant to be a step-by-step review of what happened at the incident, including what went right and wrong. The PIA should include an in-depth look at the procedures and resources used at the incident to identify whether the equipment was adequate or whether new equipment and technology options should be explored and purchased. The operation's overall effectiveness should also be reviewed to determine whether policies and procedures were correct or need updating.

An informal discussion should take place prior to conducting a formal PIA. As mentioned earlier, informal discussion called incident debriefing normally takes place after every incident before any formal post-incident analysis has been scheduled. Crew members are given the opportunity to share what they did, what they saw, what they perceive went correctly, and what they perceive went wrong. The company officer also has the opportunity to identify future training needs for the crew. He or she should make notes during this informal session if a scheduled formal PIA is going to take place. The PIA should also take place in volunteer organizations that share mutual-aid or automatic aid agreements. Department leadership must not allow the PIA to turn into a finger-pointing session, for it is too valuable a tool.

Normally, the formal PIA takes place at some future point in time after those involved have gathered the necessary information. The crew members who were present at the incident, including the command staff, should be invited to the scheduled PIA. After the PIA has been completed, there should be some type of follow-up concerned with procedural changes, training needs, and equipment needs.

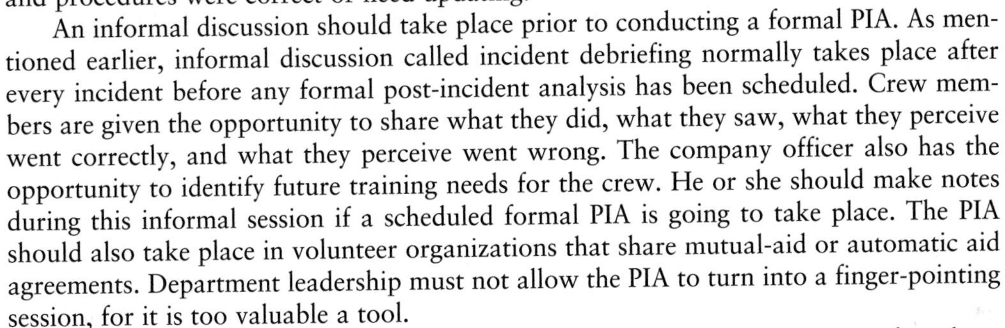

BOX 9-1: REASONS TO CONDUCT A POST-INCIDENT ANALYSIS

- As a way to capture lessons learned
- As a method to reconstruct the incident
- As a method to collect and organize incident documentation
- As a way to improve future responses
- As a method to identify training needs
- As a method to determine financial responsibility
- As a method to provide a formal investigation and report

In many cases, the incident commander will run the PIA; however, it is also a good idea to have a moderator present from the organization's administration to ensure the PIA stays on track and that everyone involved exercises diplomacy. The department SOPs and policies should be readily available during the PIA to compare actions taken and procedures followed. A deviation from the department policies and SOPs might indicate that a change to such policies or SOPs might be in order.

Some basic guidelines should be followed when conducting a PIA. The incident commander should run the PIA while the moderator focuses on analysis of the incident. In many cases, it is beneficial to include supervisory and management personnel from the dispatch center that were involved in the incident to identify whether any changes are necessary at the communications level. Radio operators can provide pertinent information about the incident. Some issues to be addressed are incident management, strategy and tactics followed, number and types of resources involved, problems, challenges, unexpected occurrences, procedures, guidelines, and training. At a minimum, the PIA should be recorded on paper using a format similar to taking the minutes of any formal meeting. The department should have a documentation template in place for personnel who are gathering data related to the PIA to use. The amassed data should be used to conduct the PIA, and then stored in a database or computerized file that is available to department personnel for review. Departments that choose to conduct quarterly or semiannual officer and/or staff meetings can use the PIA during these meetings to educate personnel within the department and on other shifts who were not present at the incident. See Box 9-2.

Generally, it is beneficial for the responders who were involved to use a plot plan of the incident location to refer to what they did, where they were located, and what specifically worked and did not work. This use of a plot plan is especially useful if a responder injury or death has occurred.

Certain liability issues exist after the completion of a formal PIA. The PIA is a useful tool departments can utilize to improve future responses. However, choosing not to improve responses can cause future failures in operations at the incident scene, which may lead to lawsuits. The State of Ohio has a statute in place that protects a peer review or quality assurance review concerning EMS organizations from public record. The statute states the following: "any discussion conducted in the course of a peer review or quality assurance program conducted on behalf of an emergency medical service organization, is not subject to discovery in a civil action and shall not be introduced into evidence in a civil action against the emergency medical service organization on whose behalf the information was generated or the discussion occurred."[1]

Unfortunately, no statute protects the fire department PIA from public record. If an incident did not go well or as planned, attorneys will have information from SOP and policy violations, disgruntled employees, and witnesses. Personnel who follow department policy and SOPs will lessen the chance of a successful lawsuit taking place.

BOX 9-2: BASIC GUIDELINES TO BE FOLLOWED WHEN CONDUCTING A POST-INCIDENT ANALYSIS

- Collect and organize all pertinent documentation.
- Interview all of the responders.
- Identify applicable laws, regulations, plans, guidelines, and training requirements.
- Draft a formal report and obtain feedback from the responders involved.
- Distribute the report to all concerned.
- Follow up on all recommendations.

The health and safety program manager and the incident safety officer(s) have a role in the PIA process. NFPA 1521, *Standard for Fire Department Safety Officer*, requires the health and safety program manager to develop procedures to ensure health and safety issues are addressed during the PIA. It is the responsibility of the incident safety officer to provide the health and safety program manager with a written report after the completion of the PIA. The health and safety program manager will use the information from the incident safety officer's report to provide a written report that includes pertinent information about the incident related to health and safety issues. It should incorporate any input obtained from the incident safety officer(s). This written report should address the following:[2]

- The incident action plan
- The incident safety officer's incident safety plan
- Issues related to the use of protective clothing and equipment
- Issues related to the personnel accountability system
- Rehabilitation procedures
- Any other issues affecting the safety and welfare of personnel at the incident scene

If the incident resulted in the injury or death of a member, the incident safety officer is also responsible for investigating the circumstances surrounding the incident, collecting and analyzing data, reconstructing the incident, and providing recommendations to the health and safety program manager or the fire chief. The incident safety officer report should focus on any violations of SOPs, poorly defined procedures, unforeseen situations, and training deficiencies. Together the incident safety officer and the health and safety program manager identify and document the lessons learned, and provide them to any group or individuals for correction. With this method, progress and changes can be tracked and reevaluated during subsequent incidents. A lessons-learned, mind-set must take the forefront in emergency services if a reduction in injuries and deaths is going to be accomplished.

#9

Critical Incident Stress Management

#1

Critical incident stress for emergency responders is as common as everyday stress is for the general public. Stress that cannot be managed effectively will lead to problems for individuals and the organization. The term *traumatic stress* often denotes the stress response, which follows traumatization, and it is usually caused by a sudden and uncontrollable event that causes disruptions in life. Stress is one of the leading causes of physical problems in society today.

As it relates to the health and safety of responders, a critical incident stress management (CISM) program must become a component of the health and safety program. The program director needs to become aware of stress-related issues and available countermeasures to effectively manage the stress associated with department members and emergency response.

Psychological trauma related to critical incident stress of emergency workers has been recognized in emergency services within the last 25 years, and its management is a relatively new field in emergency response. Emergency response professionals have been the last to be considered as vulnerable to traumatic stress. The frequency, duration, and severity of and impairment from psychological crisis are concerns with responders. Therefore, as a minimum, provisions should be made in every emergency response organization for **employee assistance programs (EAPs)**. This issue was touched on in Chapter 5, and this chapter focuses on after-incident critical incident stress management.

employee assistance programs (EAPs) ■ EAPs are intended to help employees deal with personal problems that might adversely impact their work performance, health, and well-being. EAPs generally include assessment, short-term counseling, and referral services for employees and their household members.

The health and safety considerations of stress are a major concern; many responders do not understand critical incident stress, and some are not willing to accept its effects.

Everyone under stress experiences disturbances in his or her daily routine, lifestyle, and career. Whether these disturbances are experienced as crises is but a matter of the degree. A critical incident has the ability to cause a disruption of the "steady state" of psychological processing for the responder that the mind of every individual struggles to maintain.[3]

The human body usually reacts instantly to stress. Acute and chronic stress are the types of stress in which the body reacts. Acute stress can be exciting or even thrilling. Chronic stress is not. Chronic stress occurs as events wear away on the individual day after day or year after year. Stress can be classified as good or bad, and affects each of us on a daily basis. A combination of smaller events or the single event that is of a high-stress psychological level will often affect responders. Many of these events are referred to as critical events. Over a period of time, the immune system atrophies due to the prolonged effects of stress. Some responders may feel the cumulative effects of stress to be life threatening.

Certain situations prove to be more stressful to emergency response professionals than others, such as dealing with traumatized children. In the not-too-distant past, responders were expected to deal with incident stress without open discussion or emotion. Fortunately, this is no longer the case.

The field of critical incident stress management has become a part of many emergency response organizations. The breakdown of natural defenses is most prevalent in the acute setting, such as in the street or the emergency room. In a disaster setting, emergency personnel often report that they function adequately until they come upon a child's possession such as a toy. More than any other trigger, traumatized children and death can bring to mind thoughts about life's meaninglessness and unfairness. Here are other events that may lead to critical incident stress:

- Line-of-duty deaths
- Suicide of a coworker
- Death of or injury to children
- Events involving mass casualty, terrorism, and disaster
- Events that are prolonged in nature ending in a negative outcome
- Death or injury caused by a responder
- Serious work-related injury
- Personally significant events that are powerful and overwhelming
- Multiple event incidents
- Events with a high degree of threat to personnel
- Events with excessive media interest

Responders, in many cases, have traditionally viewed the display of critical incident stress as a sign of weakness. However, critical incident stress debriefing (CISD), on-scene crisis intervention, and pre-incident education programs are all highly recommended as formal critical incident stress management interventions, which assist rescuers in coping with critical incident stress. Every organization should have procedures in place to activate the critical incident management system, which can be done when any of the listed events take place, the company officer requests activation, or the incident commander establishes a need. In certain cases, the system can even be activated prior to responders asking for intervention.

A responder may develop physical, behavioral, and emotional difficulties due to critical incident stress even if there is a system in place to deal with it. Both the immediate supervisor and the employees themselves must be trained in the signs of this stress, because many symptoms may not appear until a month or so after the event. If critical incident stress is not managed properly, employees may begin abusive behaviors such as using alcohol and drugs, or they may resign from the organization. The CISM portion of the health and safety program assists responders dealing with events associated with critical incident stress, so that they can avoid unhealthy coping behaviors.

Researchers and psychologists who specialize in job stress generally agree that those attracted to emergency work as a group are more emotionally stable than the general population and therefore less likely than the ordinary citizen to "crack" under intense pressure. Yet there are responders who report they have been emotionally and physically affected by their work at one or more emergency events. Critical incident stress attacks an individual's personal support system and basic assumptions about life in general. Even good-natured kidding, which is common in emergency services, can cause considerable anxiety.

The most well-publicized of the CISM services is the intervention technique known as **critical incident stress debriefing (CISD).** This structured group discussion, developed as a means of preventing or mitigating traumatic stress, has been widely used with emergency response personnel. In fact, CISD has grown into a virtual crisis intervention specialty. After a standard of care for it was developed, professionals recognized that the first intervention should be *pre-incident preparation.* The goals of pre-incident preparation are to set the appropriate expectancies for personnel and their families as to the nature of the crisis and trauma risk factors that they face[4] and to teach basic crisis coping skills in a proactive manner. Because the effects of a crisis influence family members in both direct and indirect ways, there needs to be a family support network and intervention in place concerning CISM.

critical incident stress debriefing (CISD) ■ This structured group discussion, usually provided 24 to 72 hours post crisis, is designed to mitigate acute symptoms, assess the need for follow-up, and if possible provide a sense of postcrisis psychological closure.

#13

In many cases, CISD teams are regional teams located to assist members of several departments or organizations. It would be difficult for an organization to have an "inside" team for CISD, because this would involve a peer member outside of the event to serve as a CISD member. Peer members of the organization, who were not at the incident scene, will be viewed as members outside of the inner circle of those who experienced the critical event. For this reason, peer members within an organization will not serve the organization or its members very well. Individuals from outside of the organization who are trained in the specialty area of CISD should be called to coordinate the CISD process. CISD is a sensitive issue that must be handled in the correct fashion or more damage than good could occur from the process, and should be handled by trained individuals who personally do not know the affected members. Total confidentiality is also an important factor with CISD, and as a minimum, a mental health professional should be involved.

The management of the organization must buy into the CISD concept in order for the system to work. The main goals of critical incident stress debriefing, which is based on sound crisis intervention and educational principles, are to mitigate the impact of the traumatic event and accelerate the normal recovery process. CISD can be used with on-site debriefing, but is commonly done after the event is over. Certain states, such as Ohio,[5] have laws in place to protect the confidentiality of critical incident stress debriefing and management. Emergency services organizations should become familiar with the specific confidentiality laws in their state pertaining to this issue and issues concerning PIAs.

On-scene CISD teams are trained to observe responders for signs and symptoms of stress and treat them appropriately. The CISD system, designed to enable the group to discuss the emotional aspects of a traumatic situation, has seven phases:

- Introduction
- Fact
- Thought
- Reaction
- Symptoms
- Teaching
- Reentry

Each phase has a specific purpose whose main goal is to facilitate maximal discussion of the traumatic incident. It is natural for emergency response professionals to discuss what happened at the event before moving on to how they feel about it. Some individuals believe from personal experience that the presence of strong emotions may interfere with the performance of their duties.

The CISD process was never meant to be used with routine incidents. If it is used too frequently, it may lose its power and effectiveness. CISD interventions appear to work best when they are provided reasonably close to the time of occurrence of the traumatic incident. The usual time frame is 24 to 72 hours after the incident. What is more important than the time frame is for the group to be ready to participate in a debriefing when it is initiated.

Because emergency response personnel tend to be cognitively biased, they typically are not able to process any of the emotional aspects of a traumatic event for the first 24 hours (which is why debriefings are generally withheld until after then). A quick defusing may be held until the entire group is ready to benefit from a debriefing. The defusing is a short-term coping mechanism that addresses immediate needs. It is informal in nature and takes place on the day of the event, often at the scene; it includes the responders who have experienced the incident before they get any sleep. The defusing explains what symptoms are normal in the short term to reassure responders. It also offers responders a telephone number should they need to talk to someone. Any effort to bring order to a chaotic situation will go far in promoting a more rapid recovery of traumatized personnel.

CISM also helps to restore the functions of the social network, which promotes recovery. The debriefing gives people a sense of direction and specific guidelines, which help them make coordinated efforts toward reaching recovery. Without crisis support services, it is likely many individuals who need such follow-up care, simply would not seek it out.

Recently, the subject of debriefing has come under fire, and the controversy about it is beyond the scope of this textbook. It is important to note that the debriefing element of CISM is just one of many types of intervention; it was never intended to stand alone or replace psychotherapy. Emergency response organizations must keep in mind that CISD represents only one component of an integrated multi-component crisis intervention system.

Today, CISD teams are being woven into the fabric of emergency services organizations. Not everyone will benefit from CISD; some individuals may not need it, whereas others require more help than a debriefing can provide. It is the safety and heath of the members of an organization that must become a priority in this critical area. CISM plays a critical role in any organization's health and safety program.

Summary

It is common for the safety and health issues after an incident to be overlooked. The department health and safety program manager must ensure the health and safety of responders is addressed after incidents and continually keeps abreast of after-incident issues as they arise. Shift commanders and the company officer have a responsibility to evaluate their personnel after an incident for signs of injury and stress. It is of the utmost importance that department safety officers be trained to assess incidents thoroughly to improve safety. The administration of the organization must be onboard with the post-incident health and safety of employees as a way to ensure a complete health and safety program is in place.

Post-incident health and safety begins at the termination of the incident. The incident commander has many decisions to make during the post-incident phase of the operation, including the demobilization of personnel and their safe return to quarters. The human component must not be ignored or taken lightly during the post-incident phase of the incident. Incident debriefing should take place after every incident that has a significant impact on personnel. In most cases, incident debriefing can be informal and take place at both the incident scene and back in quarters.

Apparatus accidents while returning to the station play a leading role in the injury and death of firefighters. It is important that the company officer pay close attention to his or her crew for signs of injury and psychological stress prior to returning to the quarters. Once the crew has returned to quarters, the company officer should continue the informal process of asking questions and observing the crew for signs of injury and critical incident stress.

A large part of post-incident safety includes the post-incident analysis (PIA) and critical incident stress management. The PIA is a two-step process that begins with an informal discussion both at the incident scene and back in quarters. Later, a formal PIA is needed to address department procedures, training, and future responses. The formal PIA should include personnel who were at the incident scene, including the command staff, and the PIA can become a progressive educational process for other members of the organization.

Critical incident stress management (CISM) has recently become a major safety and health concern for responders. It is now recognized how large a role CISM plays in the health and safety of department members and their families. Critical incident stress debriefing (CISD) teams should be available to assist members of the organization following incidents of high stress. In many cases, CISD teams are regional teams located to assist members of several departments or organizations. Company officers and department supervisors must receive training in CISM and the type of incidents that may lead to critical incident stress. Emergency response organizations must also understand the differences between CISD and CISM. It is also important to make an employee assistance program, and the means to access it, available to employees and their families.

Many health and safety issues are present during every incident; however, it is vital that the post-incident issues do not become overlooked. The post-incident issues are just as important as those that occur during every incident.

Review Questions

1. List the reasons for incident termination.
2. Describe the several phases involved with incident termination.
3. The _____________ require incident termination for hazardous materials incidents through the use of documentation.
   a. NFPA standards
   b. EPA regulations
   c. OSHA regulations
   d. NIOSH standards
4. The focus of incident debriefing should be on:
   a. Incident operations
   b. Responder safety
   c. Future training methods
   d. CISM management

5. Demobilization may focus on:
   a. Getting on-scene resources back into service
   b. Responder safety
   c. CISM management
   d. Legal requirements
6. During the demobilization process, consideration is given to:
   a. Physical condition of personnel
   b. Personal needs
   c. Adequacy of transportation
   d. All of the above
7. List the reasons why a post-incident analysis should be used.
8. It is best to allow the incident commander to conduct the post-incident analysis.
   a. True
   b. False
9. Which of the following is highly recommended as formal critical incident stress management intervention?
   a. Critical incident stress debriefing (CISD)
   b. On-scene crisis intervention
   c. Pre-incident education programs
   d. All of the above
10. Every fire service organization should have an "inside" team for CISD.
    a. True
    b. False

Case Study

On July 5, 2008, a 42-year-old male volunteer Texas fire chief was killed when he was struck by a collapsing brick parapet wall during a commercial structure fire. See Figure 9-1. The fire chief, along with four firefighters, was finishing mopping up suppression activities at a grass fire when the fire department was dispatched to a structure fire. The fire chief and two firefighters left the scene of the grass fire in a tanker and traveled to the scene of the structure fire where the fire chief began to size up the burning commercial structure while the other two firefighters traveled five blocks back to the station to obtain an engine and structural firefighting gear. The

FIGURE 9-1 Approximate location of a fire chief trapped under a brick façade collapse.

two firefighters returned to the structure fire scene with an engine parked in the street directly in front of the burning automotive repair and upholstery business. The fire chief grabbed a self-contained breathing apparatus (SCBA) from the engine and pulled a preconnected 1¾-inch hand line to the front door, assisted by a firefighter who had just arrived in her personal automobile. The fire chief worked the nozzle through the doorway (using tank water) while the other firefighters established water supply. Less than five minutes after the engine arrived on scene and shortly after water supply was established, the brick parapet wall at the front of the structure collapsed, striking the fire chief and burying him under the brick debris. Rescuers quickly uncovered the fire chief and medical treatment was started immediately. The fire chief, still conscious, was transported to a trauma hospital where he died several hours later. Key contributing factors identified in this investigation include failure to conduct a 360-degree size-up of the incident site, failure to recognize the potential collapse hazard, inadequate staffing, and inadequate fireground communications.

NIOSH had the following recommendations:

Recommendation #1: Fire departments should ensure that the incident commander conducts a complete 360-degree size-up of the incident scene, including evaluating the potential for structural collapse.

Recommendation #2: Fire departments should establish and monitor a collapse zone when conditions indicate the potential for structural collapse.

Recommendation #3: Fire departments should train all firefighting personnel in the risks and hazards related to structural collapse.

Recommendation #4: Fire departments should ensure that the incident commander maintains the role of director of fireground operations and does not become involved in firefighting efforts.

Recommendation #5: Fire departments should ensure that the incident commander conducts an initial size-up and risk assessment of the incident scene before beginning firefighting operations and continuously reevaluates the situation.

Recommendation #6: Fire departments should ensure that adequate numbers of staff are available to effectively respond to emergency incidents.

Recommendation #7: Fire departments should ensure that tactical operations are coordinated and communicated to everyone on the fireground.

Recommendation #8: Fire departments should ensure that every firefighter on the fireground has a portable radio with sufficient tactical frequencies to effectively communicate on the fireground.

Recommendation #9: Fire departments should ensure that a separate incident safety officer, independent from the incident commander, is appointed at each structural fire.

Recommendation #10: Fire departments should develop, implement, and enforce standard operating procedures (SOPs) or standard operating guidelines (SOGs) covering all aspects of structural firefighting. Periodic refresher training should also be provided to ensure firefighters know and understand departmental guidelines and procedures. Compliance with departmental SOPs/SOGs should be monitored and the appropriate action taken when noncompliance is identified. SOPs and SOGs should be applicable and common to mutual-aid departments to ensure consistency on the fireground, which can lead to a safer fireground. At the time of the incident, the department had very limited written SOGs or SOPs that covered only the fire department organizational structure.

Recommendation #11: Fire departments should be prepared to use alternative water supplies to ensure adequate water is available for fire suppression. In addition to the location and extent of the fire, factors affecting selection and placement of hose lines include the building's occupancy, construction, and size. In addition, fire load and material involved, mobility requirements, and numbers of persons available to handle the hose lines are important factors.[6] See Figure 9-2.

1. Discuss the contributing factors that led to this line-of-duty fatality.
2. Discuss how inadequate staffing played a role at this incident.

Fatal Brick Facade Collapse Concerns and Issues	**Yes**	**No**
Would a 360-degree view and size-up have produced a different outcome?		
Would better department policies and procedures have resulted in a different outcome?		
Would an incident safety officer have prevented this fatality?		
Would better radio communication have prevented this fatality?		
Was proper strategic management utilized at this incident?		
Would adequate staffing have prevented this fatality?		
Was a proper risk assessment and size-up completed at this incident?		
Would a preplan of this building have prevented this fatality?		
Would adequate training in building construction and structural collapse have prevented this fatality?		

FIGURE 9-2 Concerns and issues of a fatal brick façade collapse.

References

1. Ohio State Statutes Revised Code, Title 47, chapter 4765.12, *Guidelines for Care of Trauma Victims by Emergency Medical Service Personnel.* November 13, 2000. Retrieved February 15, 2010, from http://codes.ohio.gov/orc/4765.12; Ohio State Statute Revised Code, Title 23, chapter 2317.02, *Privileged Communication*, 2003. Retrieved February 15, 2010, from http://codes.ohio.gov/orc/2317.02
2. NFPA 1521, *Standard for Fire Department Safety Officer*, 2008. Quincy, MA: National Fire Protection Association. Retrieved February 15, 2010, from http://www.nfpa.org/aboutthecodes/AboutTheCodes.asp?DocNum=1521
3. J. T. Mitchell, and G. S. Everly, Jr., *Critical Incident Stress Management (CISM): A New Era and Standard of Care in Crisis Intervention*, 2nd ed. (p. 17). Ellicott City, MD: Chevron Publishing Corporation, 1999.
4. J. T. Mitchell, and G. S. Everly, Jr., *Critical Incident Stress Management (CISM): A New Era and Standard of Care in Crisis Intervention*, 2nd ed. (p. 17). Ellicott City, MD: Chevron Publishing Corporation, 1999.
5. Ohio State Statute Revised Code, Title 23, chapter 2317.02, *Privileged Communication*, 2003. Retrieved February 15, 2010, from http://codes.ohio.gov/orc/2317.02
6. NIOSH Fire Fighter Fatality Investigation and Prevention Program, *Volunteer Fire Chief Killed When Buried by Brick Parapet Wall Collapse—Texas*, March 12, 2009. Retrieved February 15, 2010 from http://www.cdc.gov/niosh/fire/reports/face200821.html

CHAPTER 10

Individual Responsibility with Safety—If Not You, Then Who?

KEY TERMS

incident action plan (IAP), *p. 223*

incident commander (IC), *p. 222*

recognition-primed decision making (RPD), *p. 223*

OBJECTIVES

After reading this chapter, you should be able to:

- List the individual responsibilities of responders related to the health and safety program.
- List the role and responsibilities of individual supervisors concerning the health and safety program.
- List the role and responsibilities of the organization's management concerning the health and safety program.
- List the role and responsibilities of the incident commander concerning the health and safety program.
- List the role and responsibilities of the health and safety program manager concerning the health and safety program.
- List the role and responsibilities of the health and safety committee concerning the health and safety program.
- List the role and responsibilities of the incident safety officers concerning the health and safety program.

Resource Central

For additional review and practice tests, visit **www.bradybooks.com** and click on Resource Central to access book-specific resources for this text! To access Resource Central, follow directions on the Student Access Card provided with this text. If there is no card, go to **www.bradybooks.com** and follow the Resource Central link to Buy Access from there.

It is the responsibility of individual responders to understand their relationship to the overall safety and health program. The health and safety program manager and individual supervisors must also be familiar with the responsibilities of each member of the organization. All members who are responsible for the health and safety program within the organization must recognize that health and safety is an individual endeavor that extends from the top of the organization down.

The fire service has a long and bloody history as it has held on to the belief that injuries and death are a normal part of the job. Many members of emergency response organizations adopt an attitude of "it will never happen to me." They also commonly fall into the trap of complacency, and many younger inexperienced members feel they are "bulletproof." Add to that the general public's same attitude about fires and accidents. Responders must be on their guard against this outlook in their individual attitudes. Emergency response organizations need to stress to members that the potential for injury and death applies to any member. Encouraging members to maintain a proper individual attitude will help to ensure safe operations as they comply with procedures and policies.

Emergency responders are members of a team and must work together as a team. The size of the team depends on the type of unit assignment. Individuals must internalize that their individual actions will have an impact on the team as a whole.

Individuals who do not address wellness and fitness may find themselves at any given moment in a situation in which the rest of the team will have to dedicate itself to the rescue of a team member. No one is exempt from taking an active role in the health and safety program, whether it is by exercising good safety practices, being a member of a safety committee, or making health and safety suggestions to department administration.

It is vital that new members are mentored by more experienced senior members in all areas, but none is more important than that of health and safety. Individual dedication and a proper attitude concerning health and safety will help in reducing injury and death in emergency response organizations.

First-level supervisors play a major role concerning supervision and leadership; and none is more important than their responsibility to protect the health and safety of their individual crew. The goals of the health and safety program are the same at all levels of the organization, although the responsibilities vary. Individuals need to understand their role as stakeholders in their own personal safety, health, and wellness. This chapter discusses the various levels of responsibility dedicated to the health and safety program as they relate to the current fire service culture and how to create a culture of safety within the organization. Each member of emergency services organizations has an individual responsibility with health and safety. The culture of safety is not just an issue or responsibility of the administration of an organization. Health and safety is an issue of personal and individual responsibility in which all members must take part for the health and safety program to be successful.

The Health and Safety Program

The culture of safety starts with a health and safety program. Whereas most private employers are mandated to comply with all of the OSHA regulations, fire service organizations in approximately half of the states are not covered by the OSHA regulations, except to the extent the state has adopted specific OSHA regulations (such as SCBA fit testing and two-in/two-out in structure fires). OSHA governs how organizations must take steps to protect their workers. The challenge for the fire service and other emergency response organizations is taking the necessary steps to develop the continuous theme and culture of safety.

NFPA 1500, *Standard on Fire Department Occupational Safety and Health Program*, states that the application of this standard for all agencies (public, military, private, etc.) responsible for providing fire, rescue, or other emergency services should include the minimum requirements for an occupational safety and health program.

Members must be taught that health and safety is a way of life in emergency services. Health and safety must be dynamic and encompass both on- and off-duty activities. It should include wellness and fitness programs as a large part of the culture and not just a document on a shelf. The key to success is to get the health and safety program into the hearts and minds of the individual members. Leadership and commitment on the part of the administration and officers of the organization sets the example of being actively involved in promoting health and safety. The health and safety program manager, as well as the entire staff of the organization, must recognize and keep abreast of the national safety and health programs available. The World Wide Web has become an unlimited resource for safety and health programs; it only takes time and access to a keyboard to locate these valuable resources. All members of the organization, not just the line personnel, should be encouraged to research resources available to enhance health and safety; events such as the annual National Safety Stand Down must be viewed as worthy of incorporating into the department.

Employees must be empowered so that everyone takes on the role of safety officer. Education is the key to understanding the necessity of health and safety.

#2

The culture of safety should be reflected in the department vision and mission statements. Safety should be discussed regularly at all levels within the organization. Here are just some examples: a weekly training assignment encompassing a safety-related topic, safety posters put up in buildings that house emergency response personnel and in administrative offices, a daily safety message presented at roll calls, a safety suggestion program, and computer screen savers focused on safety and health that can be shared throughout the organization. The more employees have ownership in the health and safety program, the easier it will be to obtain buy-in.

#4

Preparing for injuries and an incident action plan should be developed for all training sessions involving hands-on skills. Policy and SOP implementation should have a "why" component as part of the process to encourage buy-in from employees. Employees must understand that changes related to safety are in their best interest. Line and staff personnel should study the causes of near misses and close calls that have occurred across the nation to focus on injury and death prevention in the organization. Although injury and death will always be part of emergency response, they must never be taken lightly or accepted as inevitable; every step must be taken to reduce these statistics in the emergency services. Change within the organization never comes easy, but focusing on its benefits will help.

If Not You, Then Who?

Each responder needs to understand that he or she can make a difference in the success of the health and safety program. Without the proper attitude toward health and safety, it will not matter how many policies and SOPs are written, how much money is dedicated, or how many safety officers are appointed. The program will fall short, or it will fail. A proper attitude is critical, and the individual's lack of it is one of the most difficult obstacles to overcome. Attitude is a mind-set that affects behavior. First-level supervisors must deal with attitude every time they interact with their personnel.

#1

It is the responsibility of the first-level supervisor to ensure the team stays on track concerning a proper attitude toward health and safety. A great attitude or a poor attitude will influence not only each individual responder but the team as well. All individual responders should be encouraged to take an active role in their own health and safety as well as the health and safety program; continually make suggestions for improving health and safety; and become a mentor for younger, less experienced members.

Every emergency response department engenders its own attitude and culture due to the behavioral mind-set of its dominant members or informal leaders. Formal and informal leaders within the organization have an audience at all times, so they have the potential to influence all the other members. These leaders must take their authority seriously by instilling a safety culture within the organization.

The leadership of the organization must take a look at what is causing accidents and injuries. Skills, knowledge, attitude, training, and experienced supervision are critical components of promoting the health and safety program and the safety culture. Safety, the delivery of service, pride in the organization, brotherhood, morale, and employee retention are just a few areas of a fire service organization that can be strongly influenced by the attitudes of the individual members and leadership.

Training is an integral part of the health and safety program. Why is it that some companies train each shift whereas others do not train at all? It is worth investigating the role that attitude plays in training.

We are all influenced by our own attitudes and those of others. The impact of attitude on organizational culture is just as important as the impact of health and safety on an individual responder. Annual line-of-duty death statistics do not account for injuries or deaths caused by improper attitudes. It is often obvious when reviewing these annual reports that attitudes about safety did in fact play a role in injury and death statistics.

Attitude is always there, and emergency services leaders must first self-examine how their words and actions influence the attitudes of those around them. Organizational factors that affect attitude and the safety culture are the working environment itself, promotional practices, policies and SOPs, rules and regulations, wages and benefits, and intrinsic and extrinsic motivators. Positive attitude influences will motivate, encourage team building and cohesion, contribute to individual fulfillment, and promote the organization's safety culture. Negative attitudes will affect the team and the organization as a whole by breeding mistrust, skepticism, rumors, and an unhealthy culture.

Some individuals may have a problem with administration and the way things are being done. Negative influences within the fire station will make it difficult for first-level supervisors to provide positive leadership and take the team in the correct direction. Dealing with personnel issues that come from negative attitudes engenders a focus on individual personnel issues, not on developing a safety culture. First-level supervisors can have a positive impact with negative attitudes if they can determine their cause. Then they can help those under their command see the big picture as it relates to the safety culture.

It is also important that everyone in the organization have some knowledge of the organization's budget process. In many cases, budget cuts are viewed as an uncaring attitude on the part of the local government leaders and the fire chief. Tax revenue shortfalls that result in budget cuts certainly have a negative impact on member attitudes. However, individual members must not lose sight of the organization's vision due to money shortages. Department administration and the first-level supervisors should be able to provide members with accurate information on how budget decisions are made, so that individual attitudes are not affected negatively. A great deal of frustration can be alleviated if communication flows up and down the chain of command; then the focus can remain on the vision, goals, mission, and safety culture of the organization.

Emergency services organizations will never be able to create enough standardized operating procedures, policies, and regulations to address every possibility should systems fail and accidents or acts of nature occur. There must be a strong commitment of resources, supervision, time, and energy though, as each individual commits to being an integral part of the safety culture. Although all the risks cannot be eliminated, department efforts must be focused on where they can do the most good.

The First-Level Supervisor's Responsibilities

There are many levels of supervision and management in emergency response and fire service organizations, and this section addresses one of the most important: the first-level supervisor. The individual at this level may or may not have a formal title; but, no matter the title, the responsibilities will remain the same. See Figure 10-1.

The activities of emergency services organizations are labor intensive as they rely on their human resources to accomplish the necessary tasks. First-level supervisors, as the backbone of supervision within the organization, must take responsibility and full accountability for their personnel. It is the first-level supervisor that ensures all safety practices are followed both at the emergency scene and while the team is back at quarters. The first-level supervisor also coordinates inspections and maintenance of the station and equipment as related to safety.

Of course, the first-level supervisor is an extension of the higher levels of administration, so he or she must have the training, education, and experience to recognize and address the dangers and hazards associated with the job. Among the first-level supervisor's important roles is to ensure that policies, SOPs, rules, and regulations are followed and that department personnel continually train to reduce the potential of injury and death. The first-level supervisor expresses team safety concerns to the administration and must encourage personnel to take an active role concerning their safety at all times.

FIGURE 10-1 The many responsibilities of the first-level supervisor are concerned with the safety of the crew.
Courtesy of Lt. Joel Granata.

Participation breeds ownership, and ownership instills pride. Many younger firefighters expect their supervisors to look out for their safety, involving the new members in the process by asking for their suggestions and comments. This is a much different working environment than in the past. First-level supervisors must learn to listen to younger employees' concerns about safety.

Results are achieved concerning improving the safety culture when team members believe their contributions to the organization are appreciated. The first-level supervisor must apply his or her experience and knowledge, set an example, provide leadership, define parameters and expectations, and take responsibility for the crew, being continually involved with safe operations and maintaining awareness.

The minute the new employee or volunteer walks in the door, the first-level supervisor must promote and cultivate the safety culture. He or she cannot unrealistically expect new employees just to fall in line when it comes to the safety culture, because they will not know what is required of them.

First-level supervisors should ask themselves two questions: "Am I doing the right thing?" and "Is my obligation as a first-level supervisor to the individual involved or to the organization?" The short answer to the second question is to both. Answering these two questions correctly will enable the first-level supervisor to focus on the organization's safety culture. Unfortunately, many organizations tolerate employees' minor safety infractions. When the infraction is of a serious nature, the first-level supervisor is the one who must step in and take it to the next level.

It is important that first-level supervisors do the right thing concerning the organization and the employee. Yet, in many cases, what the organization and the employees perceive as right can be as different as night and day. It is the job of the first-level supervisor to keep employees on track concerning the safety culture and to make decisions, not write formal policies, based on what benefits the safety of the organization and the employees. Face-to-face supervision is essential in promoting safety, even though it can sometimes be uncomfortable.

Interaction with the team is the key to success. The first-level supervisor must deal with employees in a direct fashion so that it is clear what he or she thinks and expects. As the first-level supervisor, just because you don't agree with a policy, SOP, rule, or directive doesn't mean you don't have to enforce it. You are the front person for the organization and the safety culture; the attitude of the team and their safety reflects your attitude and standards.

Encourage employees so that they will feel a deep sense of worth and ownership of the organization. The first-level supervisor must never overlook what makes people tick, nor be afraid to go against a negative firehouse culture. As the most influential individual in the training of department personnel, the first-level supervisor ensures that the team is prepared for the next event and that its members go home safely at the end of the shift. After all, training may be the only opportunity the crew has to work together outside of the emergency event. Because training is directly related to the safety of personnel, the first-level supervisor must never approach it in an unenthusiastic or lackluster way. Attitude alone will make or break any training session and ultimately affect the safety of personnel.

The first-level supervisor must make every effort to have a positive effect on the crew's skills, knowledge, abilities, and attitude. For example, certain first-level supervisors believe all training sessions must be formal and lengthy, yet many times training and safety can be accomplished in a short time frame without going out of service. Short training sessions pay large dividends that will improve the crew's ability to do the job safely and effectively. According to NFPA 1021, *Standard for Fire Officer Professional Qualifications*, it is desirable to have first-level supervisors certified as instructors as part of their duties. Even if the first-level supervisor does not provide formal training every shift, his or her responsibility is to ensure the crew operates safely. For example,

FIGURE 10-2 Short training sessions are an effective way to ensure proficiency. *Courtesy of Joe Bruni.*

a short training session could involve providing a copy of the SOP on accountability and having the members go over it on their own as a way to reinforce this important area. See Figure 10-2.

First-level supervisors must always remember that they are the number-one influence on the crew members, who are counting on these leaders to keep them safe. First-level supervisors must learn to think outside of the box and recognize that every incident is a perfect opportunity to administer training or conduct an incident review. First-level supervisors who are surprised by any incompetence in their crew should rethink their goals and responsibilities. Whether first-level supervisors are providing impromptu or formal training, they must ensure that all personnel get something out of it. Safety is nothing more than making certain the crew does it right every time. By being prepared to coach and not criticize crew members, the safety culture should easily fall into place.

The Organization's Staff and Management

The levels of staff and management depend upon the type of organization. For the purpose of this textbook, the senior management will be referred to as the staff.

#6

When it comes to the safety culture, the organization's staff must exercise leadership and commitment. The health and safety program is the responsibility of the organization's staff, which should ensure the program contains a wellness component. A health and safety program is of little value if its documentation just sits on the shelf in the fire chief or staff members' office. The staff needs to set an example in being actively involved in promoting safety by filling many roles in the health and safety program. The staff can do this in the following ways: by reinforcing the values and mission of the organization, by discussing safety on a regular basis at staff meetings, by wise control of the budget resources, and with final approval of organizational policies and procedures.

FIGURE 10-3 Example of an organizational safety policy.

The Mayor has asked the fire and rescue department to institute an effective safety program. Therefore, the ________ Fire Department has committed to establish and maintain a safe and healthy environment for our employees and the public through the implementation of an effective Departmental Safety Plan. Our plan will provide a systematic way to minimize incident frequency and severity, contribute to positive morale, effectively maintain city resources, and increase operational efficiency.

The following concepts are integral to the success of the Departmental Safety Plan. All ________ Fire Department employees are responsible for applying these concepts to each task and must work together as a team to achieve effective implementation.

1. The safety of employees will receive high priority in achieving objectives and goals.
2. Safe work methods always take precedence over expediency.
3. Incident prevention and efficient performance go hand in hand.
4. Safety must be integrated into every job task.
5. Safety is a team effort. Managers and supervisors are accountable for the proper training and supervision of their employees. Employees are responsible for following the rules and for working in a safe manner at all times.
6. Every effort will be made to reduce the frequency of losses that result in injury and suffering, property damage, and interruption of our service to the public.

I encourage each ________ Fire Department employee to play an active role in the safety program.

Thank you all for your contributions of service to the public.

Fire Chief ________ Date ________

The safety component should be reflected in the organization's policies and SOPs. It is the staff that must give the health and safety program support administratively and financially. Employees need to understand the "why" as it relates to safety, and the staff must ensure employees are continually apprised of why changes to policies and SOPs have taken place.

A people-oriented safety management program should result in the development of a culture of safety, as employees are encouraged to understand the staff is acting with their best interests in mind. The staff should encourage employee ownership of the health and safety program. Emergency services employees must learn not to accept injury and death as part of the culture. According to NFPA 1500, *Standard on Fire Department Occupational Health and Safety Program*, it is the staff's responsibility to adopt an official written departmental occupational safety and health policy that identifies specific goals and objectives for the prevention and elimination of accidents and occupational injuries, exposures to communicable disease, illnesses, and fatalities. See Figure 10-3.

incident commander (IC) ■ The IC is the person responsible for all aspects of an emergency response, including quickly developing incident objectives, managing all incident operations, applying resources, as well as responsibility for all persons involved. The IC sets priorities and defines the organization of the incident response teams and the overall incident action plan.

Incident Commanders

Many injuries and fatalities in the fire service occur because of a lack of adequate supervision by the **incident commander (IC)**. The NFPA standards require the use of an incident management system (IMS) at all incidents. OSHA requires the IMS at hazardous materials incidents. If an organization has accepted federal grant monies, it must adopt and train personnel in the National Incident Management System (NIMS). As the top of the IMS structure, the incident commander is responsible for the overall incident. The IC establishes immediate priorities, especially the safety of responders, other emergency workers, bystanders, and people involved in the incident. The purpose of an IMS is to stabilize the incident by ensuring life safety and managing resources efficiently and cost effectively. The IC has a great deal of responsibility to maintain the safety culture at all times by ensuring that adequate safety measures are in place. Based on the hazardous substances and/or conditions present, the individual in charge of the IMS needs to implement appropriate emergency operations and make certain that the personal protective equipment worn is appropriate for the hazards to be encountered.

The IC is also responsible for developing the strategic goals and assigning individuals the tactics to meet those goals. The risks must be balanced with the strategic goals, keeping safety in mind at all times. A risk-versus-benefit analysis must be completed quickly during the beginning stages of an incident and continually reevaluated throughout the course of the incident. The roles of the IMS participants will also vary depending on the incident and may even change during a particular incident.

Another key aspect of an IMS is the development of an **incident action plan (IAP).** The IAP initially dictates whether an offensive or defensive strategy will be in place at the incident. Incident commanders must base their decisions on experience, risk assessment, and knowledge. They should have a good base of knowledge in a breadth of areas, especially field knowledge, in order to operate effectively and safely as an IC. The incident commander must process a great deal of information during the initial stages, and, much of it may not be accurate. The IC needs to continually update the information coming in and base all decisions on life safety.

incident action plan (IAP) ■ An IAP formally documents incident goals, operational period objectives, and the response strategy defined by incident command during response planning. It contains general tactics to achieve goals and objectives within the overall strategy, while providing important information on event and response parameters.

The incident commander plays a major role in the safety culture by ensuring adequate health and safety measures are in place at all times during the incident and providing leadership during response. The effective IC must know which measures are effective and which are not. This knowledge is gained from past incidents and attending the post-incident analysis of his or her and other incident commanders' responses. Lessons learned can always be applied to the next incident, especially in the area of safety. The decision-making procedure of emergency responders, including the IC, at incident scenes is not the step-by-step process commonly taught in textbooks and classroom settings.

The **recognition-primed decision making (RPD)** process is what is actually being used at the incident scene. Emergency response personnel learn the step-by-step process during the beginning stage and in the classroom setting; however, the incident scene does not afford the time or the opportunity to use this process. Emergency response personnel including incident commanders arrive on scene and begin to take control by basing their incident decisions on their training or what has worked in the past at the same type of incident. RPD decision making may be rapid, but it is prone to serious failure in unusual or misidentified circumstances.

recognition-primed decision making (RPD) ■ This is a model of how people make quick, effective decisions when faced with complex situations. The decision maker is assumed to generate a possible course of action, compare it to the constraints imposed by the situation, and select the first course of action that is not rejected.

When it comes to RPD thinking, the difference of being experienced or inexperienced plays a major factor in the decision-making process. Experienced personnel will generally be able to come up with a quicker decision because the situation may match a situation they have encountered before. Inexperienced individuals will run several possibilities through their minds and tend to use the first possibility they believe will work. Unfortunately, for the inexperienced IC, this trial-and-error approach could compromise personnel safety in situations that are foreign to the inexperienced IC. RPD thinking causes an individual to select different courses of actions to take; if one does not work, he or she proceeds to the next until the first effective course of action is found. Unfortunately, inexperienced decision makers are less likely to choose the most proficient course of action first.

At times, RPD thinking can cause a compromise involving safety with inexperienced personnel both at the IC level and at the lower levels of emergency responder.[1] In essence, RPD thinking can either save lives or cost lives. Therefore, inexperienced incident commanders should continually practice incident simulation exercises as a way to ensure safety is addressed in a nonemergency setting. See Figure 10-4.

FIGURE 10-4 RPD thinking can either save lives or cost lives. *Courtesy of Larry McGevna.*

The Health and Safety Program Manager

After the enactment of the OSHA Act of 1970, the fire service continued to view itself as a macho organization that accepted and embraced risks. Many leaders involved with NFPA 1500 began to send the message to the U.S. fire service that the fire service must take care of its own. Members of the fire service and other emergency services organizations must understand that the only ones who truly care about "us" are "us." No amount of regulations and standards can reduce the high injury and death rate in the fire service. The only way to truly see a reduction in injury and death is for the fire service organizations to create individual safety cultures within their own organizations. See Figure 10-5.

It is the fire chief who has the responsibility to have a health and safety program in his or her organization. NFPA 1521, *Standard for Fire Department Safety Officer*, and NFPA 1500, *Standard on Fire Department Occupational Safety and Health Program*, state the department must have a health and safety officer to manage the health and safety program. The individual assigned the role of health and safety program manager should be a member of the organization's staff or executive administration. He or she oversees the program and all of its components. The goal of NFPA 1500 is to outline the minimum requirements and specify the safety requirements of the occupational health and safety program. Certain members of emergency services organizations mistakenly believe they do not have to follow OSHA regulations because they do not operate within a state that has its own OSHA plan. Many also believe their organizations do not have to follow the NFPA consensus standards because they are not law. The question to ask is: "What would you consider as reasonable to protect workers' health and safety under the general duty clause in your jurisdiction?" More than likely, the answer would be the NFPA 1500 standard.

The health and safety program is really more of a system than a program, and its components include policies, SOPs, and training to achieve its goals. Depending on the size of the organization, the health and safety program manager may have a staff of assistants or in smaller organizations may also serve as the department training officer or training chief. See Box 10-1.

FIGURE 10-5 Firefighters must learn to accept but not embrace risk.
Courtesy of District Chief Mike Zamparelli.

BOX 10-1: RESPONSIBILITIES OF THE HEALTH AND SAFETY PROGRAM MANAGER

- Serves as chair of the occupational health and safety program
- Serves as the risk manager for the department
- Develops and coordinates a confidential record-keeping system related to health and safety
- Coordinates an accident investigation program including apparatus accidents and firefighter fatality and injury investigations
- Acts on recommendations from individuals and the safety committee
- Implements the department risk management plan
- Is involved with the development of health and safety policies
- May function as the department infection control officer or serve as a liaison with medical control personnel
- Is a liaison with the department physician concerning health-related issues
- Is a liaison with the workers' compensation provider
- Assists in incident scene safety by filling the role of incident safety officer or providing other assignments deemed necessary by the incident commander, such as those within the incident command structure
- Participates in and develops health and safety programs
- May perform facility inspections for unsafe conditions
- Conducts research on apparatus and equipment concerning safety-related issues
- May develop specifications on apparatus and equipment in conjunction with a representative group from the organization
- Attends and participates in post-incident analysis
- Has input in department SOPs
- Maintains an awareness of trends in emergency services health and safety including standards, regulations, and court decisions

In order to be effective, the department health and safety program manager must have sufficient education and experience concerning risk management, safety and health issues, cost–benefit evaluation, infection control, and incident operations. The department health and safety program manager must also meet the requirements of Fire Officer Level I found in NFPA 1021, *Standard for Fire Officer Professional Qualifications*. He or she assesses which programs are most important and decides what actions need to be taken to meet the intent of the NFPA 1500 standard. The qualifications of the health and safety program manager include:

- Maintain knowledge of current principles and techniques of occupational health and safety management systems.
- Maintain knowledge of NFPA 1500 and other current laws, codes, and standards regulating occupational health and safety.
- Maintain knowledge of hazards related to diseases, illnesses, and injuries related to emergency and nonemergency operations.
- Maintain knowledge of current health maintenance and physical fitness issues.
- Maintain knowledge of infection control practices and procedures required in NFPA 1581.
- Maintain knowledge of practices and procedures for live fire training evolutions related to NFPA 1403.
- Maintain knowledge and awareness of the work of safety organizations, standards-making organizations, and regulatory agencies as they relate to unsafe practices and existing hazardous conditions that affect the policies and procedures of the department.

The department health and safety program manager must also be able to effectively answer the following questions: How can we establish a priority action plan? Whose authority, permission, or support will be needed? Are there roadblocks getting in the way? What resources are necessary and required? What is the best way to get started?

The Safety Committee

The emergency services organization should have an occupational safety and health committee that serves in an advisory capacity to the fire chief or director. This committee should be comprised of the health and safety program manager or safety officer, individual members from the bargaining unit if applicable, management, and other resources. Other resources could include personnel from risk management, workers' compensation, and the legal department that represents the department. The safety committee can be a valuable and committed resource for the health and safety program director and other members of the organization. Its purpose is to review safety matters, conduct research, and develop safety recommendations. In some organizations, the safety committee may assist with the health and safety program manager's review of accidents and injuries.

#9

The safety committee may also contribute its efforts to reduce accidents and injuries. Employee and union involvement help to ensure buy-in with the health and safety program, as well as development of the safety culture within the organization. It is important that employees feel involved in the area of safety recommendations and procedure and policy development.

The safety committee typically meets semiannually or annually; meeting minutes should be taken and retained for members of the organization. Sometimes it may be necessary for the group to meet more often. In the State of Florida, the safety committee is required to meet every quarter. It can be one of the most effective groups within the organization concerning the health and safety of personnel, with the potential to be the keystone of personnel safety. See Figure 10-6.

FIGURE 10-6 The safety committee should be comprised of members at various levels of the organization. *Courtesy of Joe Bruni.*

In many cases where applicable, a bargaining unit may be required by the collective bargaining agreement to establish a health and safety committee within the organization. Yet even when no contractual agreement requires a health and safety committee, one should be in place and have representation from all ranks. The lone establishment of a health and safety program and committee is not enough. The health and safety committee must evaluate the effectiveness of the health and safety program, or the program is destined to fail. If possible, the health and safety committee's chairperson should not be the health and safety program manager; however, many organizations utilize the health and safety program manager in this role because of this individual's knowledge and experience.

#16

The health and safety committee should establish and embrace several core elements: first and foremost, strong organizational support and leadership, participation and commitment, specific goals and vision, and a continuous need for improvement. Recommendations from the health and safety committee should be forwarded to the staff and management of the organization to ensure action on these issues. It will be beneficial if members of the health and safety committee are the types of personnel who network with other agencies in emergency services. The health and safety committee and the knowledge of its members are well suited to spot trends in emergency services and help implement positive change in areas such as training, SOPs, policies, and equipment.

The health and safety committee can also serve as a powerful research group for the organization by providing historical information concerning injuries, injury reports, and past recommendations. There are many obstacles to firefighter safety, and the health and safety committee has the responsibility to work to overcome these obstacles. One of the common obstacles is the organization's operating budget, yet budget considerations should not dictate the health and safety issues the health and safety committee must address. If an issue is important, the health and safety committee must discover a way to make it happen.

The Incident Safety Officer

It is now standard for organizations to have two types of safety officers: the department health and safety officer or program manager, and the incident safety officer. This section covers the role and responsibilities of the incident safety officer (ISO).

The role and function of the ISO greatly contributes to the overall health and safety program and the safety culture. In many cases, the size of the organization dictates who will serve as the ISO. Some organizations assign one or more individuals from within a safety division to be in the dedicated position of ISO. Other organizations may choose to utilize first-level supervisors at each specific incident. Some departments utilize dedicated units and personnel that respond to incidents to handle the many safety-related roles such as incident safety, rapid intervention team, entry control officer, and accountability.

Each department must ensure the individual serving in the incident safety role possesses the necessary skills and knowledge. ISOs need to have formal training in the core elements of incident scene safety as well as a great deal of experience related to the incident at hand, and they should not be assigned to any other duties. Incidents other than typical everyday type of events—such as hazardous materials, technical rescue, and water-related incidents—necessitate a higher level of incident scene safety officer.

The incident itself and the expertise of the individual should dictate who will be assigned as the ISO. NFPA 1521, *Standard for Fire Department Safety Officer*, states the responsibilities and the requirements related to the ISO. This NFPA standard also requires the appointment of an ISO when the need arises. In addition, NFPA 1021, *Standard for Fire Officer Professional Qualifications*, requires the ISO to have the education and experience to serve as the ISO for the given incident. A thorough knowledge of safety and health concerns, hazards, and the organization's accountability system should dictate who is assigned to the safety officer role at every type of incident. A safety officer assigned to a fire incident should be very familiar with building construction, fire behavior and growth, and rehabilitation procedures.

#3

In order to possess a thorough understanding of operational risk management, ISOs must study pre-incident risk management. Such research will lead to a better understanding of the immediate risks faced at the incident scene. ISOs must have developed the ability and experience to predict and forecast events at the incident scene.

Just as with incident commanders, ISOs must be able to quickly assess the situation and predict where the incident is going in order to keep personnel out of harm's way. The effective way to do this is by using a 360-degree view of the incident, a review of preplans, and an evaluation of strategy and tactics. Much like the incident commander, the ISO should talk to occupants and neighbors concerning the incident, known hazards, and events leading up to the incident. See Figure 10-7.

Operations should always be assessed from the safety standpoint. Every incident varies from the next; however, the role of the ISO is to evaluate the risks versus the benefits, and make a coordinated effort to ensure the scene that response personnel work within is properly managed from the safety standpoint. The ISO may assist the incident commander by attempting to talk to occupants, property owners, and neighbors about the situation faced by response personnel. In many instances, people located at the incident scene can offer detailed and valuable information. The 360-degree walk around the structure or incident also enables the ISO to make contact with the general public located at the scene. The ISO reports directly to the IC and must have the authority to instantly stop any unsafe acts occurring on the emergency scene. It should be a top priority for the ISO to be familiar with the incident action plan (IAP), so he or she can be better aware of the type of operations taking place.

FIGURE 10-7 The incident safety officer should be experienced and knowledgeable.
Courtesy of Lt. Joel Granata.

The incident action plan defines the strategic goals, tactical objectives, risk management, support necessary, and safety of members. Along with becoming familiar with the incident action plan, the ISO should evaluate operation personnel in the area of personal protective gear, personnel accountability, and freelancing. Any unsafe action that is stopped by the ISO should be reported to the IC as quickly as possible. Occupational health and safety programs throughout the fire service have made improvements in the health, safety, and welfare of responders; however, each individual member shares the responsibility to comply with the particular department's health and safety program. Although several members within each emergency response organization have definitive roles regarding the health and safety of members, the ISO and IC play the leading roles at the incident scene.

The ISO is a part of the incident commander's command staff as the ISO works to ensure the incident command system minimizes the risk to responders. Whereas the IC has overall responsibility for responder safety, the ISO has direct responsibility related to responder safety. An effective ISO utilizes a worksheet at the incident scene to remind him- or herself about all safety issues. Several types of incident worksheets should be available to the ISO, depending upon the type of situation faced. For example, a worksheet for a structure fire would differ from a worksheet involving vehicle extrication, technical rescue, or a hazardous materials incident. See Figure 10-8.

Worksheets can also become part of a post-incident analysis. An individual who has the knowledge of safety, hazards, and the organization's safety procedures should fill the ISO position. The IC must be able to rely on the ISO's expertise and make use of this valuable individual at the incident scene. The ISO must keep in mind that many times responders focus on the work at hand and not on their own safety. The only job of the ISO is to focus on safety during emergency operations. The duties and qualifications of the incident safety officer are clearly defined in NFPA 1521. See Box 10-2.

FIGURE 10-8 Incident safety officer checklist for a hazardous materials incident.

SAFETY OFFICER CHECKLIST

Safety Officer: __________ Date: __________

Duty Checklist:

__________ Report to command and establish communications with the incident commander.
__________ Obtain a briefing and review incident action plan.
__________ Identify hazardous and unsafe situations associated with the incident.
__________ Assist the incident commander in developing the safety plan.
__________ Brief team members and additional safety officers as necessary regarding hazards of the emergency.
__________ Assist in determining the zones of operation (hot, warm, cold).
__________ Ensure that the team members are wearing the correct protective clothing.
__________ Ensure the backup team is in place.
__________ Ensure decon is in place.
__________ Ensure incident action plan is adequate.
__________ Conduct operations and safety briefing for all personnel.
__________ Ensure that an adequate medical evaluation has been made of entry and backup teams and findings are recorded.
__________ Ensure emergency medical treatment and transport capability is in position and operational.
__________ Ensure that SCBA air time is recorded and tracked.
__________ Exercise your authority to stop and/or prevent any and all unsafe acts.
__________ Provide update to IC and hazmat team leader.
__________ Monitor activities until operations have been terminated.
__________ When ordered, secure operations and forward all necessary reports/logs to the incident commander and hazmat team leader.

BOX 10-2: DUTIES OF THE INCIDENT SAFETY OFFICER

- The ISO and assistant ISOs must be identified clearly on the scene. The initial ISO must make the IC aware of the need for additional ISOs.
- The ISO must monitor responder safety and make sure all actions fall within the incident action plan. The ISO has the authority to terminate or suspend operations that fall outside the incident action plan or risk management plan.
- The ISO must ensure command establishes rehabilitation for responders.
- The ISO must monitor the scene and provide the IC with reports on conditions, hazards, and risks.
- The ISO must ensure the department accountability is in place and used.
- The ISO must review the incident action plan and provide the incident commander with a risk assessment based on the plan.
- The ISO must ensure that collapse zones, hot zones, and safety zones are known to all members operating at the incident scene.
- The ISO must monitor vehicle traffic near the incident to ensure the safety of responders, and may cause apparatus to reposition to provide a shield.
- The ISO must assist with the safe establishment of landing zones when helicopters are used at incident scenes.
- The ISO must ensure rapid intervention teams are in place and advise the IC of potential building collapse, rapid fire progression, fire extension, and access and egress points for personnel.
- At the EMS incident, the ISO must ensure infection control measures are in place and that rehabilitation and critical incident stress debriefing are in place when needed.

(continued)

- At the scene of hazardous materials and special operations incidents, the ISO must attend planning sessions to provide safety input. The ISO must ensure a safety briefing for responders is conducted. The ISO will develop and distribute a safety incident action plan and ensure the hot, warm, and cold (decontamination) zones are clearly marked. The ISO will meet with the IC to ensure rehabilitation, accountability, rapid intervention, and provisions for feeding and hygiene are in place for incidents of long duration.
- The ISO will communicate information concerning responders who become ill or are injured on scene. The ISO will ensure an accident investigation is conducted and request assistance from the department health and safety officer when needed.
- The ISO will prepare a written report concerning the post-incident analysis that includes information about the incident from an occupational health and safety perspective.
- The ISO must monitor radio communications and ensure that any barriers that would affect responder safety are addressed.

Source: FEMA.

The qualifications of the ISO include:

- Meet the requirements of Fire Officer Level I.
- Possess the knowledge, skills, and abilities to manage incident scene safety.
- Maintain knowledge of safety and health hazards in emergency operations.
- Maintain knowledge of building construction.
- Maintain knowledge of fire science and behavior concerning hostile fire events.
- Maintain knowledge of the department's accountability system.
- Maintain knowledge of incident scene rehabilitation strategies.

CHAPTER REVIEW

Summary

Every member involved in emergency response organizations has a responsibility to the overall safety and the health program. The program will not be a success without dedication from top management down to each individual responder. The responder must learn the negative consequences of becoming complacent and adopting the mentality of "it will never happen to me."

Many members within the organization play a specific role regarding health and safety. These include the health and safety program manager, incident commanders, and the incident safety officers. The individual at the center of the health and safety program—the health and safety program manager—must be experienced and knowledgeable in many areas in order to be effective. The most crucial of these is risk management, which is also an area of importance for incident commanders and incident safety officers.

It is vital for each emergency services organization to have a safety committee in place that provides input concerning health and safety matters. Its purpose is to review safety matters, conduct research, and develop safety recommendations. It is also important that employees feel involved in this top priority area of health and safety. The safety committee can also be instrumental in injury and accident investigations, as well as SOP and policy development concerning health and safety matters. The safety committee should have representation from all levels of the organization, including management. The incident safety officer responds to the scene, becomes part of the command staff, and reviews the overall scene and the incident action plan in order to make informed decisions and recommendations to the incident commander. Many safety-related issues exist at each type of incident, and the incident safety officer should use a checklist specific to the type of incident. The appointment of a qualified incident safety officer at incident scenes displays the organization's commitment to health and safety.

Review Questions

1. Name one of the beneficial reasons for having a department health and safety committee.
2. Name a positive attribute of an individual responder.
3. What is the most important role of the first-level supervisor?
   a. To provide crew training
   b. To protect the health and safety of the crew
   c. To become a member of the department health and safety committee
   d. To meet the requirements to become an incident safety officer
4. List two things that an incident safety officer might look for at an incident.
5. What is the purpose of the health and safety committee?
   a. Review safety matters
   b. Conduct research
   c. Develop safety recommendations
   d. All of the above
6. A culture of safety starts with:
   a. Individual responders
   b. Incident safety officers
   c. Incident commanders
   d. A health and safety program
7. Which organization governs how emergency services organizations must take steps to protect their workers?
   a. NIOSH
   b. NFPA
   c. OSHA
   d. CDC
8. Health and safety must be dynamic and it must encompass only on-duty activities.
   a. True
   b. False
9. What will make it easier to develop the health and safety program and the culture of safety?
10. Name the most important area concerning health and safety.

Case Study

On January 25, 2009, two male career firefighters, age 28 (Victim 1) and age 45 (Victim 2), died after falling from an elevated aerial platform during a training exercise in Texas. The firefighters were participating in the exercise to familiarize fire department personnel with a newly purchased 95-foot mid-mount aerial platform truck. A group of four firefighters were standing in the aerial platform, which was raised to the roof of an eight-story dormitory building at a local college. The platform became stuck on the concrete parapet wall at the top of the building. During attempts to free the platform, the top edge of the parapet wall gave way and the aerial ladder sprung back from the top of the building, and then began to whip violently back and forth. Two of the four firefighters standing in the platform were ejected from the platform by the motion. They fell approximately 83 feet to the ground and died from their injuries.

Key contributing factors identified in this investigation include the firefighters being unfamiliar with the controls of the newly purchased aerial platform truck, training in a "high-risk" scenario before becoming familiar with new equipment, failure to use fall restraints, the design of the platform railing and integrated doors, and the location of the lifting eyes underneath the platform, which contributed to the platform snagging on the building's parapet wall.

Occupational injuries and fatalities are often the result of one or more contributing factors or key events in a larger sequence of events that ultimately result in the injury or fatality. NIOSH investigators identified the following items as key contributing factors in this incident that ultimately led to the fatalities:

- Unfamiliarity with the controls of a new fire apparatus
- Training in a high-risk situation without adequate familiarization with the fire apparatus
- No fall restraint devices in use during training at height
- Design of the lifting eyes (one of which snagged the parapet wall) and platform doors, which sprung outward during the incident.

NIOSH had the following recommendations:

Recommendation #1: Fire departments should ensure firefighters are fully familiar with new equipment before training under "high risk-scenarios." Training on the proper method of operating an aerial platform should be done in low-risk settings. In this incident, the fire department initiated additional training following a structure fire so that firefighters could gain experience in setting up and operating the newly purchased aerial platform apparatus. A number of firefighters interviewed by NIOSH investigators reported jerky and unsteady movement of the ladder and platform while operating the controls. The use of a spotter stationed on the roof of the dormitory may have aided in guiding the platform operator while the platform approached the building.

Recommendation #2: Fire departments should ensure fall protection is used whenever firefighters and other personnel are working in elevated aerial platforms. NFPA 1901, *Standard for Automotive Fire Apparatus,* 2003 edition, requires the apparatus to be equipped with four (4) ladder belts before the apparatus is placed in service. The use of safety belts is required by the NFPA 1500 standard. The most obvious was a "Warning" box located in the aerial safety section, which stated, "Do not allow personnel on the end of a moving aerial unless they are secured to the aerial with a personal protective safety belt."

Recommendation #3: Fire departments should follow standard operating procedures (SOPs) for training, including the designation of a safety officer. In this incident, the entire duty shift reported to the college dormitory for hands-on training to familiarize firefighters with the operation of the new ladder truck. The fire department was in the process of writing a new standard operating procedure (SOP) for the ladder truck, but it had not been completed. Hands-on training provided by the manufacturer prior to the incident was conducted without the use of fall protection. Firefighters interviewed by NIOSH reported that informal practice with the new aerial platform was also conducted without the use of fall protection. A designated safety officer at the training exercise may have positively influenced the use of fall protection during the training exercise.

Recommendation #4: Fire departments should ensure SOPs covering the operation and use of fire apparatus (including aerial platform apparatus) are developed and followed during training exercises as well as in fire suppression activities. In this incident, the fire department was in the process of developing a new SOP to cover the operation and use of the newly purchased aerial platform apparatus.

Recommendation #5: Fire apparatus manufacturers should provide fall protection belts with all aerial ladder and platform apparatus, and ensure that fall protection is used in manufacturer-provided training. In this incident, the newly purchased fire apparatus was delivered without fall protection included. Fall protection belts from the original ladder truck were available, and new fall protection belts for the new platform apparatus were on order but had not been delivered to the fire department at the time of the incident.

Recommendation #6: Fire apparatus manufacturers should ensure that aerial platforms and other aerial devices are designed to reduce or eliminate the potential for snagging on buildings or other elevated surfaces. By necessity, aerial platforms should be designed to reduce or eliminate the potential to snag or catch on buildings or other structures as the aerial platform is being maneuvered.

Recommendation #7: Fire apparatus manufacturers should ensure that aerial platform doors or gates are designed to prevent opening in the outward direction. During this incident, both platform doors were sprung outward past their intended stopping point, resulting in the two victims falling through the openings in the platform guardrail. Wider retaining strips at the top and/or additional retaining strips below the door latch may have served to keep the doors from moving past their intended stopping point. The platform doors were equipped with simple spring-loaded latches. The use of a different type of door latch, such as one requiring a turning or twisting motion to open the latch, may also have prevented the doors from opening inadvertently.[2] See Figure 10-9.

1. Discuss the contributing factors that led to the fatalities in this incident.
2. Discuss how the design of the apparatus contributed to this incident.

Fatal Aerial Incident Concerns and Issues	Yes	No
Did the department provide adequate training with this apparatus?		
Should the apparatus have been set up over a building during training?		
Was there an adequate incident management system in place?		
Should this training event have been considered a high-risk event?		
Would the use of the proper safety equipment have prevented this fatality?		
Was a safety officer assigned to this training event?		
Did the lack of recognition of the hazards play a role in these fatalities?		
Should an SOP have been established prior to training with new apparatus?		
Should the manufacturer have provided this type of training?		
Should the manufacturer have provided fall protection for this apparatus?		

FIGURE 10-9 Concerns and issues concerning fatal aerial apparatus incident.

References

1. G. Klein, "Recognition-Primed Decisions," in *Advances in Man-Machine Systems Research*, vol. 5, ed. W. B. Rouse (pp. 47–92) (Greenwich, CT: JAI Press, 1989).
2. NIOSH Fire Fighter Fatality Investigation and Prevention Program, *Two Career Fire Fighters Die After Falling from Elevated Aerial Platform—Texas*, July 13, 2009. Retrieved February 15, 2010, from http://www.cdc.gov/niosh/fire/reports/face200906.html

CHAPTER 11

Safety Program Appraisal—Making It Happen—If Not Now, Then When?

KEY TERMS

evaluation process, *p. 237*
formative evaluation, *p. 238*
impact evaluation, *p. 240*
outcome evaluation, *p. 240*
paradigm shift, *p. 242*
process evaluation, *p. 238*

OBJECTIVES

After reading this chapter, you should be able to:

- Describe the reasons why an evaluation of the health and safety program is so important.
- List the several methods used in the evaluation process.
- Describe the evaluation process.
- Describe the differences between the various methods used in the evaluation process.
- Describe who is responsible for the evaluation process.
- Describe the frequency of evaluation, and list the factors that will affect the frequency of evaluation.

ResourceCentral

For additional review and practice tests, visit **www.bradybooks.com** and click on Resource Central to access book-specific resources for this text! To access Resource Central, follow directions on the Student Access Card provided with this text. If there is no card, go to **www.bradybooks.com** and follow the Resource Central link to Buy Access from there.

The purpose of the appraisal process for a health and safety program is to determine its effectiveness. According to NFPA 1500, *Standard on Fire Department Occupational Safety and Health Program*, a department needs to evaluate the effectiveness of the occupational safety and health program at least once every three years. The appraisal process assesses the goals and objectives of the health and safety program. An evaluation questionnaire should be developed and distributed throughout the organization. Of course, there should be time constraints associated with a health and safety program evaluation. The time required for the evaluation should not drag on for months. The fastest way to complete an evaluation is to have the administrative staff come together during the evaluation process, and use this opportunity as a brainstorming session to formulate the health and safety plan. If the health and safety plan is developed by the health and safety program manager as a lone individual, the administration will undoubtedly have questions which will have to be addressed slowing down the process toward implementation. The organization should utilize strategic planning to determine its present condition and compare this to where it would like to be in the near future. First, individuals involved in the evaluation process need to thoroughly understand what process the organization will utilize, and an effective system is required to measure results. Once these are in place, the organization can move forward with the evaluation process. See Figure 11–1.

The following questions are designed to help you evaluate the effectiveness of your current written safety and health program.

Written Safety and Health Program

1. Does your agency/institution have a *current written* safety and health program that addresses the following elements?
   - a. Management, commitment, and leadership? Yes ____ No ____
   - b. Safety performance standards for managers and supervisors? Yes ____ No ____
   - c. Employee involvement/safety committees? Yes ____ No ____
   - d. Written safety rules/procedures? Yes ____ No ____
   - e. Safety inspections/hazard assessments? Yes ____ No ____
   - f. Loss prevention and control techniques? Yes ____ No ____
   - g. Regulatory compliance activities? Yes ____ No ____
   - h. Safety and health training? Yes ____ No ____
   - i. Accident reporting and investigation? Yes ____ No ____
   - j. Safety and health promotion activities? Yes ____ No ____
   - k. Return to Work Policy? Yes ____ No ____
2. Are all managers, supervisors, and employees aware of the program? Yes ____ No ____

Comments/Action Steps:

____

____

____

____

FIGURE 11–1 Safety and health written program evaluation questionnaire.

Making It Happen—The Appraisal Process

The **evaluation process** normally involves an in-depth appraisal over the course of time. Evaluation of a safety program consists of four methods: formative evaluation, process evaluation, impact evaluation, and outcome evaluation. The evaluation process of the safety program consists of the planning phase as well as following the program through to completion. Each of the elements of the development and implementation of the program plan has a corresponding level of evaluation. Defining the goals, objectives, and strategies of a program determines what is measured at each level of evaluation. Many tools can be used to perform an evaluation of the health and safety program. One such example is the evaluation checklist developed by the Florida State Fire College in compliance with the OSHA regulations and the Florida Administrative Code.

evaluation process ■ This is a systematic method of collecting, analyzing, and using information to answer basic questions about projects, policies, and programs.

See the Fire Department Health and Safety Self-Evaluation Compliance Checklist for more safety initiatives information.

The important issue is to compare a department's actual outcomes with the expected outcomes. A questionnaire can be distributed department-wide to evaluate the effectiveness of the health and safety program. The categories within the health and safety questionnaire may include information about the following:

- The health and safety written program
- Managers
- Supervisors
- Health and safety program manager
- Employees
- Safety committees
- Safety and health inspections
- Regulatory compliance
- Safety and health training
- Illness/injury prevention activities
- Accident reporting and investigation
- Emergency preparedness

The health and safety committee can be the most effective when it comes to health and safety program evaluation. In many fire service organizations, the health and safety committee includes both labor and management representatives. There may even be a provision in the collective bargaining agreement regarding the health and safety committee. Safety committees are well suited to spot trends as they review accident or injury reports, research safety complaints, or network with others in the fire service. The safety committee's collective knowledge may prove instrumental in evaluating the health and safety program. When an emergency response organization discovers that its health and safety is not what it should be, the health and safety committee makes recommendations for positive change. An effective and credible safety committee normally keeps accurate records, which will be an effective tool for program evaluation. The safety committee also serves as a source of historical information, which may include lessons learned by other agencies that have conducted their own internal research following injuries or line-of-duty deaths.

The safety committee can also function as a powerful research group when it comes to program evaluation. Safety committee members provide insight into historical practices and know where to find relevant information. Accurate and detailed records will help with any historical research. During times of change, the safety committee will be viewed as subject matter experts as it relates to program evaluation.

There are many reasons to evaluate health and safety program effectiveness:

- The organization must have a clear understanding of the program's effectiveness.
- The response to the program's effectiveness must be determined from the organization's members.
- It is a successful way to implement changes within the program related to where it stands as compared to where it needs to be.

FIGURE 11–2 The safety program manager performing an annual evaluation of the safety program. *Courtesy of Joe Bruni.*

The evaluation process should involve two methods: process evaluation and outcome evaluation. See Figure 11–2.

FORMATIVE EVALUATION

formative evaluation ■ The purpose of this type of evaluation is to improve programs. It can be contrasted with other types of evaluation that have other purposes.

Procedures, activities, and materials become the focus of **formative evaluation**, which ensures that they will work as planned. The formative evaluation offers the opportunity to solve any unexpected problems that may arise early in the safety program. It requires documentation of the program including its plan, goal, objectives, target groups, control or comparison groups, program strategies and organization, management, and program implementation. Pilot testing of procedures, activities, and materials is done for feasibility and acceptability.

Achieving consistent quality is the goal of this stage of evaluation. Formative evaluation is important when the safety program must be modified, altered, or adapted for use with a different target population in a new location, because there is no guarantee that a program that is successful under one set of circumstances will translate its success to other circumstances. Once a safety program is implemented, continuing formative and/or process evaluation will provide feedback to the health and safety program manager, as well as an evaluation of the consistency, integrity, and quality of the program throughout its existence. The formative evaluation phase could also include research into line-of-duty deaths done by the NIOSH Fire Fighter Fatality Investigation and Prevention Program or at OSHA.

process evaluation ■ This evaluation focuses on how a program was implemented and operates. It identifies the procedures undertaken and the decisions made in developing the program. It describes how the program operates, the services it delivers, and the functions it carries out.

PROCESS EVALUATION

A **process evaluation** focuses on what services were provided to whom and how. Its purpose is to describe how the health and safety program was implemented, who was involved, and what problems were experienced. It is useful for monitoring program implementation; for identifying changes needed to make the program operate as planned; and, generally, for program improvement. Process evaluation is basically an analysis of the health and safety program's procedures. The process evaluation should be able to determine how well the program accomplished what it was set up to do. For example, training designed to reduce or eliminate certain types of injuries should have a component to examine whether the training had the desired effect. The process evaluation should answer the following questions:

- What area of the program appears to be the most effective?
- What area of the program appears to be the least effective?
- What personnel did the program affect?
- To what extent were they affected?
- Are improvements occurring as expected?

The training evaluation considers whether what is being accomplished is right versus being easy. In most cases, it is the first-level supervisor or company officer's job to verify the level of knowledge and performance of his or her personnel. From the perspective of the supervisor, the evaluation concerning training should be a combination of the supervisor's perspective and the organization's standards. The first-level supervisor determines the necessary critical tasks in order to evaluate whether training has brought about desired improvements. First-level supervisors must understand their obligation is to do the job correctly as it relates to training their personnel. They should evaluate employee strengths and weaknesses, and establish an environment that creates motivation. Each individual fire department should consider the Insurance Services Office (ISO) rating schedule as it relates to training. A review of the ISO schedule helps an organization use process evaluation to determine whether it is meeting the necessary benchmarks in the training area. The key to doing so is to document and keep the necessary training records. Capturing an organization's training records enables those doing the process evaluation to determine whether the necessary training has been delivered as planned. NFPA 1500 also requires each organization to maintain each member's training records. The process evaluation determines whether the training records indicate dates, subjects covered, satisfactory completion, and any certifications achieved.

As it relates to training, the process evaluation should be able to determine whether the organization is administering and utilizing the types of standard training evolutions required by NFPA 1500. The health and safety program manager's responsibility is to provide training in safety procedures related to operations to all members of the organization. The process evaluation will determine the effectiveness of these efforts and assist those in charge of training to assess whether recommendations concerning the investigation of accidents, injuries, deaths, illnesses, and exposures are being addressed.

Process evaluation determines whether the program has been implemented or delivered as planned. It also measures the following: the reach of the program to the intended group, implementation of activities, participant satisfaction, performance of materials and other components, and the program's ongoing quality assurance. It is not uncommon to find that changes in program delivery occur over time, so that materials and activities are no longer the same as at the beginning of a program, for example, changes in staff and settings and in availability of certain materials.

The ways to achieve and perform process evaluation are varied but include written and practical testing concerning procedures and policies, observations at both the incident scene and during downtime at the station, and supervisor feedback. Results should be measured against the program objectives. The health and safety program manager must review the results of program evaluation to decide on necessary changes within the organization. If an evaluation of measures shows low response to program activity, then additional evaluation may identify the problem and help provide a solution before evaluating the impact of the program.

IMPACT EVALUATION

impact evaluation ■ It assesses the changes in the well-being of individuals that can be attributed to a particular intervention, such as a project, program, or policy.

Impact evaluation assesses changes in risk factors and measures achievement of program objectives. Impact evaluation also involves the assessment of immediate program effects on risk factors for the health and safety problem. In the case of a safety program, impact evaluation measures the effect of the program on known and measurable risk factors for injury. Therefore, any changes in knowledge, skills, and attitudes of a specific target group, level of injury hazards, the environment, and relevant policy indicate the immediate effect of the program. If the effects are not in the desired direction, then the program is not progressing as planned. It should then be reviewed and modified concerning additional process evaluation of the modifications.

The methods for measuring change range from simple local surveys to full-scale investigations. A straightforward questionnaire could be used to ascertain knowledge and attitudes, with results sent to the target group and compared for any changes. If participants or target group members transfer their new knowledge into modification of other risk factors for injury, then this demonstrates attitudinal change and an intervention's ongoing safety effect on preventing future injury.

OUTCOME EVALUATION

outcome evaluation ■ This type of evaluation measures the effects of an intervention. Typically, the emphasis is on the measurement of desired intended effects, but sometimes the impact on possible negative effects is also measured.

Outcome evaluation involves the assessment of longer-term effects of a program focused on its goals after the program has been in place for a while. The outcome evaluation examines what the program has accomplished.

In many cases, outcome evaluation is concerned with increasing a specific competency, although its normal focus is on increasing safety and effectiveness. In regard to an issue in the fire service such as knee injuries, an outcome evaluation would be used at the end of a certain period of time, most likely a year or two, to determine what effect the program may have had. Did it actually reduce knee injuries? Several areas should be assessed and analyzed:

- Current injury rates and severities
- Assessing the changes that have occurred in performance, behavior, knowledge, skills, and abilities
- Analyzing the changes in safety and effectiveness
- Analyzing the response to policy and procedural changes

Outcome evaluation will measure and analyze the differences in injury rates before, during, and after the completion of the program to indicate the level of reduction of injury. Injury rates should also be compared with other local, state, and national data. The research and comparison of data were discussed in Chapter 3. Outcome evaluation may continue for years after the completion of a safety program to assess whether program effects are sustained over time. Although the measurement of injuries can come directly from statistics, the measurement of attitude and knowledge requires a cognitive assessment.

Outcome evaluation is commonly used to measure the results of a program while comparing them to the goals of the program. Limited funding for a long-term follow-up of outcome may mean that only an impact evaluation can be realistically achieved. The final stage of outcome evaluation is to compile a report documenting the evaluation results for each stage of the safety program evaluation. This document may range from a short memo to a lengthy report. A cost–benefit analysis can be used as an extension of outcome evaluation. It is important to recognize that not all safety programs require full formative, process, impact, and outcome evaluations. An outcome of injury reduction can be generally assumed if the safety program is evaluated for satisfactory implementation and penetration to the target population. If a safety program has been newly devised and there is no history of any satisfactory program effects elsewhere, then a full evaluation is essential to demonstrate its full worth. For a health and safety program in emergency services, the process and outcome evaluations should be performed as a minimum.

Limitations in funding many times do not permit a full evaluation of every aspect of the program. Realistically, it is more practical for program evaluators to focus on measures to demonstrate the success of a program. The evaluation process completed as a whole or in parts takes a comprehensive look at the health and safety program.

How Often Is Health and Safety Appraised and Evaluated?

#2

When and how often a health and safety program should be evaluated is a debatable subject. It is up to each individual organization to complete program evaluation as often as necessary. A process evaluation concerning health and safety should be an ongoing process.

An outcome evaluation should occur at the end of a set period of time. At a minimum, it should take place annually or whenever the program goals and objectives define a measured time frame. Anytime a severe injury or death occurs in emergency services, the health and safety program should be evaluated. Beyond injuries and deaths, each serious incident may uncover a need to evaluate and make necessary changes to the health and safety program. Research and development within the organization may also play a role in the appraisal process.

As new types of technology are introduced into emergency services, an analysis as to how this technology will fit into the health and safety program should be done. Any departmental changes to standards and regulations may also cause an appraisal of the health and safety program so that the organization can ensure its compliance with the changes. Program appraisal and evaluation is a requirement of NFPA 1500, *Standard on Fire Department Occupational Safety and Health Program*. The appraisal process actually starts prior to implementation of the health and safety program. The identification and analysis of risks prior to program implementation is completed through an appraisal process.

Responsibility for Appraisal—If Not Now, Then When?

Every member of an emergency response organization shares in the responsibility of health and safety program evaluation by being required to take part in the appraisal and evaluation process. During every emergency response, responders perform a process evaluation of the health and safety program, policies, and SOPs. In emergency services, the chief or director has the ultimate responsibility of ensuring the completion of the program appraisal and evaluation; however, in many cases the health and safety program manager is selected to oversee the process.

Program appraisal and evaluation may also be accomplished through the efforts of a health and safety committee, staff members, incident safety officers, and first-level supervisors. For example, the company officers or the coordinator of the department hazardous materials team can use an evaluation specific to the type of specialty team and compliant with the OSHA regulations.

Resource Central

See the Hazmat Self-Evaluation Health and Safety Compliance Checklist for more safety initiatives information.

The safety committee conducts program evaluation by examining injury statistics to determine what actually occurred, and it can review performance and behavioral changes within the organization. Members of the safety committee use their knowledge and experience to provide feedback to the organization concerning program evaluation. The incident safety officers are invaluable for providing feedback during process evaluation and input as to whether policies and procedures are being followed. They also do the

#1

following: perform incident investigation as a way to determine the contributing factors to an injury or death, or what caused damage to department property; provide information and feedback from incidents directly related to program evaluation through the post-incident analysis process; and ensure the proper use of equipment.

Department staff and administration has the responsibility to provide data and resources to the health and safety program manager and the safety committee. Data concerning incident and EMS operations should be provided by the respective administrative officers to the health and safety program manager. It is common to have the health and safety program manager act as the program evaluator, as he or she is the most familiar with the planning and implementation of the program. The first-level supervisors also provide feedback in the area of operations, behavior, and morale changes. Many times personnel's poor attitudes and behavior compromise health and safety by lowering the morale of the department. All personnel within the organization can evaluate policies and procedures to determine their effectiveness and whether they are being followed. Each individual member of the organization should be empowered to provide feedback about health and safety program evaluation. Part of this is in leadership's considering individual members' input to be invaluable. Members must know that their suggestions will be heard and considered. Individual members also have the ability to affect the attitude and knowledge of coworkers concerning the health and safety program and its effectiveness. The individual members and first-level supervisors will benefit the most from program evaluation, so they must be made aware of how important their input is to the organization as a whole.

NFPA 1500 recommends that an external evaluation done every three years will be an effective tool in the general management of the health and safety program. An individual from outside of the organization oftentimes will find areas in need of improvement that internal evaluators did not see. Outside evaluators also may suggest a **paradigm shift** within the organization that may not have been realized by internal personnel. In many cases, individuals outside the organization provide the necessary guidance in making such improvements.

paradigm shift ■ A complete change in a thinking or belief system allows the creation of a new condition previously thought impossible or unacceptable.

An external evaluation does not need to cost a great deal of money or take a large amount of resources to accomplish. In fact, many local governments have risk management and labor relations personnel who could assist with such an evaluation. Risk managers that work directly with workers' compensation claims can help in the external evaluation process. Keep in mind that risk management and workers' compensation personnel from local government both have a vested interest in the health and safety program and will be more than happy to see a reduction in injuries and claims. An individual organization may also find that the personnel of other local emergency services organizations, such as health and safety program managers, will be willing to serve as an external evaluator. Colleges and universities may also permit students to serve as evaluators for local organizations. If finances have been budgeted for health and safety program evaluation, a consultant or individual firm that specializes in this area can be contracted with to conduct an external evaluation of the health and safety program.

Summary

The purpose of the appraisal process for a health and safety program is to determine its effectiveness. NFPA 1500 requires an evaluation of the health and safety program at least once every three years. The evaluation process looks at program goals and objectives. Departments may choose from several formats to do so, including the process evaluation and the outcome evaluation. The process evaluation is an ongoing evaluation used to determine whether the program is reaching its intended participants to a certain degree. The process evaluation should also determine whether there have been changes in the areas of knowledge, behavior, attitude, and performance. The outcome evaluation takes place after the program has been in place for a predetermined time frame and actually measures the results of the program. The results of the program should be compared to the goals and objectives of the program, and then the effectiveness of the program can be determined.

The formative evaluation process ensures procedures, activities, and materials will work as planned. It offers the opportunity to solve any unexpected problems that may arise early in the safety program.

Impact evaluation assesses changes in risk factors and measures achievement of program objectives. Impact evaluation also involves the assessment of immediate program effects. Measurable risk factors for injury make up a large part of impact evaluation early in program evaluation. Impact evaluation may also include measures of change in knowledge, skills, and attitudes.

Every member of the organization will have some role to play in the evaluation process of the health and safety program. The responsibility for evaluation lies not only with the fire or EMS chief/director but also with the safety committee, staff members, incident safety officers, first-level supervisors, and department members, all of whom contribute to the evaluation process.

An important part of the evaluation process involves an external evaluation, which allows the health and safety program to be viewed from a different perspective. The evaluation process is dynamic and should be ongoing throughout each year. Changes to standards and regulations, as well as new technologies, warrant a periodic evaluation process. It should be remembered that the evaluation process actually starts prior to implementation of the health and safety program. All of the information and feedback from the evaluation process should improve and enhance the health and safety program.

Review Questions

1. List the types of program evaluation.
2. The purpose of the ____________ is to describe how the program was implemented, who was involved, and what problems were experienced.
   a. Formative evaluation
   b. Process evaluation
   c. Impact evaluation
   d. Outcome evaluation
3. State the type of evaluation that involves the assessment of immediate program effects on risk factors for the health and safety problem.
4. What is the role of the department staff in program evaluation?
5. How often should an organization complete program evaluation?
   a. Annually
   b. Biannually
   c. As often as necessary
   d. Every 3 years
6. The organization can move forward with the evaluation process after:
   a. Goals have been identified
   b. A look at where the organization is
   c. A system is in place to effectively measure results
   d. All of the above
7. State one way the safety committee can assist with program evaluation.

8. Who in the organization is responsible for health and safety program evaluation?
   a. The administrative staff of the organization
   b. The fire chief
   c. The health and safety committee
   d. Everyone in the organization
9. State what becomes the focus of formative evaluation.
10. A process evaluation focuses on:
    a. Services that were provided
    b. Procedures, activities, and materials
    c. Provided training
    d. Goals, objectives, and target groups

Case Study

On February 10, 2010, at approximately 1543 hours, a volunteer fire chief (the victim) was crushed between two fire trucks while one of the trucks was being backed into the fire station. See Figure 11–3. The chief died from his injuries at the scene.

The chief and a firefighter had traveled to the fire station to respond with a tanker truck to a mutual-aid incident. The chief and the firefighter arrived at the station, and the chief pulled Pumper 611 out of the way to allow the firefighter to pull Tanker 614 out of the station. The chief then backed Pumper 611 back into the station. *Note:* Two pumpers and one tanker occupied the two-bay station; one of the pumpers had to be moved to access and remove the tanker, which was stored between the two pumpers. The chief and the firefighter were the only department personnel that responded to the mutual-aid call.

The firefighter responded to the mutual-aid call with Tanker 614. Shortly after leaving the station, the response was canceled and Tanker 614 was directed to return to the station. The firefighter stated during interviews that when he returned to the station he observed the chief get into Pumper 611 and pull it out of the station. The firefighter backed Tanker 614 into the station, then walked over to Pumper 611 and got into the driver's seat to back it into the station. The firefighter told the NIOSH investigator that he had trouble seeing the inside of the station bay due to poor lighting and glare from the sun. The firefighter stated that when he was backing Pumper 611 into the station he observed, by line of sight, the chief standing outside the station next to a main door. The firefighter realized that he was too close to Tanker 614 (parked in the middle of the two station bays), and he pulled the pumper forward to straighten up and then backed into the station. He did not have a spotter helping him to back in, and there was no verbal communication between the chief and himself. He then got out of Pumper 611 and came around the front of the truck and yelled for the chief. He did not get a response. He then walked between Pumper 611 and Tanker 614 and found the chief lying on the bay floor beside the pump panel of Tanker 614, unresponsive. The firefighter called 911 and started CPR on the victim.

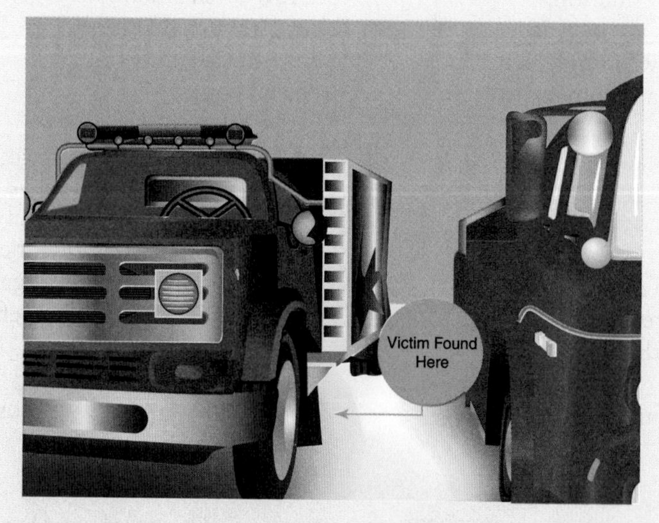

FIGURE 11–3 Victim location in a fatality backing incident.

NIOSH had the following recommendations:

Recommendation #1: Fire departments should ensure that standard operating procedures are developed, implemented, and enforced on safe backing of fire apparatus and include adequate training to ensure firefighter comprehension.

Recommendation #2: Fire departments should ensure that apparatus storage bays are well lit and have adequate room to store apparatus.

Recommendation #3: Fire departments should implement proper procedures for inspection, use, and maintenance of safety equipment used to assist in the backing of fire apparatus to ensure the equipment functions properly when needed.

Recommendation #4: Fire departments should ensure that fire apparatus drivers have unobstructed views from their rear-view mirrors.

Recommendation #5: Fire departments should consider replacing fire apparatus more than 25 years old.

Recommendation #6: Fire departments should be aware of programs that provide assistance in obtaining alternative funding for replacing or purchasing fire apparatus and equipment and for facility modification.[1] See Figure 11–4.

Fatal Backing Incident Concerns and Issues	Yes	No
Would an adequate SOP have prevented this fatality?		
Would better communication have prevented this incident?		
Was there adequate lighting and room for apparatus provided in the fire station?		
Would proper safety equipment have prevented this incident?		
Would the use of the proper safety equipment have prevented this fatality?		
Did the apparatus operator have an adequate, unobstructed view?		
Did the lack of recognition of the hazards play a role in these fatalities?		
Would a weekly inspection of apparatus have prevented this incident?		
Was the apparatus operator properly trained?		
Was there adequate room provided in this fire station for the safe operation of a spotter?		

FIGURE 11–4 Fatal backing incident concerns and issues.

1. Describe what actions could have been taken to prevent this incident.
2. What safety measures could have been in place to prevent this incident?

Reference

1. NIOSH Fire Fighter Fatality Investigation and Prevention Program, *Volunteer Fire Chief Dies After Being Crushed Between Two Fire Trucks—Kansas,* August 23, 2010. Retrieved March 12, 2011, from http://www.cdc.gov/niosh/fire/reports/face201007.html

CHAPTER 12

Data Collection and Information Management

OSHA's Form 300 (Rev. 01/2004)

Log of Work-Related Injuries an

You must record information about every work-related death and about every work-related injury or illness that i
days away from work, or medical treatment beyond first aid. You must also record significant work-related injuri
care professional. You must also record work-related injuries and illnesses that meet any of the specific recordi
use two lines for a single case if you need to. You must complete an Injury and Illness Incident Report (OSHA F
form. If you're not sure whether a case is recordable, call your local OSHA office for help.

Identify the person			Describe the case	
(A) Case no.	(B) Employee's name	(C) Job title (e.g., Welder)	(D) Date of injury or onset of illness	(E) Where the event occurre (e.g., Loading dock north end
			/ month/day	
			/ month/day	
			/ month/day	
			/ month/day	
			/ month/day	
			/ month/day	

KEY TERMS

Fire Fighter Fatality Investigation and Prevention Program (FFFIPP), *p. 255*
Health Insurance Portability and Accountability Act of 1996 (HIPAA), *p. 249*
hits, *p. 258*
learning management system (LMS), *p. 259*
links, *p. 258*
OSHA 300 log, *p. 253*
OSHA 300-A log, *p. 253*
OSHA 301 incident report, *p. 253*
search engines, *p. 258*

OBJECTIVES

After reading this chapter, you should be able to:

- Describe why data collection and the analysis of information are important to the health and safety program.
- Identify the type of data that should be collected.
- Describe the type of data that should be collected for external agencies and organizations.
- Describe why it is important to make an internal health and safety report available throughout the organization.
- Describe why the World Wide Web is an important tool for the health and safety program manager and the organization.

> **Resource Central**
>
> For additional review and practice tests, visit **www.bradybooks.com** and click on Resource Central to access book-specific resources for this text! To access Resource Central, follow directions on the Student Access Card provided with this text. If there is no card, go to **www.bradybooks.com** and follow the Resource Central link to Buy Access from there.

Data collection and information management are a necessary component of performing an evaluation of the health and safety program. The analysis of data is also used for other components of the health and safety program, such as establishing goals and objectives. The data and information collected by many organizations are not initially in a user-friendly format.

Data must be put into a useful and retrievable format that is compatible with the intended application. In most cases, computer software makes it easier to gather data and evaluate information. It is the responsibility of the department administration to be committed to provide accurate and sufficient data to be used within the organization. Data collected at the local, state, and national levels relate to trends within similar organizations, which is also useful. After data and information have been collected and analyzed, they can be distributed throughout the organization in a published health and safety report.

Change within emergency services organizations is occurring at a much faster rate than ever before. Much of this rapid change can be attributed to advancing technology, and the access available to it that emergency services organizations have has never been greater. The World Wide Web has become an important tool for every employee involved in emergency services. The health and safety program manager making use of the Internet will find an endless amount of information at his or her fingertips about not only health and safety but also standards and regulations, standard operating procedures, and policies from other departments and organizations.

The Importance of Data Collection and Information Management

OSHA and state labor departments require data collection in a number of areas and jurisdictions. In many cases, noncompliance concerning mandatory data reporting to federal and state agencies will be associated with a fine. It is vitally important for emergency response organizations to have data and information available for analysis and evaluation. Data collection, record keeping, and quality analysis enable an organization to protect itself and its employees legally. Data collection and information management not only provide support for the organization's management, vision, mission, and long-range planning but also aid the emergency response organization with its public education and information needs. See Figure 12–1.

The proactive organization pursues efficient data management with an electronic information management and filing system. Injury reports and data regarding workers' compensation claims enable the organization to help predict future premium costs. The greater the number of workers' compensation claims, the higher the cost to the organization's budget. A high number of injuries that prevent individuals from working on the line will normally equate to a higher cost in overtime to the department. Emergency response organizations now have the ability electronically to simply and efficiently update and submit the required

Month | Year
Payroll # | Name | Shift | Date
Date of Birth | SS Number | Incident #
Incident Date | Address of Incident
Description of Incident/Task Being Performed:

Exposure Information:
☐ Bloodborne ☐ Airborne

Disease Exposed To	At Pt. Time	Exposed Before

Type of Exposure	Area Exposed
☐ Droplet Spread	☐ Hands
☐ Direct Contact	☐ Mouth
☐ Needle Stick	☐ Face/Eyes
☐ Blood or Body Fluids	☐ Nose
☐ Other	☐ Other

Previous Immunizations (date)
HBV
MMR
DPT

Medical Facilty:
Name:
Follow-up Care:
Date of Follow-Up:

Reporting Process: Supervisor Notified:

Protective Barrier Used: Gloves | Mask | Eyeshield | Gown | Other

Additional Info

☐ Were proper protection barriers used?
☐ Was there knowledge of the disease before entering?
☐ Could exposure have been prevented under the circumstances?
☐ Did affected personnel request a CISD?
☐ Was this a True Exposure?

Recommended PEP ☐
Recieved PEP ☐
PEP Meds Given ☐
Employee Given Counseling ☐

Investigating Officer | Description
Date

Source Patient Information

Patient Name | Patient Sex: ○ Female ○ Male | Patient Age | Source Patient Blood Draw HIV (rapid test), HBV, HCV: ○ Yes ○ No

FIGURE 12–1 Communicable disease exposure report. *Courtesy of St. Petersburg Fire/Rescue.*

information annually. Emergency services organizations must conduct research to determine which federal and state agencies require electronic submittal of information.

Information can help protect employees, and it must be available to all members of the organization. Emergency response organizations can utilize data and information in a variety of ways. The important aspects are to:

- Compile accurate information.
- Have the data and information in a format that can be used effectively.
- Relate the compiled information effectively and appropriately.

One of the most fundamental aspects of record keeping and data collection is consistency, which becomes especially important in report information and comparative data. Reports must include the same information for each record, and so a set format should be used to compile and present information and data concerning trends and comparisons. There is a vast array of records that should be kept by every emergency response organization. Certain basic information should be captured in each report and data collection system.

Individual personnel records are one of the most important areas of data collection. Records should cover an individual's service from appointment to retirement. Performance reviews, training accomplishments, certifications, commendations, promotions, and disciplinary actions are vital information that needs to be maintained.

Most fire department organizations use some type of run report to gather and compile information. Many departments use the National Fire Incident Reporting System (NFIRS), which contains most of the standard information on the response and nature of the incident. The NFIRS represents the world's largest national database of fire incident information, which is updated annually. It is designed to capture the nation's fire problem, as well as its detailed characteristics and trends. The NFIRS has two objectives: to help state and local governments develop fire reporting and analysis capability for their own use, and to obtain data that can be used to more accurately assess and subsequently combat the fire problem at a national level. The standard NFIRS package includes incident and casualty forms. The NFIRS 5 Alive software now permits individual departments to evaluate fractile reports that provide the counts for incident performance over an increasing length of time. Basically the count of incidents is converted into percentages, and the fractile reporting can be displayed using graphs. Fractile analysis can be very useful in establishing an individual organization's performance. Risk analysis is also a component of NFIRS 5 Alive that will benefit a department's accreditation study. Recording incident history and response time performance for each risk location is an effective way to display a community's risk on a map.

#7

NFPA 1401, *Recommended Practice for Fire Service Training Reports and Records*, outlines what type of information should be contained in training records. Such information may prove extremely important in areas such as workers' compensation claims and civil suits against the department. A lack of sound information can lead to liability and lawsuits.

Medical information and data are confidential and must not be accessed by the organization's administration or staff. Federal privacy standards protect patients' medical records and other health-related information. The standard that provides patient privacy protection is known as the **Health Insurance Portability and Accountability Act of 1996 (HIPAA)**, which prevents an emergency response organization from gaining access to an employee's medical records and information. The emergency response organization must keep this in mind as it directly relates to injury reporting and other privacy issues. In most cases, legal advice should be sought prior to the health and safety program's evaluation. The department policies and SOPs in place must comply with the Health Insurance Portability and Accountability Act of 1996.

Health Insurance Portability and Accountability Act of 1996 (HIPAA) ■ The Office for Civil Rights enforces the HIPAA Privacy Rule, which protects the privacy of individually identifiable health information, and the confidentiality provisions of the Patient Safety Rule, which protect identifiable information being used to analyze patient safety events and improve patient safety.

Data Collection Within the Organization

The task of specifying what type of data should be collected related to the health and safety program can be approached in various ways: the organization must first determine what data and information are required and then identify how to gather the data. The organization must also consider what information is available and from what sources. What can be obtained from these sources must also be taken into consideration prior to data collection. Typically a few of the common sources will provide much of the needed data. Emergency response organizations collect much of the same types of data. The collection of data is commonly known as a management information system (MIS). Concerning the health and safety program, the use of computers is highly beneficial for gathering and analyzing data, because it allows the organization to set up files, maintain the files, and provide specialized analysis. Several common data types fall in this category:

- Personnel records
- Injury and accident reports
- Training records
- Drug-testing reports
- Exposure reports

In the past, the data required for these reports were captured using some sort of standardized report or form. In the modern emergency response organization, it is common to gather such data in written form to be transferred into a computerized database. In this fashion, the data can be retrieved and then analyzed at a future point in time. In many cases, the written reports and forms can be destroyed after their data have been entered into a computer database. Keep in mind there may be laws both governing the length of time to keep records and requiring that hard copies of records be kept.

Injury and accident reports are the most common type of record generated by emergency services organizations. Medical records have recently become a subject of debate between labor and management in many departments. Because individual employee medical records are private information between the department physician and the employee, the individual department administration does not have the right to review an employee's entry or annual physical results without written permission from the individual employee. See Figure 12–2.

It should be required that fire department physicians are knowledgeable about occupational medicine as it relates to the job of a firefighter. The standard related to performing a fire department physical is NFPA 1582, *Standard on Comprehensive Occupational Medical Program for Fire Departments*. The medical requirements in this standard are applicable to fire department candidates and members whose job descriptions as defined by the authority having jurisdiction (AHJ) are outlined in NFPA 1001, *Standard for Fire Fighter Professional Qualifications*. The only issue the department administration should be concerned with is a written document from the department physician stating whether an employee is either fit or unfit for duty after having completed a medical evaluation. The NFPA 1582 standard addresses medical conditions that affect the ability to safely perform essential job tasks by categorizing these medical conditions as category A and category B. Employees who fall into the A category shall not be certified as meeting the medical requirements of the standard. Those with B category medical conditions shall be certified as meeting the medical requirements of the standard only if they can perform the essential job tasks without posing a significant safety and health risk to themselves, members, or civilians.

#6

Data Collection from Outside the Organization

Sources of data collected from outside the organization include other emergency services organizations, socioeconomic statistics and demographics, national databases, and insurance information. The information required may or may not all be the same for local and state agencies as that required internally for the individual organization. Many of the forms and records used internally by individual organizations usually provide adequate background and information for the external organizations that collect,

FIGURE 12–2 Injury report. *Courtesy of St. Petersburg Fire/Rescue.*

REPORT OF AN INJURY TO AN EMPLOYEE

This is a mandatory report to be completed in every case of an ON-THE-JOB INJURY* to a City employee regardless if medical treatment is needed or not. It is to be completed in DETAIL and a signed copy of this report must be sent to the Workers' Compensation Office within 24 hours of the time the injury occurred. The Workers' Compensation Office reserves the right to investigate any claim to determine compensability. Contact the Workers' Compensation Office immediately if any injury/accident reported by the employee is questionable.

***"Arising out of and in the course of employment."**

This section is to be completed by the SUPERVISOR in charge of the injured employee.

Department: ______________________ **Division:** ______________________

Acct #: ______________________ **Payroll #:** ______________________

Employee's Name: ____

Date of Birth: ______________________ **Soc. Sec. #:** ______________________

Home Address: ____

Telephone #: ______________________ **Date Employed:** ______________________

Occupation: ______________________ **Classification #:** ______________________

Supv. name: ______________________ **Supv. Tel. ext.:** ______________________

Hourly rate of pay: $__________ **Full time** __________ **Part time** __________

Number of hours worked per week: __________ **Current scheduled days off:** __________

Date of accident: ______________________ **Time of accident:** ______________________

Date injury reported: ______________________ **Time injury reported:** ______________________

Injury reported to: ______________________ **Telephone ext.:** ______________________

Witness: ______________________ **Address:** ______________________

Witness: ______________________ **Address:** ______________________

Place of accident: ____

Employer's Premises: ____

Yes __________ **No** __________

Supervisor's description of accident (State in detail what the employee was doing; name the object, substance, or exposure that directly injured the employee; whether the employee was struck, fell, etc.; and describe the injury and indicate part of body affected):

____

Supervisor's recommendations that would prevent similar accidents in the future:

____

Employee **Did not need medical treatment:** __________

Sent to ______________ **hospital:** __________

Other: ____

ATTACH FIRST DOCTOR's SLIP (Medical Disposition Report)

(Employee is responsible for turning in all medical disposition reports to the department immediately after appointment.)

After reviewing the medical disposition report, I have determined that work is available within the employee's restrictions (if any):

__________ Yes, returned to work on: __________ No, first day off: __________

______________________ ______________________

Supervisor's Signature Date

(continued)

THIS SECTION TO BE COMPLETED BY THE INJURED EMPLOYEE

Do you concur with **Supervisor's** description of the accident?

Yes ______________ No ____________

If no, **Employee's statement:**

____

I understand that any medical treatment sought related to this injury must be authorized by the Workers' Compensation Office and that I must attend all scheduled appointments so that I will not complicate my injury or prolong my recovery.

I acknowledge receipt of the "Important Workers' Compensation Information for Florida's Workers" brochure.

ANY PERSON WHO, KNOWINGLY AND WITH INTENT TO INJURE, DEFRAUD, OR DECEIVE ANY EMPLOYER OR EMPLOYEE, INSURANCE COMPANY, OR SELF-INSURED PROGRAM, FILES A STATEMENT OF CLAIM CONTAINING ANY FALSE OR MISLEADING INFORMATION COMMITS INSURANCE FRAUD, PUNISHABLE AS PROVIDED IN S.817.234. I HAVE REVIEWED, UNDERSTAND, AND ACKNOWLEDGE THE ABOVE STATEMENT.

______________________ ______________

Employee's Signature Date

MANAGER'S COMMENTS:

__________ Recommended for On-Duty Injury Pay. Disqualifying negligence was not involved.

__________ Not recommended for On-Duty Injury Pay. Employee did negligently:

____

__________ Disciplinary action is not indicated.

__________ Disciplinary action is being initiated toward:

____

__________ State action being taken to prevent ALL employees from receiving a similar injury:

____

______________________ ______________

Manager's Signature Date

SAFETY OFFICER'S REVIEW:

__________ Received and reviewed

Action needed:

______________________ ______________

Safety Officer's Signature Date

DIRECTOR'S COMMENTS:

__________ On-Duty Injury Pay is approved.

__________ On-Duty Injury Pay is not approved because:

____

______________________ ______________

Director's Signature Date

FIGURE 12-2 *(continued)*

#7

compile, and publish data; for example, NFPA, IAFF, IAFC, USFA, NIOSH, NFIRS, and the Firefighter Near-Miss System, to name just a few. The types of information that most emergency services organizations gather and store is common in nature and type. Of course, differences will exist between paid and volunteer organizations related to personnel, financial data, and the extent of services offered by the individual organization.

WORKERS' COMPENSATION

Injury data wanted and collected by workers' compensation are considered external to the organization. There are laws, and many organizations have rules in place, dictating the time frame in which workers' compensation must be notified of injuries. Workers' compensation may be a division within a city department such as risk management. Workers' compensation will also provide a summary of claims and injuries to the emergency response organization over the course of time. The injury report data should be broken down and compiled into a usable format in order for the health and safety program manager to view the type of injuries as well as the activities in which they have occurred. The organization can use such data to move forward with prevention measures and education of its members. Injury reports and data can provide useful information concerning health and safety program evaluation. Workers' compensation claims offer information about lost time and treatment. If the number of claims is increasing, an analysis and evaluation should be done to determine whether this is due to specific events or to what the health and safety program deems reevaluation with needed changes. Injuries occur in several areas in emergency services. The most common injuries occur at the incident scene; however, there are injuries that occur at the quarters of fire departments and EMS organizations. Injuries also occur during training events. The goal is to evaluate where and why injuries occur to determine if changes are needed to the health and safety program. The data gathered from workers' compensation can be used to analyze the type and frequency of specific injuries for each individual department.

NATIONAL FIRE PROTECTION ASSOCIATION (NFPA)

The NFPA has been publishing injury report data and an annual fatality report since 1974. In order for the NFPA's data collection program to be effective, organizations must participate in it. The NFPA injury reports are designed to sample and predict the nation's fire service injuries. The NFPA gathers and compiles the injury report data from the U.S. Fire Administration's National Fire Incident Reporting System (NFIRS). The statistics from the injury report data continue to demonstrate that firefighting presents substantial risk of personal injury to the members of the fire service. National estimates of total fireground injuries are made based on data reported by fire departments to the NFPA in its Annual Fire Experience Survey. Detailed firefighter casualty information is also based on data reported by fire departments participating in the current version of NFIRS.

OCCUPATIONAL SAFETY AND HEALTH ADMINISTRATION (OSHA)

OSHA does not apply to every fire department in the United States; however, there are reporting requirements in the states in which OSHA does apply. Where certain OSHA regulations have been adopted as state law, all of the federal OSHA regulations do not apply. States and local jurisdictions that have adopted an OSHA regulation have adopted these regulations as law. An example would be the two-in-two out regulation that most fire departments adhere to. Record-keeping and reporting may also be a requirement. It is important emergency services organizations research the requirements in their specific state. The OSHA record-keeping requirements do not apply to every fire department in the nation; however, there are such requirements in the states in which OSHA does apply, and states that have adopted specific OSHA regulations must comply with the OSHA regulations and requirements. One of those requirements is the OSHA regulation that requires reporting of injuries and illnesses.

The OSHA 1904 regulation requires that the **OSHA 300 log** and **OSHA 301 incident report** be completed within seven calendar days of a reported injury or illness. The entries on the OSHA 300 log must be summarized on the **OSHA 300-A log** at the end of the year.

OSHA 300 log ■ This OSHA log of work-related injuries and illnesses notes the extent and severity of each case.

OSHA 301 incident report ■ The OSHA injury and illness incident report must be completed within seven calendar days after receiving information that a recordable work-related injury or illness has occurred.

OSHA 300-A log ■ This is a summary of work-related injuries and illnesses. Employees, former employees, and their representatives have the right to review the OSHA Form 300 in its entirety.

The injured individuals' privacy is taken into account concerning the OSHA regulations. See Figure 12–3.

The OSHA 3110 regulation is titled Access to Employee Exposure and Medical Records (Title 29 of the Code of Federal Regulations [CFR], Part 1910.1020). Employees who believe they may have been exposed to toxic substances or harmful physical agents in the workplace have a right to obtain relevant exposure and medical records. Designated employee representatives—an individual or organization to which an employee has given written permission to exercise a right of access—may be permitted access to such employee records in very specific circumstances. An individual may access employee exposure records that show the monitoring of an employee's own exposure to a toxic substance or harmful physical agent. If the employer does not have specific records on an individual employee, the employee is permitted to access the records of other employees who have engaged in similar work or been under similar conditions. In addition, the employee may have access to compilations of data or statistical studies. Before permitting employee access to the analysis, the employer is required to remove identifiers such as an employee's name, address, social security number, and job title.

OSHA 1904 offers the provisions to which the health and safety program manager can refer. Guidelines exist on the types of injuries that must be reported. For example, an injury must be reported if it results in the following:

- Death
- Days lost from work
- Restricted work or transfer to another job
- Medical treatment beyond first aid
- Loss of consciousness

Cases must also be considered that may meet the general reporting criteria if they involve a significant injury or illness diagnosed by a physician or other health care provider even

OSHA's Form 300 (Rev. 01/2004)

Log of Work-Related Injuries and Illnesses

Attention: This form contains information relating to employee health and must be used in a manner that protects the confidentiality of employees to the extent possible while the information is being used for occupational safety and health purposes.

Year 20__ __

U.S. Department of Labor
Occupational Safety and Health Administration

Form approved OMB no. 1218-0176

You must record information about every work-related death and about every work-related injury or illness that involves loss of consciousness, restricted work activity or job transfer, days away from work, or medical treatment beyond first aid. You must also record significant work-related injuries and illnesses that are diagnosed by a physician or licensed health care professional. You must also record work-related injuries and illnesses that meet any of the specific recording criteria listed in 29 CFR Part 1904.8 through 1904.12. Feel free to use two lines for a single case if you need to. You must complete an Injury and Illness Incident Report (OSHA Form 301) or equivalent form for each injury or illness recorded on this form. If you're not sure whether a case is recordable, call your local OSHA office for help.

Establishment name ______________

City ______________ State ______________

Identify the person			Describe the case			Classify the case — CHECK ONLY ONE box for each case based on the most serious outcome for that case:				Enter the number of days the injured or ill worker was:		Check the "Injury" column or choose one type of illness:					
								Remained at Work				(M)					
(A) Case no.	(B) Employee's name	(C) Job title (e.g., Welder)	(D) Date of injury or onset of illness	(E) Where the event occurred (e.g., Loading dock north end)	(F) Describe injury or illness, parts of body affected, and object/substance that directly injured or made person ill (e.g., Second degree burns on right forearm from acetylene torch)	Death (G)	Days away from work (H)	Job transfer or restriction (I)	Other record-able cases (J)	Away from work (K)	On job transfer or restriction (L)	Injury (1)	Skin disorder (2)	Respiratory condition (3)	Poisoning (4)	Hearing loss (5)	All other illnesses (6)
___	___	___	___/___ month/day	___	___	☐	☐	☐	☐	___ days	___ days	☐	☐	☐	☐	☐	☐
___	___	___	___/___ month/day	___	___	☐	☐	☐	☐	___ days	___ days	☐	☐	☐	☐	☐	☐
___	___	___	___/___ month/day	___	___	☐	☐	☐	☐	___ days	___ days	☐	☐	☐	☐	☐	☐
___	___	___	___/___ month/day	___	___	☐	☐	☐	☐	___ days	___ days	☐	☐	☐	☐	☐	☐
___	___	___	___/___ month/day	___	___	☐	☐	☐	☐	___ days	___ days	☐	☐	☐	☐	☐	☐
___	___	___	___/___ month/day	___	___	☐	☐	☐	☐	___ days	___ days	☐	☐	☐	☐	☐	☐
___	___	___	___/___ month/day	___	___	☐	☐	☐	☐	___ days	___ days	☐	☐	☐	☐	☐	☐
___	___	___	___/___ month/day	___	___	☐	☐	☐	☐	___ days	___ days	☐	☐	☐	☐	☐	☐
___	___	___	___/___ month/day	___	___	☐	☐	☐	☐	___ days	___ days	☐	☐	☐	☐	☐	☐
___	___	___	___/___ month/day	___	___	☐	☐	☐	☐	___ days	___ days	☐	☐	☐	☐	☐	☐
___	___	___	___/___ month/day	___	___	☐	☐	☐	☐	___ days	___ days	☐	☐	☐	☐	☐	☐
___	___	___	___/___ month/day	___	___	☐	☐	☐	☐	___ days	___ days	☐	☐	☐	☐	☐	☐
___	___	___	___/___ month/day	___	___	☐	☐	☐	☐	___ days	___ days	☐	☐	☐	☐	☐	☐

FIGURE 12–3 OSHA 300 log of work-related injuries and illnesses. *Source: OSHA, OSHA Forms for Recording Work-Related Injuries and Illnesses*, p. 7. Retrieved May 26, 2011, from http://www.osha.gov/recordkeeping/new-osha300form1-1-04.pdf

if it does not result in anything listed above. The state department of labor should be contacted to determine state reporting requirements for emergency response organizations that do not fall under the OSHA guidelines.

FIREFIGHTER AUTOPSY PROTOCOL

The Firefighter Autopsy Protocol was originally published in 1994 and has had many subsequent revisions. Its purpose is to provide guidelines for medical personnel performing autopsies on firefighters who have suffered a line-of-duty death. This protocol does the following: effectively serves the forensic professional and provides a consistent basis for examining firefighter deaths at a more in-depth level, has a direct bearing on the health and safety of firefighters, and offers guidance in performing an autopsy if a non-line-of-duty death may be linked to a line-of-duty exposure. It is hoped that through the Firefighter Autopsy Protocol general autopsy procedures will be supplemented with additional analyses to ascertain specific causes and mechanisms of death. The goal is to add to the area of knowledge of firefighter fatalities. It is hoped that the use of the Firefighter Autopsy Protocol will prevent future firefighter fatalities. The protocol's recommended procedures are intended to address the complexity involved between the firefighter and the hazardous work environment in which they work and the duties they perform. The protocol is intended to provide guidance in the special considerations applicable to a firefighter autopsy that must go beyond the practices of the standard autopsy. The Firefighter Autopsy Protocol can be accessed through the U.S. Fire Administration at USFA .dhs.gov.

UNITED STATES FIRE ADMINISTRATION (USFA)

The NFIRS reporting system is used by the USFA to gather data concerning health and safety. NFIRS has the capability of capturing and reporting firefighter fatalities and casualties; however, because an organization's participation in the NFIRS system is voluntary, the data are incomplete. NFIRS only provides data on injuries and causalities on those departments that participate in the NFIRS program. The data and information entered into the system are compiled into an annual report concerning the nation as a whole for those fire agencies that use the NFIRS system for reporting.

INTERNATIONAL ASSOCIATION OF FIREFIGHTERS (IAFF)

The IAFF also collects data associated with annual injuries, exposures, and fatalities. The IAFF's data focus on career fire departments that have an affiliation with or are members of the IAFF, including career fire departments in both the United States and Canada. A recent report published by the IAFF also considers EMS-related incidents more than it has in the past.

NATIONAL INSTITUTE FOR OCCUPATIONAL SAFETY AND HEALTH (NIOSH)

In 1997 NIOSH began tracking firefighter line-of-duty deaths (LODDs). The project, named the **Fire Fighter Fatality Investigation and Prevention Program (FFFIPP)**, performs independent investigations of firefighter fatalities in the United States. Its goals are (1) to better define the characteristics of line-of-duty deaths among firefighters, (2) to develop recommendations for the prevention of deaths and injuries, and (3) to disseminate prevention strategies to the fire service. Congress funded NIOSH to implement FFFIPP because it recognized the need for further efforts to address the continuing national problem of occupational firefighter fatalities—an estimated 105 each year. This program was assigned to NIOSH as a result of fire service requests to

Fire Fighter Fatality Investigation and Prevention Program (FFFIPP) ■ NIOSH conducts investigations of firefighter line-of-duty deaths to formulate recommendations for preventing future deaths and injuries. The program does not seek to determine fault or place blame on fire departments or individual firefighters, but to learn from these tragic events and prevent future similar events.

Congress to have firefighter line-of-duty deaths investigated and causes reported in an attempt to decrease deaths and injuries.

NIOSH is not an enforcement agency for the fire service. Rather it is a research and educational organization that has limited investigative authority to support the research mission, but no powers and no mandate to conduct "official fact-finding" investigations of specific incidents. The current program effort is directed toward discovering and making the fire service aware of information concerning firefighter fatalities that will help to prevent future fatalities. This process needs to evolve to produce usable data that can impact the problems that lead to firefighter deaths and injuries. Attempts to carry out the FFFIPP's mission are divided between two teams: one to investigate cardiovascular and medically related fatalities, and the other to investigate traumatic fatalities. See Box 12-1.

BOX 12-1: DESCRIPTION OF THE NIOSH FIRE FIGHTER FATALITY INVESTIGATION AND PREVENTION PROGRAM

The United States currently depends on approximately 1.1 million firefighters to protect its citizens and property from losses caused by fire. Of these firefighters, approximately 313,000 are career and 823,000 are volunteers. The National Fire Protection Association (NFPA) and the U.S. Fire Administration (USFA) estimate that, on average, 100 firefighters die in the line of duty each year. In fiscal year 1998, Congress recognized the need for further efforts to address the continuing national problem of occupational firefighter fatalities and funded NIOSH to implement a firefighter safety initiative.

FIRE FIGHTER FATALITY INVESTIGATIONS

The NIOSH Fire Fighter Fatality Investigation and Prevention Program (FFFIPP) conducts investigations of firefighter line-of-duty deaths to formulate recommendations for preventing future deaths and injuries. The program does not seek to determine fault or place blame on fire departments or individual firefighters, but to learn from these tragic events and prevent future similar events. Investigations are prioritized using a decision flowchart available on the FFFIPP Web site. Investigation priorities will change depending on fatality data.

PROGRAM OBJECTIVES

- Better define the characteristics of line-of-duty deaths among firefighters.
- Develop recommendations for the prevention of deaths and injuries.
- Disseminate prevention strategies to the fire service.

TRAUMATIC INJURY DEATHS

The program uses the Fatality Assessment and Control Evaluation (FACE) model to conduct investigations of fireground and non-fireground fatal injuries resulting from a variety of circumstances, such as motor vehicle incidents, burns, falls, structural collapse, diving incidents, and electrocutions. NIOSH staff also conducts investigations of selected nonfatal injury events. Each investigation results in a report summarizing the incident and includes recommendations for preventing future similar events.

NIOSH staff members with respirator expertise also assist with investigations in which the function of respiratory protective equipment may have been a factor in the incident. They evaluate the performance of the self-contained breathing apparatus (SCBA) as a system and conduct evaluations of SCBA maintenance programs upon request.

CARDIOVASCULAR DISEASE (CVD) DEATHS

NFPA data show that heart attacks are the most common cause of firefighter line-of-duty deaths. NIOSH investigations of these fatalities include assessing the contribution of personal and workplace factors. Personal factors consist of identifying individual risk factors for coronary artery disease. The workplace evaluation includes the following assessments:

- Estimating the acute physical demands placed upon the firefighter
- Estimating the firefighter's acute exposure to hazardous chemicals

(continued)

- Assessing efforts by the fire department to screen for coronary artery disease
- Assessing efforts by the fire department to develop fitness and wellness programs

INFORMATION DISSEMINATION

The Fire Fighter Fatality Investigation and Prevention Program posts all investigative reports on the NIOSH Web site and notifies a list of subscribers of each posting. In addition, printed copies of related publications are available.

WHAT TO EXPECT DURING A NIOSH INVESTIGATION

NIOSH is notified of a line-of-duty death in a number of ways, including by the U.S. Fire Administration (USFA), a fire department representative, the International Association of Fire Fighters (IAFF), or the state fire marshal's office. NIOSH conducts investigations of both career and volunteer firefighter line-of-duty deaths.

Once notified of a fatality, a NIOSH representative contacts the fire department. NIOSH investigators then review all applicable documents (e.g., department standard operating procedures; dispatch records; training records for the victim, incident commander, and officers; the victim's medical records, where applicable; coroner/medical examiner's reports; death certificates; blueprints of the structure; police reports; photographs; and video). Additionally, investigators interview fire department personnel and firefighters who were on the scene at the time of the incident. NIOSH may work closely with other investigating agencies. When needed, NIOSH enlists the assistance of other experts, such as those in motor vehicle incident reconstruction or fire growth modeling.

Once the investigation is completed, NIOSH summarizes the sequence of events related to the incident and prepares a draft report. Each department, union (if present), or family (where applicable) will have the opportunity to review this portion of the report in draft form to ensure it is technically accurate. The report is then finalized with the addition of recommendations for preventing future deaths and injuries under similar circumstances. In selected cases, NIOSH also enlists the assistance of subject matter experts to review complete draft reports. Once the fire department, union (if present), and family (where applicable) have received the final copy of the NIOSH incident report, it is made available to the public through the Fire Fighter Fatality Investigation and Prevention Program Web site.

WHOM DO I CONTACT FOR FURTHER INFORMATION?

If you have any questions regarding the NIOSH Fire Fighter Fatality Investigation and Prevention Program, please contact the NIOSH Division of Safety Research at:

National Institute for Occupational Safety and Health
Division of Safety Research
Surveillance and Field Investigations Branch
1095 Willowdale Road, M/S 4020
Morgantown, WV 26505-2888
Phone: (304) 285-5894

Source: NIOSH, from http://www.cdc.gov/niosh/fire/

The Annual Health and Safety Report

The division of the emergency services organization responsible for health and safety is involved in every aspect of the organization. The health and safety division is normally in charge of investigating injuries and deaths of members, inspecting and maintaining standards for personal protective equipment, responding to safety inquiries, and promoting the overall health and wellness of personnel. Larger fire departments may have a division dedicated to the health and safety of members, whereas smaller departments may have an individual assigned to and responsible for health and safety instead of having a division. In order to publish a health and safety report on an annual basis, a data collection and retrieval system must be put in place. After data collection is completed, some sort of analysis will be necessary. The collected data must be put into a usable format so members of the organization can easily understand the information, which should be published in an annual report to be made available to the administrative staff of the city or county and the elected officials. The annual report is very useful as a resource, and as a method to support the organization's applying needed resources to the health and safety program. The annual report should contain the following information:

- Introduction
- Emergency activity
- Fire prevention information
- Emergency medical services
- Training information
- Emergency management information
- The strategic plan
- Operations information
- Administration and budget information
- Health and safety information
- A summary and conclusion

The fire chief or director should be prepared to present the information contained in the annual report if called upon. The health and safety program manager should also be prepared to present or assist the fire chief or director in his or her presentation as it relates to the health and safety of the organization. Health and safety information should be a vital component as it relates to the annual report.

Use of the World Wide Web to Assess Health and Safety Information

The Internet, as well as other sources of electronic information, has become a valuable resource for health and safety information. The proactive health and safety program manager should make use of technology and electronic media to assist and complement the health and safety program. The Fire Department Safety Officers Association (FDSOA) is one such source dedicated to both the incident safety officer and the health and safety program manager. Most updates and changes in information can quickly be made available throughout the organization by using computers. Not all organizations have the same level of technology available; however, the organizations that do can increase the quality and amount of information in a very rapid fashion. The Internet can also provide data and information from other organizations including local, state, and federal agencies. Access to annual information and data is accessible from several organizations, such as the USFA, the NFPA, and NIOSH, to name just a few. It is important that organizations have access to the Internet.

Many emergency services organizations do not provide Internet access to their members for fear of the members' accessing inappropriate Web sites. If an organization provides training and signed documentation to its members concerning Internet use and puts security measures in place, this fear of access can be greatly reduced. Additional security measures include the monitoring of members' access.

The benefit of Internet use for emergency services organizations is through the use of **search engines** to search for Web pages that have the words or phrases being searched for within the document or Web page. A health and safety program manager who wants to conduct a search for hazardous materials training information will have volumes of information available at his or her fingertips. The search engine will display a search as **hits**, and the user can select any of the Web pages displayed on the list by clicking on them. Once accessed, many Web pages also have **links** that enable the user to access additional Web pages to related topics. Much of the information available on the Internet is useful, timely, and directly related to an organization's health and safety program. Not all information is factual or correct, so members of an organization that use the Internet

search engines ▪ Web search engines are tools designed to search for information on the World Wide Web. The search results are usually presented in a list and are commonly called hits. The information may consist of Web pages, images, information, and other types of files.

hits ▪ A request for a file is made by a user-agent. User-agents include Web browsers and search engine indexing programs.

links ▪ These are used on Internet pages so that each page may be directly tied to a page on another Internet site.

must keep this in mind when conducting research through the Internet. Discovered information that may seem questionable concerning validity should be further researched at an in-depth level to ensure accuracy.

In many cases, an overwhelming amount of information is available when completing a search. In these cases, users may narrow their search by being more specific with their words or phrases. Many of the organizations and governmental agencies listed earlier in the chapter can easily be located on the World Wide Web. Literally hundreds of Web sites related to the health and safety of emergency services, as well as Web sites to be used for training purposes, are available.

Health and Safety Training and Information Management

There are many methods for organizations to choose from to deliver training related to health and safety. Over the course of time, it has been difficult to have a quality assurance method in place that ensures the training is actually completed unless the training has taken place in the classroom setting. One of the newest methods to deliver and track health and safety training is through a **learning management system (LMS)**.

learning management system (LMS) ▪ An LMS is a software application that automates the administration, documentation, tracking, and reporting of training events.

LMSs range from systems for managing training/educational records to software for distributing courses over the Internet and offering features for online collaboration. There are more and more emergency services organizations purchasing LMSs to automate record keeping and data management, as well as registration and delivery of online courses and training. A Web-based learning management system provides online delivery and assessment, as well as a database for data collection. LMSs are used by regulated industries such as emergency services organizations as a method to ensure compliance. LMS is also used to enhance and support classroom and hands-on practical teaching.

Emergency services organizations that are striving to comply with current regulations and standards are discovering that an LMS is a valuable delivery and assessment tool, and an effective LMS will meet unique needs. Two of the most important aspects of an LMS are the recording of data from learners and providing reports to the organization's administration. The LMS provides an "anytime, anywhere" learning environment that is of great value to emergency services organizations. Responders are able to access the educational and training material during the course of their shift without going out of service. Responders whose learning process has been interrupted during the course of their shift can come back and pick up where they left off at before they were interrupted.

CHAPTER **REVIEW**

Summary

Data collection and the management of information are two of the most important aspects of the health and safety program. Data collection is a necessary component of establishing goals and objectives; however, data must be put into a useful and retrievable format that is compatible with the intended application. Data collection and information management provide valuable feedback that will aid in the evaluation of programs. It is important to collect data not only at the local level but also at the state and national levels. A consistent means should be established in which to gather and evaluate data.

Change within emergency services organizations is occurring at a much faster rate than ever before. Much of this rapid change can be attributed to advancing technology. Once data have been analyzed and put into a usable format, an annual report should be compiled and distributed throughout the organization that answers these questions: Where are we now? Where are we going? How do we get there?

The Internet has become a valuable tool allowing an organization to communicate and gain access to information at a much faster rate than ever before. Many local, state, and national agencies have information available on the Internet related to health and safety. Such information provided through the Internet can be used to assist the health and safety program manager and the safety committee in program design, training, and evaluation. There is no limit to the access of information and new technologies available.

The health and safety program manager must learn to focus on tools that provide both quality assurance and risk management. Data collection and assessment are the keys to effective health and safety program management. The health and safety program manager must stay proactive and up to date with current technology to remain as effective as possible in the area of information management.

Review Questions

1. Name the method that is most useful in collecting and analyzing data concerning emergency services organizations.
2. What governmental agencies require data collection in many areas and jurisdictions?
3. Describe the qualities that collected data should have.
4. Injury and accident reports are considered external data.
   a. True
   b. False
5. Describe why the annual health and safety report is a very useful tool.
6. Describe the keys to effective health and safety program management.
7. Who is normally responsible for investigating injuries and/or deaths of members?
   a. The fire chief
   b. The health and safety division
   c. The training division
   d. The local union
8. OSHA requirements are in many cases enforced at the:
   a. Local level
   b. County level
   c. State level
   d. None of the above
9. Describe a necessary component of performing an evaluation of the health and safety program.
10. Describe a learning management system.

Case Study

On February 9, 2007, a 29-year-old female, career probationary firefighter died while participating in a live-fire training evolution at an acquired structure. The victim's class was conducting a live-fire training drill as required by the department's training protocol for its NFPA 1001, *Standard for Fire Fighter Professional*

Qualifications, Fire Fighter I. The victim was part of a four-person engine company, led by an adjunct instructor, that made the initial attack on a training fire in a vacant, condemned, three-story, end-unit townhouse. The scenario called for the victim's crew to enter the front of the townhouse and proceed to the third floor to find and extinguish any fire there. They were to bypass any fire on the second floor so that the second-due engine could practice suppression on that floor. The victim's crew encountered heavy fire on the second floor and third floor stairwell as they proceeded to the third floor. The victim, operating the nozzle, and the adjunct instructor attempted to fight fire on the third floor, but conditions made it untenable. The adjunct instructor was able to exit through a window located on the third floor landing followed by a firefighter who was backing up the victim on the hose line. However, the victim got stuck attempting to exit the window, which was 41 inches above the floor. The victim became unresponsive as the adjunct instructor and other firefighters attempted to free her from the window. After she had been freed, she was transported to a local trauma center where she was pronounced dead.

The victim was a probationary firefighter who had not yet completed the last requirement for NFPA 1001, *Standard for Fire Fighter Professional Qualifications*, Level I certification, which was live-fire training. She reportedly had experienced mask claustrophobia while wearing a face piece in previous training evolutions. It was also reported that the victim did not meet the minimum required time to complete the physical ability test developed by the fire department when she was hired. She reportedly was retested just prior to the incident and produced a slower time than on the initial test. Documentation was not provided to the investigative team regarding her physical ability testing.

The officer in charge (OIC) was a lieutenant who had received training through the State Fire and Rescue Institute on topics such as Instructor I and II and Fire Skills for Instructors. He had over 25 years of firefighting experience.

The safety officer was a district chief who was in charge of the training academy. He had received training through the State Fire and Rescue Institute on topics such as Leadership and Supervision, Fire Command, and Methods of Instructor Level II—Teaching and Program Development Techniques. He had over 20 years of firefighting experience.

None of the adjunct instructors that participated in this training evolution had any training as a fire instructor. The adjunct instructor for the victim's squad had over 9 years of firefighting experience. This was the first time he had acted as a training instructor.

The department was using an acquired structure for its live-fire training, which was a three-story, end-unit townhouse of ordinary construction that came to a 45-degree angle at the A/D corner. It had three bedrooms and approximately 1,200 square feet of living area. The building was designated a vacant structure by the city in June 1999 and then condemned in January 2004.

Several weeks prior to the day of the incident, the structure had been used to practice forcible entry and ventilation operations. Portions of the walls and ceilings had been removed from the structure during those exercises.

NIOSH had the following recommendations:

Recommendation #1: Fire departments should conduct live-fire training exercises in accordance with the most recent edition of NFPA 1403, *Standard on Live Fire Training Evolutions*.

Recommendation #2: Fire departments should ensure all training and education, including live-fire training, is conducted under the direct supervision of a qualified instructor(s) who meets the requirements of NFPA 1041, *Standard for Fire Service Instructor Professional Qualifications*.

Recommendation #3: Fire departments should provide the Training Academy and Safety Division with adequate resources, personnel, and equipment to accomplish their training mission safely.

Recommendation #4: Fire departments should screen recruits to ensure they meet the physical performance requirements as established by the fire department prior to entering a training program to become a firefighter.

Recommendation #5: Fire departments should develop and maintain a comprehensive respiratory protection program that complies with NFPA 1404, *Standard for Fire Service Respiratory Protection Training*.

Recommendation #6: Fire departments should ensure all recruits meet the requirements of NFPA 1582, *Standard on Comprehensive Occupational Medical Program for Fire Departments*, prior to entering the training program.

Recommendation #7: Fire departments should develop inspection criteria to ensure that all protective ensembles meet the requirements of NFPA 1851, *Standard on Selection, Care, and Maintenance of Protective Ensembles for Structural Fire Fighting and Proximity Fire Fighting*.

Recommendation #8: Fire departments should ensure coordinated communication between the instructor in charge and the live-fire training participants.

Recommendation #9: Fire departments should utilize the incident command system and a personnel accountability system, for all incidents, including live-fire training exercises, that meets the requirements of NFPA 1561, *Standard on Emergency Services Incident Management System*.

Recommendation #10: Fire departments should create a training atmosphere that is free from intimidation and conducive to learning.

Recommendation #11: States should develop a permitting procedure for live-fire training to be conducted at acquired structures and also ensure that all the requirements of NFPA 1403 have been met before issuing the permit.[1] See Figure 12–4.

1. Describe some of the contributing factors with this line-of-duty death.
2. Describe how communication may have played a role in this incident.

Maryland Training Fire Fatality Concerns and Issues	Yes	No
Should the interior crews have been permitted to pass active fire on the second floor?		
Should the victim have been in a live-fire situation after experiencing claustrophobia with a face piece?		
Was there an adequate incident management system in place?		
Was the victim's physical condition questionable to perform as a firefighter in this situation?		
Did the incident commander continuously evaluate changing conditions?		
Was crew integrity maintained throughout this event?		
Did the lack of recognition of the fire conditions by the instructors play a role at this event?		
Was the PPE used for this training adequate?		
Would adequate firefighter survival training have played a role at this fire?		
Was personal protective gear worn properly at this event?		
Did intimidation play a role at this event?		
Was NFPA 1403 followed at this event?		

FIGURE 12–4 Maryland training fire fatality concerns and issues.

Reference

1. NIOSH Fire Fighter Fatality Investigation and Prevention Program, *Career Probationary Fire Fighter Dies While Participating in a Live-Fire Training Evolution at an Acquired Structure—Maryland,* December 8, 2008. Retrieved May 26, 2011, from http://www.cdc.gov/niosh/fire/reports/face200709.html

CHAPTER 13

Specific Issues Concerning Occupational Safety Services

KEY TERMS

Age Discrimination in Employment Act of 1967 (ADEA), *p. 266*

Americans with Disabilities Act (ADA), *p. 265*

change agent, *p. 272*

tort liability, *p. 267*

OBJECTIVES

After reading this chapter, you should be able to:

- Describe the ethical, legal, and financial issues concerning the health and safety program.
- Discuss why proper ethics are important in emergency services.
- Describe the type of funding available for the health and safety program.
- Describe the adjustments that must be made to the health and safety program as they relate to a diversified workforce.
- Describe the implications of future trends and technologies as they relate to the health and safety program.

ResourceCentral

For additional review and practice tests, visit **www.bradybooks.com** and click on Resource Central to access book-specific resources for this text! To access Resource Central, follow directions on the Student Access Card provided with this text. If there is no card, go to **www.bradybooks.com** and follow the Resource Central link to Buy Access from there.

Many of the topics discussed in this chapter are being considered by today's proactive emergency services organizations and may also be regarded as future trends in emergency services. Hopefully the discussion of these issues will cause many in other organizations to stand up and take notice of what is to come. A great deal of research must be done in these areas in order to implement the necessary changes in the field of emergency services. Everyone agrees that emergency services organizations will need to address many challenges in the near future. Yet change does not come easily in the emergency services field. An emergency services organization that chooses the status quo will find itself falling farther and farther behind those that strive to be proactive. Much of the future changes will revolve around trends and changing technology, as well as financial issues. The reduction in funding for emergency services in recent times will cause continued challenges and force emergency services organizations to reinvent how they provide service and how they will continue to conduct business.

#1

Ethics in Emergency Services

Ethical leadership continues to be a hot topic in society in general, and it is extremely important in emergency services organizations. Proper ethics can be described as "a set of behavior rules that allow proper growth and development."

Over the years, fire department officers have been caught up in serious ethical issues concerning safety. For example, fire department officers were terminated after a probationary firefighter was killed in the line of duty while live-fire training was being conducted. The department health and safety program manager is concerned with the ethical issues involving the health and safety program. The leading ethical priorities focused on health and safety concerns are to provide the safest work environment, encourage supervisors and leaders to ensure members are operating as safely as possible, and ensure members are protected with the safest possible protective and safety equipment. Unfortunately, in emergency services there is seldom a completely safe working environment, and control over the workplace is limited as it relates to the incident scene; however, when it comes to facilities, equipment, and operations, the organization has an ethical responsibility to provide as safe a workplace as humanly possible to ensure the safety of workers. It is important for every emergency services organization to recognize the value of supervisor input related to safety issues. Supervisors and employees often will provide valuable input to the administration of the emergency services organization. This input should be evaluated and used in every area of the safety environment.

#4

Because life in emergency services is a great deal more complicated than it ever has been, so are its ethical issues. Certain individuals and committees are directly involved in emergency services investigations of accidents, fatalities, and management of records and data. Leadership must continually stress to involved individuals the importance of not sharing specific sensitive information with other members of the organization. Those who have been exposed to such information must be trusted to keep it confidential. The sharing of confidential information could possibly do irreparable damage to not only the health and safety program and but also, quite possibly, members within the organization. Those involved with the health and safety program must continually be encouraged to keep the ethical standards of the organization in mind at all times. Both department culture and policy must encourage individuals within the organization to do the right thing when faced with circumstances that present them with more than one right way or, worse, when there is no clear right way at all.

Emergency services personnel must continually be educated in ethical standards as they relate to the health and safety program. In emergency services it is not uncommon for members of the organization to become offended when one of our own fails to follow ethical standards in doing his or her duty; however, the proper perspective is that most emergency services personnel effectively meet ethical challenges and perform their duties satisfactorily. As a collective organization, members must strive to have a significant impact on the ethical behavior of coworkers and be continually reminded that one of the highest virtues of professionalism in emergency services is abiding by an endorsed code of ethics. Members should have a very clear understanding of what constitutes acceptable and unacceptable behavior. Unfortunately, ethical issues in emergency services can be very complex, and in most organizations, the code of ethics is informal and simple in nature, although the IAFC has published a formal code of ethics. The basic code of ethics in emergency services should include respect for people, not causing harm, being truthful, and following laws to the letter. It is the custom of emergency services to employ individuals with impeccable integrity.

Court and Law Issues

Litigation against fire departments and other emergency services organizations has increased during recent times. Although the legal actions affect many areas of emergency response organizations, many cases pertain directly to health and safety. For example, the safety program concerning blood-borne pathogens may be brought into question after an employee suffers an exposure from a needlestick, or an organization may be brought into court for a case related to inadequate safety equipment. Although many of the investigative reports on incidents provide insight to the nation as a whole, these same reports could be used by a plaintiff in court proceedings against an emergency response organization. A large number of line-of-duty deaths result in lawsuits against the local governments responsible for providing emergency services. A number of fire service organizations cannot find the funding to fully comply with many of the NFPA consensus standards, even though they are not required by law to comply. Noncompliance with the consensus standards will many times result in legal action should injury or death occur. When specific issues arise, the legal department or legal representatives that represent the department should be contacted for direction and advice. Many of the common issues emergency response organizations face in court include, but are not limited to, liability, employment policy, employee physicals, personal injury, internal investigations, company officer operations, personnel issues, and risk management.

Americans with Disabilities Act

The **Americans with Disabilities Act (ADA)** Title I of 1990 took effect July 26, 1992, and was amended by Congress in January 2009. The amended ADA act broadened its protection of employees and overturned a series of decisions made by the Supreme Court. The act prohibits private employers, state and local governments, employment agencies, and labor unions from discriminating against qualified individuals with disabilities in job application procedures; hiring; firing; advancement; compensation; job training; and other terms, conditions, and privileges of employment. The ADA requirements and court rulings are under continual changes. It is recommended emergency service organizations keep abreast of the most recent updates and changes.

Americans with Disabilities Act (ADA) ▪ The ADA prohibits private employers, state and local governments, employment agencies, and labor unions from discriminating against qualified individuals with disabilities in job application procedures, hiring, firing, advancement, compensation, job training, and other terms, conditions, and privileges of employment.

The ADA covers employers with 15 or more employees, including state and local governments, and also applies to employment agencies and labor organizations. To be protected by the ADA, one must have a disability or have a relationship or association with an individual with a disability. An individual with a disability is defined by the ADA as a person who has a physical or mental impairment that substantially limits one or more major life activities, a person who has a history or record of such an impairment, or a person who is perceived by others as having such an impairment. The ADA does not specifically name all of the impairments that are covered.

The health and safety program manager's responsibility is to become familiar with the provisions contained in Title I of the ADA. One area of concern to emergency response organizations is medical examinations and disability. Job applicants may no longer be questioned about the existence, nature, or severity of a disability. The law restricts questions that can be asked about an applicant's disability before a job offer is made, and it requires employers to make reasonable accommodation to the known physical or mental limitations of otherwise qualified individuals, unless doing so results in undue hardship. Applicants may be asked about their ability to perform specific job functions. A job offer may be conditional based on the results of a medical examination, but only if the examination is required for all employees who are gaining employment in similar jobs. Medical examinations of employees must be job related and consistent with the needs of the employer.

Beginning in 1999, the United States Supreme Court issued a series of decisions limiting the scope of the ADA. As a result of those court decisions, persons with certain types of impairments that have been mitigated by corrective measures, such as corrective eyeglasses for myopia and medication for high blood pressure, are not considered *disabled* under the ADA. Since 2009, the Supreme Court decision was made that an impairment is not a disability covered by the ADA unless it severely restricts a person from doing activities that are of central importance to most people's daily lives. These Supreme Court decisions have limited the people who can claim protection of the federal ADA; however, this does not eliminate the need for the department health and safety program manager to keep abreast of ADA guidelines concerning emergency services–related employment decisions. Separate disability protections exist under the laws of specific states, and some of these laws afford greater protection than that given by the ADA.

Under the amended ADA act, the terms *major life activities* and *substantially limited* are better defined as an effort to determine whether an individual falls under the guidelines and definitions of the ADA. The amended ADA act provides reasonable accommodations that are required only for individuals who can demonstrate they have an impairment that substantially limits a major life activity or a record of such impairment. Accommodations do not need to be provided to an individual who is regarded as having only an impairment.

Age Discrimination in Employment Act

Age Discrimination in Employment Act of 1967 (ADEA) ■ The ADEA protects individuals who are 40 years of age or older from employment discrimination based on age. The ADEA's protections apply to both employees and job applicants.

The **Age Discrimination in Employment Act of 1967 (ADEA)** is a federal regulation with which the health and safety program manager, labor relations manager, and human resources manager should all be well versed. The ADEA is meant to protect individuals who are 40 years of age or older from employment discrimination. An employer may not discriminate based on age with respect to any term, condition, or privilege of employment including, but not limited to, hiring, firing, promotion, layoff, compensation, benefits, job assignment, and training. The ADEA applies to employers with 20 or more employees, including the federal, state, and local governments, and also encompasses employment agencies and labor organizations. A section of the ADEA applies to state laws on the hiring of firefighters and law enforcement officers.

Liability

Liability is one of the leading issues for the health and safety program manager and all members of emergency response organizations. In cases concerning injury and death, the liability of the health and safety program will come into question. Liability issues have become a major concern for every emergency response organization and are at the forefront of the everyday issues that public officials face. Public officials, including emergency response personnel, can be found personally liable for compensatory or punitive monetary damages.

Numerous states protect firefighters and EMS personnel from liability due to state statutes, many of which require proof of willful or wanton misconduct. Willful misconduct is considered as doing something that is wrong, or intentionally failing to do that which should be done. The circumstances must also disclose that the defendant knew or should have known that such conduct would probably cause injury to the plaintiff. As a general rule, every person may be presumed to intend the natural and probable consequences of his or her actions. Willful misconduct implies an intentional disregard of a clear duty or of a definite rule of conduct, a purpose not to discharge such duty, or the performance of wrongful acts with knowledge of the likelihood of resulting injury. Knowledge of surrounding circumstances and existing conditions is essential; actual ill will or intent to injure need not be present.

Wanton misconduct must be done under such surrounding circumstances that the party doing the act or failing to act must be aware, from his or her knowledge of such circumstances and conditions, that such conduct will probably result in injury. Wanton misconduct implies a failure to use any care for the plaintiff and an indifference to the consequences, when the probability that harm would result from such failure is great, and such probability is known, or ought to have been known, to the defendant. Moreover, many of the states and courts within the states have defined "willful and wanton misconduct" as "behavior demonstrating a deliberate or reckless disregard for the safety of others."

Tort liability is defined as a legal obligation stemming from a civil wrong or injury for which a court remedy is justified. A tort liability arises from a combination of a direct violation of a person's rights and the transgression of a public obligation causing damage or a private wrongdoing. A tort is an injury to another person, group, or to property that is compensable under the law. Categories of torts include negligence, gross negligence, and intentional wrongdoing, with negligence being the most common. A tort can occur when a person or group acts, or fails to act, without right and harms another directly or indirectly. To give rise to a legal claim in negligence, an act (or inaction) must satisfy four elements:

tort liability ■ Defined as a legal obligation stemming from a civil wrong or injury for which a court remedy is justified. A tort liability arises because of a combination of a direct violation of a person's rights and the transgression of a public obligation causing damage or a private wrongdoing.

- There must be a legal duty of care to another person.
- There must be a breach of that duty.
- The claimant must have suffered damages.
- The damages must have been proximately caused by the breach of duty.

Four types of torts are applicable to the health and safety program:

1. ***Intentional liability.*** This is any intentional acts that are reasonably foreseeable to cause harm to an individual, and that do so. There is normally a violation of the law or regulations, and harm results.
2. ***Strict liability.*** This is a legal doctrine that makes some persons responsible for damages their actions or products cause, regardless of any "fault" on their part. There can also be a violation of law or regulations, even if the violation is unintentional.
3. ***Negligent liability.*** This is the failure to exercise the required amount of care to prevent injury to others. A person or group fails to do what a reasonable and prudent person would have done under the same or similar circumstances.
4. ***Warrant liability.*** The promised level of service is not delivered and harm results. A liability is normally associated with the warranty of a product.

It is easy to see the types of liability that will affect the health and safety program. If an emergency response organization does not adhere to the applicable OSHA 1910.120 regulation, the organization may be guilty of an intentional or strict liability. For example, a firefighter injured at a hazardous materials incident directly related to the lack of utilizing the incident command system would have a case of liability against the department.

The issues of liability that emergency services organizations face are ever changing. Each court decision brings changes in emergency services operations, and the organization's administration and the health and safety program manager must keep on top of liability and legal issues. They need to get up close and personal with the legal department or legal staff that represents the emergency response organization.

Economic and Budgetary Issues

It should come as no surprise that health and safety programs cost money and take considerable resources to have in place. The costs associated with operating a career fire department have escalated, as have the costs of administering a health and safety program. The administration of any fire service organization must realize that the cost of not having a health and safety program in place will be more than that of having such a program in place. For example, funding the IAFF/IAFC's Fire Service Joint Labor Management Wellness/Fitness Initiative will reduce costs associated with injuries and unhealthy lifestyles. The cost of funding this program and other types of wellness programs serves to reduce the cost of workers' compensation claims.

Many emergency services organizations find common ways to fund a program; for example, they may use monies from the organization's annual budget or seek out and obtain grant funding. A cost–benefit analysis should be completed prior to attempting to obtain money and resources from either the annual budget or grant funding.

Of course, taxation and fire service fees are the most common ways for an emergency services organization to obtain funding. The health and safety program manager should be familiar with budgeting and the associated process, using persuasion with the department's administration as well as with the officials in charge of financing the budget of the organization as it relates to the health and safety program. The health and safety program manager must work hard to convince the administrative staff what parts of the program are absolutely necessary. Conducting cost–benefit analysis prior to budget formulation is an effective method of encouraging the organization's administration to buy in to the program. It may also be necessary for the health and safety program manager to present a cost–benefit analysis to the governmental administration outside the emergency services organization.

The health and safety program manager must keep in mind that, aside from new health and safety program components, the health and safety program has annual ongoing costs, which will vary from organization to organization. A proactive health and safety program manager will anticipate and calculate the ongoing as well as new costs associated with the program.

Recently, grant monies that are related to the health and safety program have become available for purchases of capital items. Then the organization must come up with future funding after the grant has been exhausted.

GRANT FUNDING

In recent years, the available grant funding has come from federal, state, and even some local funding sources. Many local governments at the city level have individuals or committees there that specialize in grant writing; they are familiar with ways to streamline the process.

The health and safety program manager should focus research efforts on all possible areas of available funding and grants. The *Federal Register* publishes federal grant programs that can be accessed and applied for. In a down economy, emergency services organizations will be faced with possible budgetary cuts and a reduction in resources, which could easily threaten the health and safety program; therefore, the health and safety program manager should work at becoming an expert in the area of resources related to grant funding.

Some emergency services organizations have received grant funding related to personnel staffing (SAFER Grant) and technology such as teleconferencing (Assistance to Firefighters Grant) for efficient communications among various county, city, and local

BOX 13-1: COMMON EMERGENCY SERVICES GRANT PROGRAMS

Assistance to Firefighters Grant Program (AFGP). Administered by the Department of Homeland Security's Center for Domestic Preparedness, this program assists rural, urban, and suburban fire departments throughout the United States. The grant funding is used by the nation's firefighters to increase the effectiveness of firefighting operations, to improve firefighter health and safety programs, and to establish or expand fire prevention and safety programs. AFGP, originally known as the Fire Investment and Response Enhancement (FIRE) Act, was moved from FEMA to the Department of Homeland Security in 2004.

Volunteer Fire Assistance (VFA). Formerly known as the Rural Community Fire Protection Program (RCFP), the purpose of the VFA program is to provide federal financial, technical, and other assistance to state foresters and other appropriate officials to organize, train, and equip fire departments in rural areas and communities to prevent and suppress wildland fires. A rural community is defined as a community with a population of 10,000 or less.

Department of Justice. Grant funding is available for terrorism preparedness equipment and training.

Hazardous Materials Emergency Preparedness (HMEP). This grant funding provides financial and technical assistance, as well as national direction and guidance to enhance state, territorial, tribal, and local hazardous materials emergency planning and training. The HMEP grant program distributes fees collected from shippers and carriers of hazardous materials to emergency responders for hazardous materials training and to local emergency planning committees (LEPCs) for hazardous materials planning.

departments. Through teleconferencing, communications is available during disaster operations at the emergency scene, which helps to enhance safety.

Grant applications must show how the funding will be distributed and how the department plans to continue the program after the grant funding has been exhausted. Certain grants require a local matching of funds, whereas others do not. The health and safety program manager is responsible to find the available source of grant funding and to oversee the process. See Box 13-1.

#10

Safety Considerations Involved with Diversified Generations

It should come as no surprise that diversity and a generational gap in the workforce exist. Such diversity will continue to have repercussions for emergency response organizations, including the following:

- The average worker will continue working past normal retirement age.
- The population growth will become sluggish.
- More women will seek careers and wish to work.
- The number of white males in the workforce will diminish.
- The number of minority workers will increase.

The repercussions associated with the changing workforce do not necessarily cause concerns for the health and safety program or its manager; however, issues will arise regarding communications and diversity.

Physical fitness is a perennial issue in emergency services organizations. Physical fitness should become a top priority in emergency services organizations, particularly the fire service. As many emergency services employees are forced to work longer until retirement, the corresponding rise in injuries and cardiac-related injuries will become more of an issue in the near future. Emergency services workers may become a hazard to themselves and the organization through the aging process. For those who choose or are forced to work longer, physical fitness and wellness should become a top priority to the individual member and the organization as a whole. A joint wellness-fitness initiative created by the IAFF and the IAFC has a physical abilities test for entry into the fire service called the candidate physical abilities test (CPAT). CPAT consists of eight timed, performance-based events. Information concerning CPAT is available from both the IAFF and the IAFC, or can be found from an Internet search. The portability of CPAT is possible; however, each agency that administers the CPAT test must be licensed by the IAFF. A review of the CPAT test between agencies is advisable if one agency is willing to accept the CPAT results of a candidate from another agency to ensure the test is being administered as intended by the IAFF/IAFC joint wellness-fitness initiative. CPAT can be administered only in strict compliance with the licensing agreement from the IAFF. The use of CPAT for candidate physical abilities testing without a license or the misuse of the CPAT program is in direct violation of the IAFF copyright of the program.

#6

The entry of women into emergency services has produced a diversified workforce, and emergency services has adapted well. Considering the history of emergency services, the entry of women has been relatively new. All emergency services organizations should have policies in place addressing pregnancy. During gestation, women should not be exposed to hazardous substances nor required to perform any function that could harm the baby. Many organizations place pregnant women on restricted or modified duty during their pregnancy. Organizations that do not offer such choices place women on sick leave or permit them to use vacation time.

The diversified workforce also means there are individuals of varying size and weight. Such circumstances should cause the health and safety program manager, possibly with an apparatus committee, to study equipment and apparatus design to determine the type and size of apparatus that will best suit the organization. They must be sure to take the needs of the diversified workforce into consideration. Equipment should be mounted and apparatus designed so the organization's employees can have access with relative ease. Properly designed apparatus helps to reduce injuries and aid in achieving the health and safety program's goals.

Consideration should also be given to equipment purchases. PPE and SCBA masks will have to be fit tested for all employees, and the organization must have varying sizes of PPE in stock to fit all employees. Personnel should be able to comfortably wear PPE without its adding undue weight or causing unnecessary heat stress. Improperly fitting PPE will affect job performance and cause the potential for injury and added stress. The preceding examples are just a few concerning a diversified workforce in emergency services and are not meant to be all-inclusive. What is most important is that the health and safety program manager should make every effort to keep abreast of issues that may seem small in nature, but have the potential to cause employee injury and stress.

It must be recognized there are several generations of workers making up the ranks of emergency services organizations. Although these generational differences do not have a direct negative influence on the health and safety program, issues may arise related to conveying the health and safety message and program to individuals of different generations. The various generations in the workforce will view authority figures and supervisors differently. Individuals also view training from an individual and generational perspective. The generations found in emergency services organizations can be made up of traditionalists, baby boomers, and those from Generation X and Generation Y. A perceived decline in work ethic is the prime issue causing associated generational problems in the workforce today. Although this is a debatable issue, managers and supervisors must keep in mind that perception is reality to employees.

It is common for both younger and older workers to complain about the lack of respect from individuals of various generations. Research suggests that older generations value authority more than younger generations do, and older workers in fire service organizations display command-and-control leadership and personality styles similar to military organizations. Although members of younger generations may not always be impressed with titles and rank in emergency services, many were raised with the philosophy that it is all right to ask questions and interact with superiors. Although these actions are not signs of disrespect, supervisors must ensure their subordinates follow directives, policies, and procedures related to the health and safety program. It is common for younger workers to believe that respect for authority must be earned; however, they also must understand that following orders concerning health and safety is not a choice based on the lack of respect for authority. Personnel must understand and follow directives concerning health and safety, and not question directives based on a lack of respect of certain officers or individuals.

It is now known that members of different generations have different learning styles. Whereas members of older generations do not always require feedback associated with learning, younger members do. For this reason, supervisors and older peers should make every attempt to communicate to younger members about why certain procedures, practices, and policies are in place regarding health and safety issues. As it relates to the health and safety program, members of various generations will differ in their perceived training needs. It would serve any organization well to provide training and instruction in this important area of generational differences as a way to ensure issues related to health and safety are addressed in the most effective way across generational lines.

The Future Developments and Tools in Emergency Services

The proactive health and safety program manager must understand that change has always played a large part in emergency services. Change is inevitable in emergency services, and the health and safety program manager's ability to cope with and manage change in an effective and meaningful manner will enhance his or her position.

Consider the changes in technology in the emergency services field that have occurred over the course of time. The health and safety program manager should face the future with anticipation and embrace the possibilities of creative thinking. The health and safety program manager should try to spot trends early concerning the health and safety of members and be willing to test new ideas and adopt the ones that work.

Unfortunately, change and new technology sometimes bring expensive equipment and technology that is difficult to afford. In order for an emergency services organization to accept change and new technology, these resources must be useful and adaptable to many emergency services organizations. A method or technology must be in place for an extended period of time before many in the fire service will accept the change it has brought. New technology and equipment are initially expensive; however, over the course of time the associated costs usually go down as demand rises.

The health and safety program manager must understand that emergency services organizations are a reflection of society. Developments in technology, tools, and methods concerning other areas of society will find their way into emergency services organizations whether they embrace them or not. Keep in mind the SCBA as it is known today in the fire service was not originally developed for the fire service but for the mining industry. The same could be said of the thermal imaging camera and portable radios, both of which were developed for other industries. Just a few short years ago, few fire departments had thermal imaging cameras. The costs have been reduced with advancing technology, and several departments have purchased them using available grant money. Many departments now have cameras on each emergency response apparatus, and in the future probably every firefighter will carry one. See Figure 13–1.

#3

#7

The health and safety program manager must understand that trends and patterns involve examination of both the past and the present of emergency services. The health

FIGURE 13–1 A firefighter uses a thermal imaging camera to increase health and safety. *Courtesy of Joe Bruni.*

and safety program manager with a heightened awareness need only look at advancing trends and patterns in society. The effective health and safety program manager must embrace being a **change agent**; however, he or she must resist the temptation to embrace every latest "craze" until its consequences can be properly evaluated. The health and safety program manager must understand that as a positive change agent, many within the organization will assess the impact of any change, sometimes in a resistive manner, especially when it comes to changes addressing health and safety.

change agent ■ This individual is recruited prior to implementation of a change; must be representative of the user population, understand the reasoning behind the change, and help to communicate the excitement, possibilities, and details of the change to others within the organization.

One of the areas that must also be addressed concerning developments, tools, and technology is the ability to make use of better data management. This area has been somewhat lacking in recent years until advancing technologies have enabled organizations to track and keep records in a much easier fashion than ever before. In order for the health and safety program manager to be effective, his or her focus must go beyond emergency response operations. The program manager must stay focused on issues and concerns affecting health and safety in a broader scope. The list in this area will be endless, and must go beyond issues that will arise beyond the emergency incident scene. Everyday operations must be evaluated continually for health and safety issues.

#8

Summary

Coverage of the topic of health and safety would not be complete without discussing ethical, legal, and financial issues as well as the importance of an increasingly diverse workforce, technologies, and future trends that will affect the health and safety program. Research in these areas will be necessary in order to implement change in the field of emergency services.

One constant that will remain in emergency services organizations is ethics. Individuals involved with the health and safety program will be held to a higher standard concerning ethics. The health and safety program manager must maintain a heightened awareness of legal concepts; however, legal counsel can best address issues such as negligence. The health and safety program manager should have input into the budget process and convince the administrative staff what parts of the program are absolutely necessary. Conducting a cost–benefit analysis is necessary to obtain the necessary budget dollars. The annual budget is the most common funding source; however, external sources are also available such as grant funding from numerous sources.

Diversity in the workforce will continue to increase in the future and will impact the health and safety program. Diversity will also bring implications for emergency services organizations; for example, minorities will become the larger share of new candidates into the emergency services field.

Future technologies and trends will also affect the health and safety program. Change has always played a large part in emergency services concerning trends and technology. Change is inevitable in emergency services, and the health and safety program manager's ability to cope with and manage change in an effective and meaningful manner will enhance his or her position. The effective health and safety program manager must embrace being a change agent. The focus of the health and safety program manager is continued support and improvement of the health and safety program.

One of the areas that must also be addressed concerning trends and technology is the ability to make use of better data management. In order for the health and safety program manager to be effective, his or her focus must go beyond emergency response operations.

Review Questions

1. Describe the common funding sources for the health and safety program.
2. Describe the leading ethical priority focused on health and safety concerns.
3. To give rise to a legal claim in negligence:
   a. There must be a legal duty of care to another person
   b. There must be a breach of that duty
   c. The claimant must have suffered damages
   d. All of the above
4. Describe the four types of tort liability.
5. Describe three concerns and issues that may or will affect the health and safety program in the future.
6. _____________ should be concerned with certain ethical issues involving the health and safety program.
7. Providing as safe a workplace as possible as a way to ensure the safety of workers is what type of issue?
   a. An ethical issue
   b. A logistical issue
   c. A training issue
   d. An operations issue
8. Describe 8 of the issues common to most emergency response organizations involving liability.
9. The Americans with Disabilities Act covers employers with __________ or more employees.
   a. 10
   b. 15
   c. 20
   d. 25
10. Many states protect firefighters and EMS personnel from liability due to:
    a. State statutes
    b. Local laws
    c. County laws
    d. Federal laws

Case Study

On February 19, 2009, a 49-year-old male volunteer lieutenant and a 26-year-old male firefighter were fatally injured while combating a mobile home fire. They arrived on scene to find a camper fully involved with fire and flames impinging on an adjacent mobile home. The occupants of the camper and mobile home escaped without injury, prior to the fire department's arrival. The victims entered the mobile home through the front door with a charged 1½-inch hose line. Within 5 to 10 minutes of them entering, the pump operator sounded the evacuation alarm when he noticed his tank water was low. The victims did not evacuate from the structure. Firefighters on scene attempted to contact the victims via radio and by yelling into the mobile home. The fire chief and a firefighter tugged on the 1½-inch hose line several times with no response. They then pulled on the hose line and it came freely from the mobile home. Fire conditions were primarily contained to the one side of the structure, while several attempts were made to locate the victims by firefighters entering through the front door and noninvolved side of the mobile home. The victims were eventually discovered in the front room of the mobile home—several feet from the front door they had entered through. Their face pieces were not on when they were found. The victims were pronounced dead on scene.

NIOSH had the following recommendations:

- Recommendation #1: Ensure that firefighters use their self-contained breathing apparatus (SCBA) during all stages of a fire due to the potential exposure and health affects of fire-produced toxins.
- Recommendation #2: Ensure that all SCBAs are equipped with an integrated personal alert safety system (PASS) device.
- Recommendation #3: Ensure that all firefighters are equipped with a means to communicate with fireground personnel before entering a structure fire.
- Recommendation #4: Ensure that the incident commander (IC) does not become involved with firefighting activities.
- Recommendation #5: Ensure that the incident commander (IC) maintains close accountability for all personnel operating on the fireground and that procedures and training for the use of a personnel accountability report (PAR) are in place.
- Recommendation #6: Ensure that a properly trained incident safety officer (ISO) is appointed at all structure fires.
- Recommendation #7: Ensure that a rapid intervention team (RIT) is established and available to immediately respond to emergency rescue incidents.
- Recommendation #8: Ensure that hose line operations are properly coordinated so as not to impede search-and-rescue operations.
- Recommendation #9: Develop, implement, and enforce written standard operating procedures (SOPs) for fireground operations.
- Recommendation #10: Ensure that all firefighters properly wear their department-issued turnout gear and personal protective equipment (PPE) during fire suppression activities.
- Recommendation #11: Develop and maintain a comprehensive respiratory protection program.
- Recommendation #12: Ensure that firefighters are aware of the dangers involved in fighting mobile home fires.
- Recommendation #13: Ensure that policies and procedures for proper inspection, use, and maintenance of self-contained breathing apparatus (SCBA) are implemented to ensure they function properly when needed.[1]

1. Describe the alternative tactics that could have been used at this incident to prevent the line-of-duty deaths.
2. Describe how the lack of communications played a role at this incident.

Reference

1. NIOSH Fire Fighter Fatality Investigation and Prevention Program, *Fire Fighter Trainee Suffers Fatal Exertional Heat Stroke During Physical Fitness Training—Texas,* June 2010. Retrieved September 13, 2010, from http://www.cdc.gov/niosh/fire/reports/face200917.html

CHAPTER 14 Lessons Learned from Incidents

KEY TERM

risk-versus-benefit analysis, *p. 276*

OBJECTIVES

After reading this chapter, you should be able to:

- Describe the focus and purpose of the post-incident analysis.
- Describe the benefits of the post-incident analysis.
- List the problems and threats commonly encountered at the large-scale incident.
- List the reasons why structural firefighting personal protective equipment is not well suited for non-fire-related environments.
- List the reasons why an incident management system is a benefit to the health and safety of operating forces.
- List the reasons why everyone at the incident scene should complete a risk-versus-benefit analysis.
- Describe why incident commanders must have adequate resources and properly trained personnel at the incident scene.
- Describe why an ongoing apparatus operator training program is beneficial to the health and safety program.

Resource**Central**

For additional review and practice tests, visit **www.bradybooks.com** and click on Resource Central to access book-specific resources for this text! To access Resource Central, follow directions on the Student Access Card provided with this text. If there is no card, go to **www.bradybooks.com** and follow the Resource Central link to Buy Access from there.

The use of case histories in emergency services plays a valuable role in ensuring occupational health and safety. Coupled with incident investigation, post-incident analysis has evolved from what was once called a post-incident critique to what is now referred to as a post-incident analysis (PIA). In the past, post-incident critiques sometimes turned into finger-pointing sessions, which served no real benefit to emergency services members. Instead, the focus and purpose of the PIA is to emphasize what was done correctly at the incident scene and the subsequent lack of injuries and fatalities. NFPA 1500, *Standard on Fire Department Occupational Safety and Health Program*, contains a section in Chapter 8 on post-incident analysis and "lessons learned from investigating incidents," which discusses using case histories in the fire service—in other words, what lessons can be learned from investigating and analyzing incidents. NFPA 1500 states that the department safety officer should be part of the post-incident analysis process. The PIA should involve a basic review of the conditions present at the incident scene, the actions taken by members, and the effect of the conditions and actions on the safety and health of members.

As a result of this review, the PIA should also identify any action necessary to update any safety and health program elements or components to improve the health and welfare of department members. The department needs to develop a standardized action plan during or shortly after the PIA that includes the needed changes and the responsibilities, dates, and details of such actions.

Information gained from the investigation of incidents has played a major role in assisting the NFPA and other standards-making organizations with the development of and revisions to their health and safety standards. There are always lessons to be learned by analyzing incidents, as each incident contains aspects that can be reviewed and used to educate emergency responders with the hope that tragedies will not be repeated.

Some emergency response personnel do not understand the limitations of or place much faith in their personal protective equipment (PPE). Proper training is crucial so that personnel are thoroughly familiar with their issued PPE, because without it, personnel are bound to exceed the limitations of PPE in certain situations. Those who have had proper training will aid in improving the health and safety program. See Figure 14-1.

The lack of an effective incident management system (IMS) at an incident scene has been identified through the NIOSH Fire Fighter Fatality Investigation and Prevention Program as one of the causes of firefighter injury and death. Incident commanders must receive adequate training and continuously conduct a **risk-versus-benefit analysis** when determining whether the response will be offensive or defensive. The first arriving officer conducts the initial size-up to make an assessment of the conditions and to assist in planning the suppression strategy.

risk-versus-benefit analysis ■ This is the comparison of the risk of a situation to its related benefits.

FIGURE 14-1 Responders must understand the limitations of their PPE.
Courtesy of Joe Bruni.

The incident commander must perform a risk analysis to determine what hazards are present, what the risks to personnel are, how the risks can be eliminated or reduced, and the benefits to be gained from interior or offensive operations. Because the injury and death rate in emergency services remains relatively unchanged, we can only wonder to what extent the emergency services has learned these lessons. Incident analysis along with preparation is the key to protecting the health and safety of emergency responders.

Post-Incident Analysis

Post-incident analysis (PIA) is the reconstruction of an incident to assess the events that took place, the methods used to control the incident, and how the actions of emergency personnel contributed to the eventual outcome. In many cases, it is advisable to review the PIA with legal counsel because the PIA may be discoverable in litigation in certain states. The PIA's main purpose is to reinforce personnel actions and departmental procedures that are effective, and to give management insight into how to improve the department's operations. The health and safety program manager must learn to use both internal and external PIAs. The benefits of a PIA include:

- A comprehensive analytical record of an incident from which to evaluate departmental procedures
- Assessment of response times and company response areas under actual conditions
- Assessment of the effects of an additional resources request
- Assessment of tools and equipment
- Assessment of safety practices and related procedures
- Assessment of training needs for department personnel
- Assessment of the department's working relationship with outside agencies and other community departments

Large-Scale Incidents

Large-scale incidents such as massive fires, civil disturbances, terrorist events, and emergency medical events are unique, and all pose challenges to emergency services organizations and impact the health and safety of responders. The lessons learned from these types of incidents will pertain only to incidents of a similar nature, scope, and magnitude. The environment and nature of work conducted by emergency responders has changed drastically in recent years. Emergency responders and their organizations now view the incidents they respond to in a new light. For example, the September 11, 2001, attacks can be compared to the Oklahoma City bombing. The major threats at these types of incidents mainly include unknown hazards (environmental, chemical, biological, as well as physical) encountered by responders and incident commanders. See Figure 14-2.

#12

Large-scale incidents involving violence have drastically increased during recent times. An incident such as mass casualties at schools (e.g., the Columbine incident), and attacks on firefighters and emergency medical responders have changed the way of doing business and the level of personal protection. The unknown hazards of these types of large-scale incidents have forced emergency response organizations to reconsider the equipment and practices they commonly use to protect themselves during the course of their work.

Incident management staff as well as responders must continually evaluate and adjust their practices accordingly in order to operate at the highest possible level of health and safety. Because it is not possible to duplicate these types of incidents during training sessions, the tabletop exercise is an effective way to prepare and train for large-scale incidents. Incidents of these types should also be analyzed and evaluated for necessary changes to standard operating procedures.

THE LIMITATIONS OF PERSONAL PROTECTIVE EQUIPMENT

One of the main issues related to the health and safety of members operating at a large-scale incident is that of personal protective equipment. As has been discussed earlier in this textbook, structural firefighting protective equipment is not intended for incidents involving chemicals and biological agents. The added weight and limitations of structural PPE can be detrimental to the health and safety of responders at long-term incidents involving search and rescue, recovery operations, and emergency medical events. Not only can the added weight of PPE negatively impact heath and safety, it can also impede

FIGURE 14-2 Workers at large-scale incidents face similar unknown problems and hazards. *Courtesy of Firefighter Gerd Schuch.*

the ability to accomplish the overall mission and mitigation of the incident. Law enforcement and EMS personnel are also not equipped to handle long-term incidents involving chemical, biological, and radioactive incidents or those incidents involving search and rescue on a rubble pile. Because the work environment at large-scale incidents has changed, the equipment and way of conducting mitigation efforts must change as well to ensure adequate health and safety measures in these types of environments.

#12

Through incident analysis, it has been discovered that certain components of the overall typical PPE ensemble and equipment performed better than others. Head protection and high-visibility vests functioned at an adequate level, whereas protective clothing and respiratory protection had serious shortcomings. Typical PPE does not offer adequate protection against biological threats and infectious disease.

The typical PPE ensemble also does not offer flexibility, allow workers to move debris from a rubble pile, or make it possible to enter confined spaces. The eye protection provided with typical PPE provides adequate impact protection, but little protection from dust hazards. During many large-scale incidents, the typical self-contained breathing apparatus (SCBA) is limited by weight and the short duration of air. Air-purifying respirators may offer adequate protection from dust hazards, but little protection from threats such as anthrax. During extended operations, equipment and PPE must offer comfortable wear during periods of demanding physical labor. It has been discovered through PIA that current technology concerning equipment and PPE requires a trade-off between being practical enough and wearable enough to allow responders to perform their tasks.

Large-scale incidents present many challenges to those responding. Emergency response organizations must address both PPE availability and PPE performance if health and safety is to remain a top priority. Providing appropriate equipment and protection is also a challenge due to the different brands of equipment used by various responder organizations. Interoperability of the various types of equipment and PPE will also be a concern due to the large amount of equipment and PPE that may be sent to the scene of a large-scale long-term incident. Matching the appropriate equipment and PPE to responders will present a difficult challenge at certain incidents.

Before an incident occurs, emergency response organizations should train their personnel on the proper selection and operation of PPE. Emergency medical personnel must also be trained and provided with the proper equipment and PPE during the large-scale incident to avoid EMS personnel working and treating victims in an area that will be hazardous to their health and safety. It is now realized that adequate on-site training must take place to protect the health and safety of workers. Emergency response organizations must also be prepared to address the PPE needs of workers from responding agencies outside the emergency response arena. Construction workers, volunteers, and other agency workers in many cases will not be familiar with PPE and equipment at the incident scene. During these instances, on-site training will have to take place in the area of respiratory protection, how to maintain the equipment, and when to change the filters.

Another issue that should receive a high priority at the large-scale incident is providing timely, reliable, and adequate health and safety information. During the large-scale incident, it is normally the unknown that kills or injures workers and responders. The health and safety information should be delivered to workers from a single reliable source to avoid confusion and misinformation. During the large-scale incident, information conflicts can be avoided by conducting a continual and ongoing risk assessment and ensuring adequate PPE standards are in place.

Information exchange affects the health and safety of those working at the large-scale incident. Keeping up with changing information can become a serious challenge for incident management staff and responders, as can be the surplus of information for responders. Consistent and better information should become a top priority at any large-scale long-term incident. The flow of consistent and adequate information will also help to motivate responders to wear their supplied PPE and could help to decrease the tendency of workers to modify or remove their PPE and equipment when it becomes uncomfortable. Of course, the dilemma of what type of equipment and PPE will be needed at the large-scale incident will continue to play a large role in the health and safety of workers. All these lessons can be considered unique in nature, and they do not compare to the typical everyday events to which the emergency services respond.

INCIDENT MANAGEMENT OF LARGE-SCALE INCIDENTS

When comparing the large-scale incident to other incidents, one of the factors they have in common is incident management. It does not matter whether the incident is as large in scope as the World Trade Center incident or as small as the typical room and contents fire, the reasons for utilizing the incident management system (IMS) remain the same: accountability, health and safety of both the responders and the public, and scene management. Even with an adequate incident management system in place at every incident, an effective PIA should be done to analyze and identify the lessons learned. The use of the IMS plays a major role in lessons learned at each incident and is a major factor contributing to the health and safety of responders and the public at large.

#2

The response to emergencies is changing since the events of September 11, 2001, as it relates to all-hazard incident management and command of emergencies. This change is in the form of a comprehensive national approach to incident management known as the National Incident Management System (NIMS). With the ever-present need to improve firefighter health and safety, the NIMS helps to ensure interoperability among varying agencies.

There is also an emergency services priority to lower the number of injuries and line-of-duty deaths. The NIMS aids emergency organizations with their efforts to improve the health and safety of responders. The benefit of NIMS is most evident at the local level, when a community as a whole prepares for and provides an integrated response to an incident, so all emergency response agencies can work together in the most efficient manner possible with the health and safety of members at the forefront of incident operations. Small or rural jurisdictions will benefit from a regional approach of the NIMS, as they may not initially have the resources to implement the NIMS; however, smaller communities will be able to pool their resources to ensure the NIMS can be in place, in turn ensuring effective and safer operations at the incident scene. A lack of an effective IMS has led to firefighter fatalities. Consequently, the lack of IMS early into an incident has led to a lack of accountability of personnel at the incident scene, directly affecting the health and safety of operating personnel.

Size-Up and Risk Management

#3

One of the major factors contributing to firefighter injuries and death is a lack of or difficulty in sizing up the incident and the scene itself. Size-up and risk management are part of an ongoing process that begins with the initial call for an emergency response and continues until mitigation of the incident has been completed.

Size-up consists of evaluating many factors at each incident. The incident commander, as well as every responder on scene, must perform a risk-versus-benefit analysis of the incident scene. The lack of an effective size-up and risk-versus-benefit analysis has led to many injuries and fatalities, and negatively affected the health and safety of responders. Proper size-up starts from the time the call is received and determines the strategy and tactics that will be used at the particular incident scene. Many firefighters and officers have not had the experience or possibly the necessary training in the areas of flashover recognition and reading smoke conditions. Due to these contributing factors, firefighters are sustaining burn injuries at a higher rate today than throughout the history of the fire service. If experienced personnel are not available during the start of initial operations, inexperienced personnel may become hard-charging gung-ho responders regardless of the conditions. See Figure 14-3.

#4

FIGURE 14-3 Company officers and incident commanders must continually perform a size-up. *Courtesy of Chief Mike Zamparelli.*

In many cases, it takes those with experience and the necessary level of training to keep personnel at the incident scene from becoming a statistic. Emergency services personnel must be taught how to recognize pre-flashover conditions so they can avoid getting caught in post-flashover conditions or smoke explosions. All personnel must learn to recognize the risk-versus-benefit profile at every incident. During post-flashover conditions, no one will have survived the conditions in the main fire area, and operations can no longer take place directly inside the fire area without placing personnel at extreme risk.

Education and knowledge in fire and smoke behavior should also be coupled with education and knowledge in building construction. It is critical for every fire department to ensure its personnel are trained in fire behavior and reading smoke. In many cases, the risk-versus-benefit analysis will be based on both fire behavior conditions and the type of construction. The decision to transition from a defensive fire operation to an offensive operation may be acceptable for a sturdy building, but unacceptable for a building of modern lightweight construction. Initial and ongoing size-up must continually be coupled with an ongoing risk-versus-benefit analysis. It does not matter whether the incident is a structural fire or a wildland fire. The ability to recognize signs of impending problems such as collapse, flashover, backdraft, or wildfire "blowup" comes with training and experience. The ability to identify that a situation is about to get worse is a recognition skill that all personnel work to perfect throughout their careers. Recognition will allow personnel, including incident commanders, time to react and remove responders and themselves from a dangerous or deadly situation. See Figure 14-4.

Resource Management

Providing adequate resources—including personnel, apparatus, and equipment—at the scene of an incident is necessary for efficient, safe operations. Sufficient standby or backup personnel are necessary to support the operations at the incident scene. Without these resources, the health and safety of personnel can become compromised. In order to manage the encountered risk at the incident scene, incident commanders must have adequate resources and properly trained personnel available in a timely fashion.

FIGURE 14-4 Company officers and incident commanders must recognize hazards to keep personnel safe. *Courtesy of Chief Mike Zamparelli.*

NFPA 1500 addresses the allocation of resources during the initial stages of an incident. In many situations, the lack of adequate resources played a major role in the outcome of the incident and the health and safety of operating forces. It is also not uncommon for initial resources to become overwhelmed during the beginning stages of an incident operation, which will have a negative impact on the health and safety of personnel. Depleted SCBA air cylinders are just one example of resources that have been overwhelmed. Firefighters who cannot continue to operate at the incident scene or make a necessary rescue provide just another example of a detrimental effect on health and safety.

#1

When an emergency responder fatality or injury occurs, there are always lessons that can be learned from the incident. Unfortunately, many of these lessons are repeated over the course of time due to a lack of awareness on the part of responders. The lessons that continue to be repeated, and hence have become all too common, include those dealing with shortcomings in incident management, personnel accountability, size-up and risk management, and resource management. Fortunately, through the use of the Internet and other sources, information and lessons learned are more available to emergency response organizations than ever before. The goal is to learn from tragic incidents in order to avoid repeating them in the future.

Vehicle Safety

In recent years, the topic of emergency vehicle operations and safety has come to the forefront in emergency services organizations. Vehicle accidents involving both response apparatus and private vehicles continue to account for a high number of injuries and fatalities of those in the emergency services annually. Next to cardiac-related events, vehicle accidents account for the most common cause of firefighter-related deaths and injuries. Most of these vehicle accidents occur while emergency workers are responding to or returning from alarms. According to the NFPA, even firefighters are more likely to die responding to or returning from a fire than actually fighting one. Vehicle crashes account for not only injuries and deaths but also apparatus out of service and liability issues.

Operational and financial hardships can be created for the emergency services organization through either an apparatus or private vehicle accident. In many cases, emergency vehicle operators are endangering the lives of civilian drivers and their passengers when they drive an emergency vehicle in an unsafe manner or too fast for conditions. In all too many incidents involving crashes, emergency vehicle operators have placed far too much faith in their emergency lights and sirens to clear traffic and intersections. Some emergency responders will risk their health and safety at the incident scene to rescue a victim, but unthinkingly place themselves and civilians at risk by aggressively responding to or returning from an incident. Many responders also place themselves at risk due to their failure to wear a seat belt while their vehicle is in motion.

It would be in the interest of the health and safety program for all emergency response organizations to develop and enforce standard operating procedures (SOPs) that require mandatory use of seat belts in all moving vehicles. Many fire department organizations have adopted the National Fire Service Seat Belt Pledge to reinforce the importance of wearing seat belts. A model seat-belt policy should include a progressive disciplinary system holding the violator and the supervisor responsible to ensure compliance with the seat-belt policy, reflecting the serious and potentially life-threatening consequences of failure to comply. The International Association of Fire Chiefs (IAFC), U.S. Fire Administration (USFA), National Volunteer Fire Council (NVFC), National Fire Protection Association (NFPA), National Institute for Occupational Safety and Health (NIOSH), and National Fallen Firefighters Foundation (NFFF) are committed to promoting firefighter seat-belt use and are supportive of this pledge. This pledge was created to honor firefighter Christopher Brian Hunton, age 27, of the Amarillo (Texas) Fire Department. In 2005, he fell out of his apparatus responding to an alarm. He died two days later from his injuries. This pledge program was created because vehicle crashes are the second leading cause of firefighter line-of-duty deaths, and not wearing seat belts is the number-one vehicle-related safety concern in the fire service.

Most vehicle accidents are preventable because the majority of crashes are caused by excessive speed, the failure to stop at traffic signals and stop signs, and the general lack of caution at intersections. These issues are not just related to emergency response apparatus but also to volunteer firefighters who respond to the fire station or directly to the incident scene in privately owned vehicles equipped with warning devices. Research has indicated a privately owned vehicle equipped with warning devices may lead to a higher number of accidents and crashes. In fact, certain jurisdictions do not permit volunteer firefighters to use warning devices on their privately owned vehicles, requiring their volunteers instead to respond in accordance with the normal flow of traffic. These jurisdictions also report no delay in response times when volunteers respond in this way. The health and safety of volunteer firefighters can be greatly enhanced through a dispatch center that differentiates between an emergency and a nonemergency response at the time of sending off volunteer firefighters. Calls that can be identified as low frequency/low severity and dispatched as nonemergency responses will help to reduce the number of vehicle crashes that occur during a lights and siren response.

Recent research has discovered that the use of emergency lights and sirens does produce a faster response time; however, the time saved does not change patient outcome on medical calls except with cardiac arrest patients and those with an airway obstruction. Every emergency response organization should review the following document: *Use of Warning Lights and Siren in Emergency Medical Vehicle Response and Patient Transport* (available online at http://www.naemsp.org/documents/UseWarnLightsSirens.pdf). The International Association of Fire Chiefs (IAFC), the International Association of Firefighters (IAFF), and the National Volunteer Fire Council (NVFC) have worked to develop the following advanced Web-based education programs:

- The *Guide to Model Policies and Procedures for Emergency Vehicle Safety*, developed by the IAFC, provides in-depth information for developing policies and procedures that support the safe and effective operation of emergency and privately owned vehicles in the fire service.

- The IAFF developed *Improving Apparatus Response and Roadway Operations Safety in the Career Fire Service*, which includes instructor and participant guides and addresses critical emergency vehicle safety issues.
- The NVFC/USFA together developed *Emergency Vehicle Safe Operations for Volunteer and Small Combination Emergency Service Organizations*, whose focus is on the specific needs of volunteer and smaller departments.

These programs address specific response issues, such as the use of seat belts, safety at intersections, the design of fire apparatus and emergency vehicles, selecting and training drivers, alcohol use and driving policies, and alternative response programs.

The health and safety program should include policies and SOPs related to vehicle safety, as well as an ongoing driver/operator education program. The developed SOPs need to incorporate guidelines for safe and prudent driver/operator training and include a detailed curriculum for the classroom and hands-on training. A large part of the ongoing driver/operator program should include recognition of road and apparatus hazards, as well as hazards that may occur during emergency operations.

It would also be beneficial for driver/operators to receive training from a state or nationally recognized agency in addition to the training provided by their individual department. A large part of the training effort concerning emergency response should center on attitude and behavior, with safety as the priority of each organization's training efforts. The proper attitude and behavior toward safety should be instilled in every apparatus operator. A commitment to driver/operator competency and accountability must also become a focus of the training efforts in this important area. Training will help to reduce accidents, injuries, and fatalities. It must not stop after driver candidates have been selected and certified as driver/operators, but be a continual process that emphasizes classroom and applied practice sessions. Instructor competency and effectiveness will help establish the foundation and framework of the program. The key is to develop driver requirements and establish basic, remedial, and continual training related to drivers and apparatus operators. In many cases, inadequate skills with apparatus operations are usually the result of inadequate training and reckless behavior. Driver/operators must be taught that excitement should not lead to impulsive behaviors. NFPA 1500 states, "Drivers of fire apparatus shall be directly responsible for the safe and prudent operation of fire department vehicles." Effective ways for emergency response organizations to comply with NFPA 1500 are to ensure driver/operator training programs are in place and provided as often as necessary to meet the requirements of NFPA 1451, 1500, and 1002. This training should incorporate specifics on intersection practices. When a department conducts initial driving and refresher driver training, it is imperative that firefighters designated as drivers understand all the principles and the myriad of potential problems that can occur while operating an emergency response apparatus.

#11

Drivers should also be required to become familiar with all of the different models of fire apparatus that they may be expected to operate. Nationally, emergency services organizations are responding to incidents at a heightened level in recent years. Many fire departments unfortunately respond to false alarms on a daily basis. It would be a direct benefit to the health and safety program for municipalities to develop and impose a false alarm ordinance that targets facilities that fail to adequately maintain their alarm systems as a way to reduce the unnecessary response to nuisance false alarms that place emergency responders in harm's way.

For years, vehicle safety in emergency services was not addressed at the level that would be considered prudent and adequate. The current heightened awareness of vehicle safety has become a priority concern and been heavily addressed with health and safety program managers across the nation.

The Post-Incident Analysis and NIOSH

The NIOSH Fire Fighter Fatality Investigation and Prevention Program (FFFIPP) has been in place for many years and can be considered an effective post-incident analysis program that is available to all fire service personnel (see Chapter 12). The program focuses on recommendations for preventing future deaths and injuries. A survey conducted by NIOSH has produced numerous valuable findings. It discovered that a large number of fire departments are aware of the FFFIPP and use its findings and recommendations in their efforts to improve firefighter safety and health. Through the survey, NIOSH has also discovered that it needs to extend its outreach to small and rural fire departments. NIOSH conducts collaborative research and policy activities with partner organizations in the fire service. Its publications—which include Line of Duty Death reports, NIOSH Alerts, Health Hazard Evaluation reports, and special documents such as NIOSH Workplace Solutions—are disseminated to fire departments through the mail, e-mail, conferences, and other venues.

The survey has revealed that the FFFIPP is only moderately known. More than half of the fire service officers are not familiar with the FFFIPP itself, particularly with the process of identifying incidents to investigate, conducting the investigation, and reporting findings. Unfortunately, two-fifths of fire departments do not disseminate information from NIOSH to front-line firefighters at all. There are many barriers keeping some fire departments from following the recommendations put out by NIOSH, but firefighter resistance does not appear to be a significant reason FFFIPP-recommended safety practices are not followed.

The reason for emergency services not following recommendations from NIOSH seems to be due more to the lack of adequate funding and resources. The kinds of fire departments that most likely follow NIOSH's safety guidelines are career fire departments in large, urban jurisdictions in the Northeast. Fire departments that have experienced a firefighter fatality are also more likely than others to implement many of the NIOSH recommendations. It should be noted that health and safety programs in numerous departments will not be as effective as they should be until organizations are able to disseminate and follow the recommendations from NIOSH and post-incident analysis.

Firefighters say that learning about specific incidents helps them develop safer work practices, thereby ensuring a successful health and safety program. Here are some of the many key implications concerning post-incident analysis and NIOSH recommendations that will help to ensure the success of an effective health and safety program:

- Small volunteer departments have the greatest challenges to following safety guidelines.
- Existing resources limit safety practices.
- Gaps in knowledge and attitudes also limit safety.
- FFFIPP investigations and LODD reports provide useful information.
- Fire departments need additional information to enhance the effectiveness of the LODD reports.
- Firefighters and fire departments need information presented in additional formats.
- FFFIPP materials need to be better marketed and distributed.
- Increasing awareness will likely improve safety practices.

A combination of both the post-incident analysis conducted by each individual organization and the NIOSH FFFIPP investigations will help personnel gain knowledge to reduce or prevent injuries and fatalities in the fire service, as well as increase the effectiveness of any department's health and safety program.

CHAPTER **REVIEW**

Summary

Coupled with incident investigation, post-incident analysis has evolved from a post-incident critique to a post-incident analysis (PIA). What was done correctly at the incident scene should be the focus of the PIA. Its main purpose is to reinforce the effective actions of personnel and departmental procedures and to give management insight into how to improve the department's operations. The benefits of the PIA are numerous, and the health and safety program manager must be aware of and use both internal and external PIAs.

Large-scale incidents can be varied as they relate to emergency services. Many large-scale incidents are new to emergency services organizations including terrorist events and large-scale disasters. Large-scale incidents involving violence have also drastically increased during recent times. Equipment and practices must now be reconsidered as they relate to the large-scale incident.

The issue of PPE must also now be reconsidered when it involves certain types of incidents. Many long-term incidents will also require the use of PPE other than the structural firefighting ensemble. Emergency response organizations must address both PPE availability and PPE performance if health and safety is to remain a top priority. Personnel must also be trained and provided with the proper and necessary equipment and PPE. It is now realized that matching the appropriate equipment and PPE to responders will present a difficult challenge at certain incidents.

When comparing the large-scale incident to other types and sizes of incidents, one of the factors they have in common is incident management. Incident management and an effective PIA should go hand in hand for each incident. The use of the IMS plays a major role in lessons learned at each incident. The lack of IMS early into an incident has led to a lack of accountability of personnel. The NIMS system helps to ensure interoperability among varying agencies in present-day emergency response organizations. The benefit of NIMS is most evident at the local level.

One of the major factors contributing to firefighter injuries and death is a lack of sizing up the incident and the scene itself. The incident commander and every responder on scene must perform a risk-versus-benefit analysis. Adequate size-up and risk-versus-benefit analysis go hand in hand with health and safety. Adequate resources must be provided at the incident scene to ensure safe and effective operations, and properly trained personnel must be at the disposal of the incident commander.

Lessons can be learned after each incident, and many shortcomings have been noted after these lessons. It is now recognized that weaknesses and deficiencies exist in incident management, personnel accountability, size-up and risk management, and resource management. The Internet provides a great deal of resources along the line of lessons learned. The goal is to learn from tragic incidents to avoid repeating them in the future.

Vehicle accidents involving both response apparatus and private vehicles continue to account for a high number of emergency responder injuries and fatalities annually. Next to cardiac-related events, vehicle accidents account for the most common cause of firefighter-related deaths and injuries. In the interest of the health and safety program, all emergency response organizations need to develop and enforce SOPs that require mandatory use of seat belts in all moving vehicles. A model seat-belt policy should include a progressive disciplinary system. Many national organizations promote firefighter seat-belt use and are supportive of the National Fire Service Seat Belt Pledge.

Web-based education programs are also available from national organizations concerning apparatus operations. These programs address many of the specific issues related to apparatus operations and apparatus selection. Along with vehicle safety and operations, municipalities would be wise to implement a false alarm ordinance, thereby reducing the number of unnecessary responses to the same location that place response personnel and the general public in harm's way.

The NIOSH Fire Fighter Fatality Investigation and Prevention Program (FFFIPP) has been in place for many years and can be considered an effective post-incident analysis program that is available to all fire service personnel. Unfortunately, the FFFIPP is only moderately known in the emergency services field. There are many barriers keeping some fire departments from following the recommendations put out by NIOSH. Health and safety programs in many departments will not be as effective as they should be until organizations are able to disseminate and follow the recommendations from NIOSH. A combination of both the post-incident analysis and the NIOSH FFFIPP investigations will help personnel gain knowledge to reduce or prevent injuries and fatalities in the fire service thus improving health and safety.

Review Questions

1. Describe the purpose of the post-incident analysis.
2. List the common threats at large-scale incidents.
3. List the reasons why structural firefighting PPE is not well suited for non-fire-related environments.
4. List the benefits of an incident management system.
5. Describe the benefits of an ongoing apparatus operator-training program.
6. NFPA 1500 states that ____________ shall be involved in the post-incident analysis process.
   a. The fire chief
   b. The department safety officer
   c. The battalion chief
   d. The operations chief
7. Describe one of the benefits of the post-incident analysis.
8. Large-scale incidents have forced emergency response organizations to reconsider the ____________ they commonly use to protect themselves during the course of their work.
   a. Equipment and practices
   b. Knowledge and skills
   c. Tools
   d. None of the above
9. Incident commanders must receive adequate training and continuously:
   a. Conduct ongoing training with personnel
   b. Conduct proper ethical behavior
   c. Conduct a risk-versus-benefit analysis
   d. None of the above
10. One of the main issues related to the health and safety of members operating at a large-scale incident is that of:
    a. Personal protective equipment
    b. A lack of experience
    c. A lack of training
    d. Skill and knowledge

Case Study

On December 22, 1999, a 49-year-old shift commander (Victim 1) and two engine operators, 39 and 29 years of age, respectively (Victim 2 and Victim 3), lost their lives while performing search-and-rescue operations at a residential structure fire. At approximately 0823 hours, the three victims and two additional firefighters cleared the scene of a motor vehicle incident. One of the firefighters (Firefighter 1), riding on Engine 3, joined the ambulance crew to transport an injured patient to the hospital. At approximately 0824 hours, Central Dispatch was notified of a structure fire with three children possibly trapped inside. At approximately 0825 hours, Central Dispatch notified the fire department, and a shift commander and an engine operator (Victim 1 and Victim 2) were dispatched to the scene in the Quint (Aerial Truck 2). At 0827 hours, Engine 3 (lieutenant and Victim 3) responded to the scene. At 0829 hours as Aerial Truck 2 approached the scene, it radioed Central Dispatch, reporting white to dark brown smoke showing from the residence, and requested six additional firefighters. Aerial Truck 2 arrived on the scene at 0830 hours. The crew of Aerial Truck 2 witnessed a woman and child trapped on the porch roof, and they were informed that three children were trapped inside the house. A police officer who was already on the scene positioned a ladder to the roof and removed the woman and child as Victim 1 proceeded into the house to perform a search-and-rescue operation. Engine 3 arrived on the scene shortly after, and the lieutenant connected a supply line to the hydrant as Victim 3 pulled the Engine into position. The lieutenant and Victim 3 stretched a 5-inch supply line and connected it to Aerial Truck 2. At approximately 0831 hours, the chief and Firefighter 1 arrived on the scene, and the chief assumed incident command (IC). Firefighter 1 pulled a 1½-inch hand line off Aerial Truck 2, through the front door, and placed it in the front room. The IC instructed Victim 2 and Victim 3 to don their protective gear and proceed into the house to assist in the search-and-rescue operations. Firefighter 1 went back to Aerial Truck 2 to gear up. At this time, one of the victims removed the first of the three children from the structure, handed the child to a police reserve officer near the front entrance of the structure, and returned to the structure to continue search-and-rescue operations. The police reserve officer transported the child to a nearby hospital. The IC charged the hand line from Aerial Truck 2 and went to the structure. At this time one of the victims removed a second child. The IC grabbed the child and began cardiopulmonary resuscitation (CPR). Due to limited personnel on the fireground, the IC directed a police officer on the scene to transport him and the child to the hospital. After donning her gear, Firefighter 1 approached the front door and noticed that the 1½-inch hand line (previously stretched) had been burned through and water was free-flowing. It is believed that the three victims were hit with a thermal blast of heat before the hand line burned through. The three victims failed to exit as 12 additional firefighters arrived on the scene through a callback method and began fire suppression and search-and-rescue operations. Victim 2 was located, removed, and transported to a nearby hospital, where he was pronounced dead. Victim 1 and Victim 3 were later found and pronounced dead on the scene.

NIOSH had the following recommendations:

Recommendation #1: Fire departments should ensure that adequate numbers of staff are available to immediately respond to emergency incidents.

Recommendation #2: Fire departments should ensure that incident command conducts an initial size-up of the incident before initiating firefighting efforts, and continually evaluates the risk versus gain during operations at an incident.

Recommendation #3: Fire departments should ensure firefighters are trained in the tactics of defensive search.

Recommendation #4: Fire departments should ensure that fire command always maintains close accountability for all personnel at the fire scene.

Recommendation #5: Fire departments should ensure that fireground communication is present through both the use of portable radios and face-to-face communications.

Recommendation #6: Fire departments should ensure that a trained rapid intervention team is established and in position immediately upon arrival.

Recommendation #7: Fire departments should ensure that firefighters wear and use PASS devices when involved in interior firefighting and other hazardous duties.[1] See Figure 14-5.

Fatality Structure Fire Concerns and Issues	Yes	No
Did interior crews perform a proper risk-versus-benefit survey?		
Should the lack of an adequate handline have prevented search-and-rescue operations?		
Was there an adequate incident management system in place?		
Was inadequate staffing an issue at this incident?		
Was an adequate size-up conducted at this event?		
Was crew integrity maintained throughout this incident?		
Did the lack of recognition of the fire conditions play a role at this event?		
Did inadequate communications play a role at this incident?		
Would adequate firefighter survival training have played a role at this fire?		
Would a rapid intervention team have made a difference at this incident?		
Did a lack of building construction recognition play a role at this incident?		

FIGURE 14-5 Fatality structure fire concerns and issues.

1. Describe how an initial size-up and report would have possibly prevented the line-of-duty death.
2. Describe how a lack of accountability played a role at this incident.

Reference

1. NIOSH Fire Fighter Fatality Investigation and Prevention Program, *Structure Fire Claims the Lives of Three Career Fire Fighters and Three Children—Iowa,* April 11, 2001. Retrieved December 8, 2009, from http://www.cdc.gov/niosh/fire/reports/face200004.html

INDEX

F

G

H

N

O

P

R

S

T

U

V

W